Open-Ended Problems

Open-Ended Problems

A Future Chemical Engineering Education Approach

J. Patrick Abulencia and
Louis Theodore

Scrivener
Publishing

WILEY

Co-published by John Wiley & Sons, Inc. Hoboken, New Jersey, and Scrivener Publishing LLC, Salem, Massachusetts.
Published simultaneously in Canada.

Limit of Liability/Disclaimer of Warranty: While the publisher and author have used their best efforts in preparing this book, they make no representations or warranties with respect to the accuracy or completeness of the contents of this book and specifically disclaim any implied warranties of merchantability or fitness for a particular purpose. No warranty may be created or extended by sales representatives or written sales materials. The advice and strategies contained herein may not be suitable for your situation. You should consult with a professional where appropriate. Neither the publisher nor author shall be liable for any loss of profit or any other commercial damages, including but not limited to special, incidental, consequential, or other damages.

For general information on our other products and services or for technical support, please contact our Customer Care Department within the United States at (800) 762-2974, outside the United States at (317) 572-3993 or fax (317) 572-4002.

Wiley also publishes its books in a variety of electronic formats. Some content that appears in print may not be available in electronic formats. For more information about Wiley products, visit our web site at www.wiley.com.

For more information about Scrivener products please visit www.scrivenerpublishing.com.

Cover design by Russell Richardson

Library of Congress Cataloging-in-Publication Data:

ISBN 978-1-118-94604-6

Printed in the United States of America

10 9 8 7 6 5 4 3 2 1

To

Nicole, who signed up for a lifetime of addressing open-ended problems with me

(J.P.A.)

To

My two long-standing Manhattan College colleagues

Dr. John Jeris

Dr. Wally Matystik

(L.T.)

Contents

Preface   xix

Acknowledgements   xxi

**Part I: IntroductionIntroduction to the Open-Ended
Problem Approach**   **1**

Part II: Chemical Engineering Topics   **13**

1   Materials Science and Engineering   **15**

   1.1   Overview   15

   1.2   Crystallography of Perfect Crystals (CPC)   17

     1.2.1   Geometry of Metallic Unit Cells   20

     1.2.2   Geometry of Ionic Unit Cells   21

     1.2.3   Packing Factors   23

     1.2.4   Directions and Planes   23

   1.3   Crystallography of Real Crystals (CRC)   25

     1.3.1   Interstitial Impurities   26

   1.4   Materials of Construction   27

   1.5   Resistivity   28

   1.6   Semiconductors   29

   1.7   Illustrative Open-Ended Problems   30

   1.8   Open-Ended Problems   34

   References   37

2   Applied Mathematics   **39**

   2.1   Overview   39

   2.2   Differentiation and Integration   41

   2.3   Simultaneous Linear Algebraic Equations   42

   2.4   Nonlinear Algebraic Equations   43

2.5   Ordinary and Partial Differential Equation   44
2.6   Optimization   45
2.7   Illustrative Open-Ended Problems   48
2.8   Open-Ended Problems   51
References   56

3   Stoichiometry   **59**
3.1   Overview   59
3.2   The Conservation Law   60
3.3   Conservation of Mass, Energy, and Momentum   62
3.4   Stoichiometry   64
3.5   Illustrative Open-Ended Problems   67
3.6   Open-Ended Problems   72
References   77

4   Thermodynamics   **79**
4.1   Overview   79
4.2   Enthalpy Effects   81
    4.2.1   Sensible Enthalpy Effects   81
    4.2.2   Chemical Reaction Enthalpy Effects   83
4.3   Second Law Calculations   84
4.4   Phase Equilibrium   86
4.5   Chemical Reaction Equilibrium   88
4.6   Illustrative Open-Ended Problems   90
4.7   Open-Ended Problems   94
References   97

5   Fluid Flow   **99**
5.1   Overview   99
5.2   Basic Laws   101
5.3   Key Fluid Flow Equations   102
    5.3.1   Reynolds Number   102
    5.3.2   Conduits   103
    5.3.3   Mechanical Energy Equation – Modified Form   103
    5.3.4   Laminar Flow Through a Circular Tube   104
    5.3.5   Turbulent Flow Through a Circular Conduit   105
    5.3.6   Two Phase Flow   106
    5.3.7   Prime Movers   107
    5.3.8   Valves and Fittings   107

5.4	Fluid-Particle Applications	108
	5.4.1 Flow Through Porous Media	109
	5.4.2 Filtration	109
	5.4.3 Fluidization	110
5.5	Illustrative Open-Ended Problems	110
5.6	Open-Ended Problems	114
	References	118

6	**Heat Transfer**	**119**
6.1	Overview	119
6.2	Conduction	121
6.3	Convection	122
6.4	Radiation	125
6.5	Condensation, Boiling, Refrigeration, and Cryogenics	126
6.6	Heat Exchangers	127
6.7	Illustrative Open-Ended Problems	129
6.8	Open-Ended Problems	134
	References	139

7	**Mass Transfer Operations**	**141**
7.1	Overview	141
7.2	Absorption	143
	7.2.1 Packing Height	144
	7.2.2 Tower Diameter	145
	7.2.3 Plate Columns	146
	7.2.4 Stripping	146
	7.2.5 Summary of Key Equations	147
7.3	Adsorption	148
	7.3.1 Adsorption Design	151
	7.3.2 Regeneration	152
7.4	Distillation	152
7.5	Other Mass Transfer Processes	158
	7.5.1 Liquid-Liquid Extraction	158
	7.5.2 Leaching	158
	7.5.3 Humidification and Drying	159
	7.5.4 Membrane Processes	159
7.6	Illustrative Open-Ended Problems	160

	7.7	Open-Ended Problems	163
		References	166

8   Chemical Reactors   **169**

	8.1	Overview	169
	8.2	Chemical Kinetics	171
	8.3	Batch Reactors	174
	8.4	Continuous Stirred Tank Reactors (CSTRs)	176
	8.5	Tubular Flow Reactors	178
	8.6	Catalytic Reactors	181
		8.6.1 Fluidized Bed Reactors	183
		8.6.2 Fixed Bed Reactors	183
	8.7	Thermal Effects	184
		8.7.1 Batch Reactors	184
		8.7.2 CSTRs	185
		8.7.3 Tubular Flow Reactions	186
	8.8	Illustrative Open-Ended Problems	187
	8.9	Open-Ended Problems	192
		References	196

9   Process Control and Instrumentation   **197**

	9.1	Overview	197
	9.2	Process Control Fundamentals	199
	9.3	Feedback Control	203
	9.4	Feedforward Control	204
	9.5	Cascade Control	205
	9.6	Alarms and Trips	206
	9.7	Illustrative Open-Ended Problems	207
	9.8	Open-Ended Problems	209
		References	212

10   Economics and Finance   **213**

	10.1	Overview	213
	10.2	Capital Costs	*216*
	10.3	Operating Costs	217
	10.4	Project Evaluation	218
	10.5	Perturbation Studies in Optimization	219
	10.6	Principles of Accounting	220

10.7　Illustrative Open-Ended Problems 221
10.8　Open-Ended Problems 225
References 230

11 Plant Design **233**
11.1　Overview 233
11.2　Preliminary Studies 235
11.3　Process Schematics 236
11.4　Material and Energy Balances 237
11.5　Equipment Design 238
11.6　Instrumentation and Controls 240
11.7　Design Approach 240
11.8　The Design Report 242
11.9　Illustrative Open-Ended Problems 243
11.10　Open-Ended Problems 246
References 250

12 Transport Phenomena **253**
12.1　Overview 253
12.2　Development of Equations 255
12.3　The Transport Equations 256
12.4　Boundary and Initial Conditions 257
12.5　Solution of Equations 258
12.6　Analogies 258
12.7　Illustrative Open-Ended Problems 262
12.8　Open-Ended Problems 264
References 267

13 Project Management **269**
13.1　Overview 269
13.2　Managing Project Activities 271
13.3　Initiating 272
13.4　Planning/Scheduling 273
13.5　Gantt Charts 275
13.6　Executing/Implementing 276
13.7　Monitoring/Controlling 277
13.8　Completion/Closing 278
13.9　Reports 279

13.10 Illustrative Open-Ended Problems   280
13.11 Open-Ended Problems   284
References   291

14 Environmental Management   **293**
14.1   Overview   293
14.2   Environmental Regulations   295
14.3   Classification, Sources, and Effects of Pollutants 296
14.4   Multimedia Concerns   297
14.5   ISO 14000   298
14.6   The Pollution Prevention Concept   299
14.7   Green Chemistry and Green Engineering   300
14.8   Sustainability   301
14.9   Illustrative Open-Ended Problems   302
14.10 Open-Ended Problems   309
References   315

15 Environmental Health and Hazard Risk Assessment   **317**
15.1   Overview   317
15.2   Safety and Accidents   319
15.3   Regulations   320
15.4   Emergency Planning and Response   321
15.5   Introduction to Environmental Risk Assessment   322
15.6   Health Risk Assessment   323
15.7   Hazard Risk Assessment   326
15.8   Illustrative Open-Ended Problems   329
15.9   Open-Ended Problems   333
References   341

16 Energy Management   **343**
16.1   Overview   343
16.2   Energy Resources   345
16.3   Energy Quantity/Availability   346
16.4   General Conservation Practices in Industry   346
16.5   General Domestic Conservation Applications   347
16.6   General Commercial Real Estate Conservation
       Applications   348
16.7   Architecture and the Role of Urban Planning   349

16.8   The U.S. Energy Policy/Independence   350
16.9   Illustrative Open-Ended Problems   352
16.10   Open-Ended Problems   355
References   361

17 **Water Management**   **363**
17.1   Overview   363
17.2   Water as a Commodity and as a Human Right   365
17.3   The Hydrologic Cycle   366
17.4   Water Usage   367
17.5   Regulatory Status   367
        17.5.1   The Safe Drinking Water Act (SDWA)   367
        17.5.2   The Clean Water Act (CWA)   369
17.6   Acid Rain   370
17.7   Treatment Processes   371
17.8   Future Concerns   372
17.9   Illustrative Open-Ended Problems   373
17.10   Open-Ended Problems   376
References   381

18 **Biochemical Engineering**   **383**
18.1   Overview   383
18.2   Enzyme and Microbial Kinetics   385
18.3   Enzyme Reaction Mechanisms   386
18.4   Effectiveness Factor   389
18.5   Design Procedures   391
        18.5.1   Design of a Batch Sterilization Unit   391
        18.5.2   Design of a Continuous Sterilization Unit   392
        18.5.3   Design of an Air Sterilizer   393
        18.5.4   Scale-Up of a Fermentation Unit   393
18.6   Illustrative Open-Ended Problems   394
18.7   Open-Ended Problems   399
References   403

19 **Probability and Statistics**   **405**
19.1   Overview   405
19.2   Probability Definitions and Interpretations   407
19.3   Introduction to Probability Distributions   408

19.4   Discrete and Continuous Probability Distributions   410

19.5   Contemporary Statistics   410

19.6   Regression Analysis (3)   411

19.7   Analysis of Variance   412

19.8   Illustrative Open-Ended Problems   413

19.9   Open-Ended Problems   418

References   425

20 Nanotechnology   **427**

20.1   Overview   427

20.2   Early History   429

20.3   Fundamentals and Basic Principles   429

20.4   Nanomaterials   430

20.5   Production Methods   431

20.6   Current Applications   432

20.7   Environmental Concerns   433

20.8   Future Prospects   434

20.9   Illustrative Open-Ended Problems   436

20.10 Open-Ended Problems   440

References   443

21 Legal Considerations   **445**

21.1   Overview   445

21.2   Intellectual Property Law   447

21.3   Contract Law   448

21.4   Tort Law   448

21.5   Patents   449

21.6   Infringement and Interferences   451

     21.6.1   Infringement   451

     21.6.2   Interferences   451

21.7   Copyrights   452

21.8   Trademarks   453

21.9   The Engineering Professional Licensing Process   454

21.10 Illustrative Open-Ended Problems   454

21.11 Open-Ended Problems   457

References   460

22 Ethics   **463**
   22.1   Overview   463
   22.2   The Present State   464
   22.3   Moral Issues   466
   22.4   Engineering Ethics   467
   22.5   Environmental Justice   468
       22.5.1   The Case For and Against Environmental Justice   469
   22.6   Illustrative Open-Ended Problems   470
   22.7   Open-Ended Problems   473
   References   480

Part III: Term Projects   **483**

23 Term Projects (2): Applied Mathematics   **485**
   23.1   Term Project 23.1   486
   23.2   Term Project 23.2   487
   References   488

24 Term Projects (2): Stoichiometry   **489**
   24.1   Term Project 24.1   490
   24.2   Chemical Plant Solid Waste   493
   Reference   493

25 Term Projects (2): Thermodynamics   **495**
   25.1   Estimating Combustion Temperatures   496
   25.2   Generating Entropy Data   496
   References   497

26 Term Projects (6): Fluid Flow   **499**
   26.1   Pressure Drop – Velocity – Mesh Size Correlation   500
   26.2   Fanning's Friction Factor: Equation Form   500
   26.3   An Improved Pressure Drop and Flooding Correlation   503
   26.4   Ventilation Model I   505
   26.5   Ventilation Model II   506
   26.6   Two – Phase Flow   506
   References   507

27 Term Projects (4): Heat Transfer **509**
27.1 Wilson's Method 510
27.2 Heat Exchanger Network I 511
27.3 Heat Exchanger Network II 513
27.4 Heat Exchanger Network III 514
References 515

28 Term Projects (5): Mass Transfer Operations **517**
28.1 An Improved Absorber Design Procedure 518
28.2 An Improved Adsorber Design Procedure 519
28.3 Multicomponent Distillation Calculations 520
28.4 A New Liquid-Liquid Extraction Process 523
28.5 Designing and Predicting the Performance of
Cooling Towers 525
References 526

29 Term Projects (2): Chemical Reactors **529**
29.1 Minimizing Volume Requirements for
CSTRs in Series I 530
29.2 Minimizing Volume Requirements for
CSTRs in Series II 531
References 531

30 Term Projects (4): Plant Design **533**
30.1 Chemical Plant Shipping Facilities 534
30.2 Plant Tank Farms 535
30.3 Chemical Plant Storage Requirements 536
30.4 Inside Battery Limits (ISBL) and Process Flow Approach 538
References 541

31 Term Projects (4): Environmental Management **543**
31.1 Dissolve The USEPA 544
31.2 Solving Your Town's Sludge Problem 547
31.3 Benzene Underground Storage Tank Leak 549
31.4 An Improved MSDS Sheet 551
References 552

32 Term Projects (4): Health and Hazard Risk Assessment **553**

32.1 Nuclear Waste Management 554

32.2 An Improved Risk Management Program 555

32.3 Bridge Rail Accident: Fault and Event Tree Analysis 557

32.4 HAZOP: Tank Car Loading Facility 558

References 560

33 Term Projects (3): Unit Operations Laboratory Design Projects **561**

33.1 Hand Pump 562

33.2 Rooftop Garden Bed 563

33.3 Hydration Station Counter 564

Reference 566

34 Term Projects (4): Miscellaneous Topics **567**

34.1 Standardizing Project Management 568

34.2 Monte Carlo Simulation: Bus Section Failures in Electrostatic Precipitators 569

34.3 Hurricane and Flooding Concerns 570

34.4 Meteorites 571

References 573

Index **575**

Preface

Chemical engineering is one of the fundamental disciplines of engineering, and contains many practical concepts that have been utilized in the past in countless real-world industrial applications. However, the profession is changing. Therefore, the authors considered writing a text that highlighted open-ended material since chemical engineers in the future will have to be innovative and creative in order to succeed in their careers. One approach to developing the chemical engineer's ability to solve unique problems is by employing open-ended problems. Although the term "open-ended problem" has come to mean different things to different people, it describes an approach to the solution of a problem and/or situation where there is usually not a unique solution. The authors of this text have applied this approach by including numerous open-ended problems in several of their courses. Although the literature is inundated with texts emphasizing theory and theoretical derivations, the goal of this book is to present the subject of open-ended problems from a pragmatic point-of-view in order to better prepare chemical engineers for the future.

This book is the result of much effort from the authors, and has gone through classroom testing. It was difficult to decide what material to include and what to omit, and every attempt was made to offer sufficient chemical engineering course material at a level that could enable chemical engineers to better cope with original and unique problems that will be encountered later in practice. It should be noted that the authors cannot claim sole authorship to all of the essay material in this text. Although much of the material has been derived from sources that both of the authors have been directly involved, every effort has been made to acknowledge material drawn from other sources.

The book opens with an Introduction (Part I) to the general subject of open-ended problems. This is followed by 22 chapters (Part II), each of which addresses a traditional chemical engineering (or chemical engineering related) topic. Each of these chapters contain a brief overview of the subject matter of concern, e.g., thermodynamics, which is followed by three open-ended problems that have been solved by the authors, employing one of the many potential possible approaches to the solution. This is then followed by approximately 30-40 open-ended problems with *no* solutions. A reference section complements the chapter's contents. Part III is concerned with term projects. Twelve chapter topics, including a total of 42 projects, are provided.

It is hoped that the book will describe the principles and applications of open-ended chemical engineering problems in a thorough and clear manner for academic, industrial, and government personnel. Upon completion of the text, the reader should have acquired not only a working knowledge of the principles of chemical engineering, but also (and more importantly) experience in solving open-ended problems. The authors strongly believe that, while understanding the traditional basic concepts is of paramount importance, this knowledge may be rendered virtually useless to future engineers if he/she cannot apply these concepts to unique real-world situations.

Last, but not least, the authors believe that this modest work will help the majority of individuals working and/or studying in the field of engineering to obtain a more complete understanding of chemical engineering. If you have come this far and read through the Preface, you have more than just a passing interest in this subject. The authors strongly suggest that you take advantage of the material available in this book and believe that it will be a worthwhile experience.

January 2015
J. Patrick Abulencia
Bronx, NY

Louis Theodore
East Williston, NY

Acknowledgements

The authors were assisted during the preparation of this book by several individuals that we wish to acknowledge. Rita D'Aquino served as our executive editor, and spent many days proofing the manuscript and preparing the index. Colleen Kavanagh and Xi Chu are two Manhattan College students who assisted with preparing the manuscript and figures. The authors are sincerely grateful for all of your contributions.

Part I

INTRODUCTION TO THE OPEN-ENDED PROBLEM APPROACH

This part is a stand-alone portion of the book, which serves the sole purpose of introducing the reader to open-ended problems and the open-ended problem approach.

The reader is constantly reminded of the need for change in the chemical engineering curriculum and, there is a need to change. The key word in the new chemical engineering curriculum will be *innovation*. It *must* be innovation if the profession is to survive. Presenting problems with "discrete" solutions thwarts the preparation of students by constraining their vitality, energy, and intellectual capabilities, and will minimize their impact on the future marketplace. Thus, failure to develop the innovative skills of future chemical engineers will adversely affect their careers. Bottom line: creativity, imagination, and (once again) innovation will be a requisite for success in the future.

Finally, the reader should note that a good part of the material presented in this Part was adapted from the earlier work by Theodore titled *Chemical Engineering: The Essential Reference*, Chapter 30 Open-Ended Problems, McGraw-Hill, New York City, NY, 2014. [1]

Overview

The phrase for success at the turn of the 20th century was: work hard and you will succeed. What was heard during the careers of both authors as educators and practitioners was the phrase: work intelligently and you will succeed. However, the key phrase for the 21st century is: be *innovative* and you will succeed. This will be the theme for the engineers and scientists of tomorrow; and, more than any other profession, it will become the key to success for future chemical engineers. For success to follow, the education of chemical engineers, in terms of the curriculum, will have to change if they are to succeed.

As noted earlier, the key word in the new chemical engineering curriculum will be *innovation*. It *must* be innovation if the profession is to survive. It will require more than possessing traditional problem-solving skills in order for the chemical engineering workforce to be appropriately educated. The authors have always advocated that one of the most important jobs of an educator is to anticipate the future.

Career paths in chemical engineering are now undergoing a change—a drastic change in the authors' opinion. The days of the need for massive numbers of chemical engineers required to size pumps, design heat exchangers, predict the performance of multi-component distillations columns is now a distinct memory. Handbook solutions are being replaced with creative, innovative action; hard work is being replaced by the need to understand software, etc. The conversion process will take time; educating and cultivating this intellectual approval for the new breed of engineers will not come overnight. But the time to start is NOW.

In terms of introduction, the cliché of the creative individual has unfortunately been aptly described throughout history- the Einsteinian wild hair, being locked in a room for days at a time, mumbling to one's self, eating sporadically, lost in a fog of conflicting thoughts, not paying attention to one's hygiene, working diligently until that times when the "light goes on" moment of discovery, etc. This chapter will provide (among other things) specific suggestions on how to develop and improve one's critical thinking abilities. [2]

Engineering is one of the noblest of professions, and the authors are extremely proud to be part of it. They are fortunate to have served as chemical engineering educators during their careers. A good part of this effort was directed to improving critical thinking skills of students in recent years. Check any engineering school's web-site and locate its mission statement. Many of these will carry the phrase "fosters creativity and innovation" among its students. But do they really? The authors hope so. But then again, how does one teach it? [2]

As a chemical engineering educator, one is required to teach traditional basic scientific and technical principles in courses like thermodynamics, heat transfer, reaction kinetics, etc., but, along with the lectures, one should include an emphasis on creativity, problem solving and *failure(s)*. These three terms are definitely interrelated. Finding solutions to problems is a creative activity. Failure comes into play since there are often solutions with high uncertainty and many or no correct answers. [2]

The remainder of this part addresses a host of topics involved with open-ended problems and approaches. The following sections are addressed:

1. General Thoughts
2. The Authors' Approach
3. Earlier Experiences
4. Developing Students' Power of Critical Thinking
5. Creativity and Brainstorming
6. Inquiring Minds
7. Final Thoughts

General Thoughts

Here are a baker's dozen general thoughts regarding the open-ended problem approach drawn from the files of one of the authors. [3]

1. Abstract reasoning is the ability to analyze information and solve problems on a complex, thought-based level.
2. Software will become increasingly more important.
3. In order to create an individual's intellectual capabilities, one has to nurture and cultivate while educating, which may take decades of effort.
4. To reach the upper levels of science and technology, one needs creativity, imagination, and innovation, which at present is not being nurtured.
5. Wealth will be generated from technological innovations.
6. The labor market is undergoing a historic change, and individuals in the future should exploit this.
7. Everything is possible for the individual who doesn't have to do it.
8. It's okay for a scheme not to work, i.e., it's okay to fail.

9. Chemical engineers in the future will be judged and rewarded by their ability to predict evolving situations and formulate concrete strategies.
10. An important part of success will be to anticipate future situations, evaluate possible outcomes, and set appropriate goals.
11. Many of the old blue-collar factory jobs will disappear.
12. The chemical engineer has to be educated to meet the challenges of this century, and this will require activities that involve creativity, artistic ability, innovation, leadership, and analysis.
13. The chemical engineer who develops good habits of problem solving early in his/her career will save considerable time and avoid many frustrations later in life.

The Authors' Approach

Here is what the authors have stressed to their students in terms of developing problem-solving skills and other creative thinking.

1. Carefully define the problem at hand.
2. Obtain all pertinent data and information.
3. Initially, generate an answer or solution.
4. Examine and evaluate as many alternatives as possible, employing "what if" scenarios [1].
5. Reflect on the above over time.
6. Consider returning to step 1 and repeat/expand the process.

The traditional methodology of solving problems has been described for decades with the following broad stepwise manner:

1. Understand the problem.
2. Devise a plan.
3. Carry out the plan.
4. Look back and (possibly) revise.

Many now believe creative thinking should be part of every student's education. Here are some ways that have proven to nudge the creative process along:

1. Break out of the one-and-only answer rut.

2. Use creative thinking techniques and games.
3. Foster creativity with assignments and projects.
4. Be careful not to punish creativity.

The above-suggested activities will ultimately help develop a critical thinker that:

1. Raises important questions and problems, formulating them clearly and precisely.
2. Gathers and assesses relevant information, using abstract ideas to interpret it effectively.
3. Comes to well-reasoned conclusions and solutions, testing them against relevant criteria and standards.
4. Thinks open mindedly within alternative systems of thought, recognizing and assessing, as need be, their assumptions, implications, and practical consequences.
5. Communicates effectively with others in figuring out solutions to complex problems.

The analysis aspect of a problem remains. It essentially has not changed. The analysis of a new problem in chemical process engineering can still be divided into four steps.

1. Consideration of the process in question.
2. Mathematical description of the process, if applicable.
3. Solution of any mathematical relationships to provide a solution.
4. Verification of the solution.

Earlier Experiences[1,4,5]

The educational literature provides frequent references to individuals, particularly engineers, and other technical fields, that have different learning styles, and in order to successfully draw on these different styles, a variety of approaches can be employed. One such approach for educators involves the use of *open-ended* problems.

The term *open-ended* has come to mean different things to different people in industry and academia. It basically describes an approach to the solution of a problem and/or situation for which there is usually not a unique solution. Three literature sources[6–8] provide sample problems that can be used when this educational tool is employed.

One of the authors of this book has applied this somewhat unique approach and has included numerous open-ended problems in several chemical engineering course offerings at Manhattan College. Student comments for a general engineering graduate course "Accident and Emergency Management" were tabulated. Student responses to the question "What aspects of this course were most beneficial to you?" are listed below:

1. "The open-ended questions gave engineers a creative license. We don't come across many of these opportunities."
2. "Open-ended questions allowed for candid discussions and viewpoints that the class may not have been otherwise exposed to."
3. "The open-ended questions gave us an opportunity to apply what we were learning in class with subjects we have already learned and gave us a better understanding of the course."
4. "Much of the knowledge that was learned in this course is applicable to everyday situations and our professional lives."
5. "Open-ended problems made me sit down and research the problem to come up with ways to solve them."
6. "I thought the open-ended problems were inventive and made me think about problems in a better way."
7. "I felt that the open-ended problems were challenging. I, like most engineers, am more comfortable with quantitative problems than qualitative."

In effect, the approach requires asking questions, to not always accept things at face value, and to select a methodology that provides the most effective and efficient solution. Those who conquer this topic have probably taken first step toward someday residing in an executive suite.

Developing Students' Power of Critical Thinking[1,9]

It has often been noted that chemical engineers are living in the middle of an information revolution. Since the term of the century, that revolution has had an effect on teaching and learning. Educators are hard-pressed to keep up with the advances in their fields. Often their attempts to keep the students informed are limited by the difficulty of making new material available.

The basic need of both educator and student is to have useful information readily accessible. Then comes the problem of how to use this information properly. The objectives of both teaching and studying such information are: to assure comprehension of the material and to integrate it with the basic tenets of the field it represents; and, to use the comprehension of the material as a vehicle for *critical thinking* and *effective argument*.

Information is valueless unless it is put to use; otherwise, it becomes mere data. For information to be used most effectively, it should be taken as an instrument for *understanding*. The process of this utilization works on a number of incremental levels. Information can be absorbed, comprehended; discussed, argued in reasoned fashion, written about, and integrated with similar and contrasting information.

The development of critical and analytical thinking is key to the understanding and use of information. It is what allows the student to discuss, and argue points of opinion and points of fact. It is the basis for the student's formation and development of independent ideas. Once formed, these ideas can be written about and integrated with both similar and contrasting information.

Creativity and Brainstorming

Chemical engineers bring mathematics and other sciences to bear on practical problems and applications, molding materials and harnessing technology for human benefit. *Creativity* is often a key component in this synthesis; it is the spark, motivating efforts to devise solutions to novel problems, design new products, and improve existing practices. In the competitive marketplace, it is a crucial asset in the bid to win the race to build better machines, decrease product delivery times, and anticipate the needs of future generations.[1,9]

One of the keys to the success of a chemical engineer or a scientist is to generate fresh approaches, process and products, i.e., they need to be creative. Gibney[9] has detailed how some schools and institutions are attempting to use certain methods that essentially share the same objective: open students' minds to their own creative potential.

Gibney [9] provides information on "The Art of Problem Definition" developed by the Rensselaer Polytechnic Institute. To stress critical thinking, they teach a seven-step methodology for creative problem development. These steps are provided below: [9]

1. Define the problem.
2. State objective.
3. Establish functions.
4. Develop specifications.
5. Generate multiple alternatives.
6. Evaluate alternatives.
7. Build.

In addition, Gibney [9] identified the phases of the creative process set forth by psychologists. They essentially break the process down into five basic stages:

1. Immersion.
2. Incubation.
3. Insight.
4. Evaluation.
5. Elaboration.

Psychologists have ultimately described the creative process as *recursive*. At any one of these stages, a person can double back, revise ideas, or gain new knowledge that reshapes his or her understanding. For this reason, being creative requires patience, discipline, and hard work.

Delia Femina [10] outlined five "secrets" regarding the creative process:

1. Creativity is ageless.
2. You don't have to be Einstein.
3. Creativity is not an eight hour job.
4. Failure is the mother of all creativity.
5. Dead men don't create.

Panitz [11] has demonstrated how *brainstorming strategies* can help engineering students generate an outpouring of ideas. Brainstorming guidelines include:

1. Carefully defining the problem upfront.
2. Allow individuals to consider the problem before the group tackles it.
3. Create a comfortable environment.
4. Record all suggestions.
5. Appoint a group member to serve as a facilitator.
6. Keep brainstorming groups small.

A checklist for change was also provided, as detailed below:

1. Adapt
2. Modify.
3. Magnify.
4. Minify.
5. Put to other uses.
6. Substitute.
7. Rearrange.
8. Reverse.
9. Combine.

Inquiring Minds

In an exceptional well-written article by Lih [12] entitled "Inquiring Minds", he commented on inquiring minds by saying "You can't transfer knowledge without them." His thoughts (which have been edited) on the inquiring or questioning process follow:

1. Inquiry is an attitude—a very important one when it comes to learning. It has a great deal to do with curiosity, dissatisfaction with the status quo, a desire to dig deeper, and having doubts about what one has been told.
2. Questioning often leads to believing—there is a saying that has been attributed to Confucius: "Tell me, I forget. Show me, I remember. Involve me, I understand." It might also be fair to add: "Answer me, I believe."
3. Effective inquiry requires determination to get to the bottom of things.
4. Effective inquire requires wisdom and judgment. This is especially true for a long-range intellectual pursuit that is at the forefront of knowledge.
5. Inquiry is the key to successful life-long learning. If one masters the art of questioning, independent learning is a breeze.
6. Questioning is good for the questionee as well. It can help clarify issues, uncover holes in an argument, correct factual and/or conceptual errors, and eventually lead to a more thoughtful outcome.

7. Teachers and leaders should model the importance of inquiry. The teacher leader must allow and encourage questions and demonstrate a personal thirst for knowledge.

Ultimately, the degree to which one succeeds (or fails) is often based in part on one's state of mind or attitude. As President Lincoln once said: "Most people are about as happy as they make their minds to be." William Jones once wrote: "The greatest discovery of my generation is that human beings can alter their lives by altering their attitude of mind." So, no matter what one does, it is in the hands of that individual to make it a meaningful, pleasurable, and positive experience. This experience will almost definitely bring success.

Final Thoughts

One of the authors [13], prior of retiring, sought consulting jobs when he was told, "We just can't figure out how to solve the problem." For example, should a heat exchanger be heated with atmospheric or superheated steam? Obviously, it would appear to be better to employ atmospheric steam. But, that may not always be the "best" approach. Tackling and solving these class of problems will only come with experience. And then there is the option of ordering a chemical reactor in assembled form rather than in sections. One would normally select the assembled option, but once again, it might require eliminating walls and/or enlarging small openings. The choice is not clear and analysis is warranted.

The traditional chemical engineering curriculum cannot be totally abandoned. It must still include material to describe the behavior of processes and the ability to design equipment. If the process or problem is complex, he/she must also be able to use approximate methods. Unfortunately, many systems with which the chemical engineer will deal with in the future do not fit simple theory.

The development of the future chemical engineer can be compared to the development of a good basketball player [14]. A basketball player must learn how to dribble and shoot. He must also develop an ability to play hard-nose defense, and he must learn the meaning of teamwork. He must also learn to take orders from his coaches. All these can be worked on and perfected individually, but the complete basketball player does not manifest until all the individual parts are put together to function as a smooth, complete unit. [14] Just as the basketball player needs to work on the individual parts of his skill, the chemical engineer still needs to study the separate operations involved in his field. Chemical engineers must study and understand the basic laws of chemistry and physics. They must know the various types of equipment and the economics involved in the over-all plant process. They must understand the unit operations of fluid flow, heat transfer and mass transfer operations, and other peripheral topics.

What about the chemical engineer sustaining his/her career? MacLean [15] recently provided some career advice that is generally universally recognized. Here is a summary of his baker's dozen (the authors have added three to his 10) pointers:

1. Pick a rigorous college.
2. Pick a future viable career path, not the current popular one.
3. EQ (emotional quotient) is as important as IQ.
4. Be really nice to everyone.
5. Choose your battles very carefully.
6. Be patient and hold your tongue.
7. Do not expect to be rewarded for hard work.
8. Understand your "customer" needs.
9. Do your fair share.
10. Maintain contact.
11. Provide praise (if appropriate) to others.
12. Luck rules; increase your odds.
13. Most importantly, develop an ability to communicate orally and in writing.

Some of the suggestions might be a good fit for some chemical engineers, and some may not apply over their entire career. Think it through.

References

1. Adapted from, L. Theodore, *Chemical Engineering: The Essential Reference*, McGraw-Hill, New York City, NY, 2014.
2. L. Theodore, *On Creative Thinking II*, Discovery, East Williston, NY, August 13, 2004.
3. Personal Notes, L. Theodore, East Williston, NY, 1995.
4. J.P. Abulencia and L. Theodore, *Fluid Flow for the Practicing Chemical Engineer*, John Wiley & Sons, Hoboken, NJ, 2009.
5. A. Flynn and L. Theodore, *An Air Pollution Control Equipment Design Course for Chemical and Environmental Engineering Students Using and Open-Ended Problem Approach*, ASEE Meeting, Rowan University, NJ, 2001.

6. A. Flynn, J. Reynolds, and L. Theodore, *Courses for Chemical and Environmental Engineering Students Using an Open-Ended Problem Approach*, AWMA Meeting, San Diego, CA, 2003.

7. L. Theodore, class notes, 1999-2003.

8. Manhattan College Center for Teaching, *Developing Students' Power of Critical Thinking*, Bronx, NY, January 1989.

9. K. Gibney, *Awakening Creativity*, ASEE Promo, Washington, DC March 1988.

10. J. Delia Femina, *Jerry's Rules*, Modern Maturity, location unknown March-April 2000.

11. B. Panitz, *Brain Storms*, ASEE Promo, Washington, DC March 1998.

12. M. Lih, *Inquiring Minds*, ASEE Promo, Washington, DC December 1998.

13. Personal notes, L. Theodore, East Williston, NY, 2010.

14. L. Theodore, *Basketball Coaching 101*, in preparation, East Williston, NY, 2014.

15. R. Maclean, *Sustaining Your Career*, EM, Pittsburgh, PA, July 2012.

Part II
CHEMICAL ENGINEERING TOPICS

Part II is concerned with subject matter of interest and concern to the practicing chemical engineer. Topics reviewed include (with accompanying Chapter Number);

1. Material Science and Engineering
2. Applied Mathematics
3. Stoichiometry
4. Thermodynamics
5. Fluid Flow
6. Heat Transfer
7. Mass Transfer Operations
8. Chemical Reactors
9. Process Control
10. Economics and Finance
11. Plant Design
12. Transport Phenomena
13. Project Management
14. Environmental Management
15. Environmental Health and Hazard Risk Assessment
16. Energy Management

17. Water Management
18. Biochemical Engineering
19. Probability and Statistics
20. Legal Considerations
21. Nanotechnology
22. Ethics

Each chapter contains three sections. The first section provides a broad (but brief) overview of the subject matter of concern. Several sections then address technical subject matter related to the topic of concern. This in turn is followed with a section that contains three *solved* open-ended problems. The chapter concludes with approximately 35-40 open-ended problems—some solutions of which are available to those who adopt the book for classroom or training purposes. A reference page compliments the presentation.

Additional details for each chapter topic is available in Theodore's "Chemical Engineering: The Essential Reference"; an extensive and comprehensive treatment of most topics is provided in Perry's "Chemical Engineers' Handbook", 8th edition (both of McGraw-Hill).

1

Materials Science and Engineering

This chapter is concerned with Materials Science and Engineering (MSE). As with all the chapters in Part II, there are sereval sections: overview several specific technical topics illustrative open-ended problems, and-open ended problems. The purpose of the first section is to introduce the reader to the subject of MSE. As one might suppose, a comprehensive treatment is not provided, although numerous references are included. The second section contains three open-ended problems; the authors' solution (there may be other solutions) is also provided. The third (and final) section contains 35 problems; *no* solutions are provided here.

1.1 Overview

This overview section is concerned—as can be noted from its title—with Materials Science and Engineering (MSE). As one might suppose, it was not possible to address all topics directly or indirectly related to MSE. Because of space limitations, only the subject of crystallography of perfect crystals (CPC) is primarily addressed. However, additional details may be obtained from the references at the end of the chapter.

Note: Those readers already familiar with the details associated with MSE may choose to bypass this Overview.

The title, *Materials Science and Engineering*, implies a double focus—one geared toward a fundamental study of the materials and their properties, and the other towards the production and use of materials for the benefit of society. This chapter is primarily concerned with the former focus.

The terms *Materials* denotes a vast areas of compiled knowledge. There is very little in all of engineering and science that does not involve materials. Obviously, the first task in preparing an abbreviated chapter in the study of materials must be the application of limits on the subject matter to be covered—a focus on specific types of materials. It is generally understood that *Materials* covers only the solid state of matter; liquids are considered only in certain cases where solid-liquid equilibrium is involved. There are many types of solids, however, and further focusing is required.

Most solids can be categorized into one of the three types: metals, plastics, or ceramics. (Ceramics are compounds of metallic and non-metallic elements such as ferrous oxide.) In this chapter, emphasis has been placed on metals because, in the opinion of the authors, this class of materials has the widest impact on all of the four major fields of engineering: chemical, civil, electrical, and mechanical. In the end, the remainder of the chapter was divided into three sections. These are briefly discussed below.

The first topic covered is *Crystallography of Perfect Crystals* (CPC). All matter is ultimately composed of atomic particles. How these particles are put together plays an extremely important role in determining a material's properties and in the various uses of that material. The purpose of this section is to provide the reader some insights into how solids (mainly metals and ionic materials) are *organized* at the atomic level and how this organization is reflected in some of the properties of the solids. Just as the organization of atoms in a solid has a critical role in determining material properties, so too does the occasional *breakdown* of this organization since there is no such thing as a *perfect* crystal. No study of crystallography would therefore be complete without a companion study of crystal imperfections, and this is the role of the CRC (*Crystallography of Real Crystals*) section. These two sections (i.e., CPC and CRC) involve a study of the makeup of the crystalline materials from an atomic standpoint, but what happens on the atomic level must obviously be reflected at the macroscopic level. The last sections discuss the behavior of materials more in terms of some directly measurable phenomena.

Finally, the reader should note that there are a host of topics that normally fall under the Materials Science and Engineering umbrella. Most of them are listed below:

1. Atomic Structure
2. Crystal Structures
3. Crystal Geometry
4. Crystalline Imperfections
5. Phase Diagrams
6. Mechanical Properties of Metals
7. Polymeric Materials
8. Diffusion in Solids
9. Engineering Alloys
10. Ceramics
11. Composite Materials
12. Corrosion
13. Electrical Properties
14. Optical Properties
15. Magnetic Properties
16. Superconductive Properties

1.2   Crystallography of Perfect Crystals (CPC)

As noted earlier, all matter is ultimately composed of atomic particles. How these particles are put together plays an extremely important role in determining a material's properties and in the various uses of that material. The purpose of the CPC (Crystallography of Perfect Crystals) section is to give the reader some insights into how solids (mainly metals and ionic materials) are organized on the atomic level and how this organization is reflected in some of the properties of the solids.

Solid materials may be either *amorphous* or *crystalline*. The word *amorphous* literally means "without form" and the atoms or molecules of solids in this category have little organization. The word crystalline implies that the component atoms, ions or molecules that make up the material are arranged spatially in an *ordered* pattern often referred to as a *crystal lattice*. In the solid state, metallic and ionic materials are almost universally found as crystals in nature; many covalent materials are crystalline as well.

There are many different types of crystal patterns or structures. For ionic and metallic materials, the main factor that determines the pattern

Table 1.1 Minimum CN radius ratios.	
CN	**(r/R)***min*
3	0.155
4	0.225
6	0.414
8	0.732
12	1.000

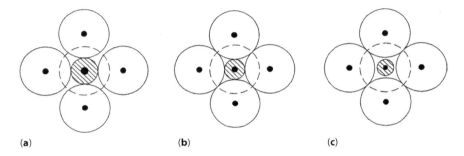

(a)   (b)   (c)

Figure 1.1 Coordination Number

or crystal type is the packing efficiency or the *packing factor*. Both metal-lic and ionic bonds are electrostatic in nature. As a result, the closer the bonded atoms are, the stronger are the bonding forces and the more sta-ble is the crystal. For purposes of this explanation, assume the atoms of a crystal to be small hard spheres tightly packed together in an ordered pattern. The metallic crystal is composed of spheres all having the same size and an ionic crystal is constructed of spheres of at least two different sizes. The packing factor is defined as the fraction of space occupied by the spheres.

The *coordination number* (CN) of an atom in a crystal is defined as the number of nearest neighbors that atom possesses. All "nearest neighbors" must be equidistant from the atom in question, which shall be referred to as the *central* atom. In the case of ionic crystals, electrical stability requires that the central and neighboring atoms be oppositely charged. In a metal, all atoms of the crystal are positively charged and are held together by an electron cloud which pervades the entire crystal.

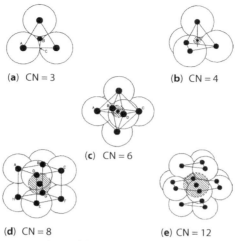

(a) CN = 3 **(b)** CN = 4 **(c)** CN = 6 **(d)** CN = 8 **(e)** CN = 12

Figure 1.2 Coordination numbers of (a) 3, (b) 4, (c) 6, (d) 8, and (e) 12.

There are five coordination numbers that occur in nature; as noted in Table 1.1, these are 3,4,6,8 and 12. The crystal pattern that a given pair of ions form depends mainly on the relative sizes of the atoms or, equivalently, on the radius ratio (r/R). In this ratio, r represents the radius of the central atom and R the radius of the neighboring atoms. The central atom is always chosen as the smaller of the two ions.

Figure 1-1 shows arrangements for a CN of six. Note that the solid circles represent atoms whose centers are in the plane of the page; the dotted circle represents two atoms whose centers are above and below the plane of the page. In 1(a), the central atom is in contact with all six neighbors simultaneously, a fact which is critical for ionic bonding. As the (r/R) ratio is decreased, the spacing between the neighboring atoms becomes smaller until the situation depicted in 1(b) is achieved. In this diagram, the (r/R) ratio is 0.414, which is the minimum ratio that is capable of supporting a CN of six. For a ratio below this minimum, it is impossible to have the central atom contacting all six neighbors at the same time. In Table 1.1, the minimum radius ratios for the five coordination numbers are presented and in Figure 1.2, the atomic arrangements for the five CNs are depicted. In each of the five diagrams of Figure 1.2, the (r/R) ratio is at the minimum for that coordination number.

The minimum radius ratios in Table 1.1 can be calculated using simple geometric and trigonometric principles. Taking the coordination number of six as an example (see Figure 1-2(c)), the three-dimensional figure obtained by joining the centers of the neighboring atoms is an octahedron.

The two-dimensional figure formed by connecting atom centers A, B, and D is an isosceles right triangle. Since each leg of the triangle is 2R and the hypothenuse is 2R+2r, the application of the Pythagorean theorem yields

$$(2R+2r)^2 = (2R)^2 + (2R)^2 \qquad (1.1)$$

or

$$2R + 2r = \sqrt{8}R \qquad (1.2)$$

which can be rearranged to give

$$\left(\frac{r}{R}\right) = \left(\frac{r}{R}\right) min = 0.414 \qquad (1.3)$$

The significance of the information contained in Table 1.1 lies in the fact that it can be used to help explain why ionic and metallic materials form the types of crystals that they do. The radius ratio of sodium chloride, for example, has been determined by x-ray diffraction to be 0.54. Since this ratio is less than 0.732, sodium chloride cannot crystallize in a pattern that requires a CN of 8; the sodium ion is simply too small to fit eight chloride ions around its periphery. This leaves coordination numbers of 6, 4 and 3 as possibilities, with 6 as the most likely prospect, since of the three, it would result in the highest packing factor.

1.2.1 Geometry of Metallic Unit Cells

The term *Bravais lattice* refers to one of 14 different patterns employed in the structure of crystals. Three of these (the only three in which the points are arranged to form a cubic pattern) are shown in Figure 1.3. These three Bravais lattices are called simple cubic (sc), body centered cubic (bcc), and face centered cubic (fcc). Although the points of the Bravais lattices are depicted as spheres in these diagrams, the points do not (at least for now) represent atoms. The *Bravais lattice* is an abstraction consisting only of a six- or eight-sided box with points placed on either inside or on the box surface. Figure 1.4 would show the hexagonal Bravais lattice if the three points that fall completely inside the eight-sided box were absent.

In order to transform the abstraction of the Bravais lattice into a real crystal structure, each point of the lattice is allowed to represent a single

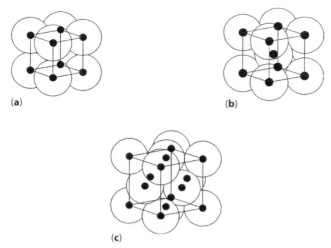

Figure 1.3 Bravais lattices: (a) simple cubic, (b) body-centered cubic, (c) face-centered cubic.

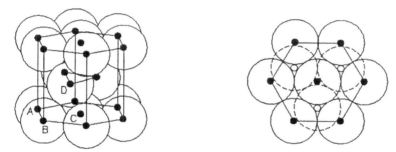

Figure 1.4 The hexagonal close-packed (hcp) unit cell.

atom or a set of atoms (like a pair of ions or a molecule). If each point of the fcc Bravais lattice shown in Figure 1.3(c) is replaced by a methane molecule, for example, the unit cell of the methane crystal (which, since methane is a gas at ambient conditions, is found only at very low temperatures) results. A unit cell can be considered the basic "building block" for the construction of the crystal, and as such must be representative of the entire crystalline material. By repeating the methane unit cell over and over again, the solid methane crystal would result.

1.2.2   Geometry of Ionic Unit Cells

In the previous subsection, it was seen that, for bcc and fcc metallic crystals, the transformation from the Bravais lattice to the unit cell is a matter of replacing a point by a single spherical atom. For the hcp metallic crystal, each Bravais lattice point is replaced by two spherical atoms. In ionic materials, a *single* ion can never replace a Bravais lattice point; all points of the Bravais lattice must be identical and must have identical neighbors. The sodium chloride structure shown in Figure 1.5, for example, could have been arrived at by replacing a point of the sc Bravais lattice by a sodium ion and an adjacent point by a chloride ion, etc. The sodium chloride structure, however, is *not* simple cubic; it is face-centered cubic. To demonstrate this, if each point of the fcc Bravais lattice pictured in Figure 1.3(c) is replaced by an ion pair, for example, one chloride and a sodium ion placed to its immediate right, the structure shown in Figure 1.5 results.

Although simple cubic *metals* do not exist in nature, simple cubic *salts* (ionic crystals) do. Cesium chloride is an example. The CsCl structure can be demonstrated by replacing each point of the sc lattice shown in Figure 1.3(a) by a cesium ion-chloride ion pair to achieve the structure shown in Figure 1.6. In this case, the chloride ions were placed at each vertex, and only one cesium ion, the one in the center of the cube, is shown. The centers of the other seven cesium ions fall outside of the cube, and are members of adjacent unit cells. In this structure, the spherical ions are in contact along the body diagonal only. Only ions of opposite charge should be touching.

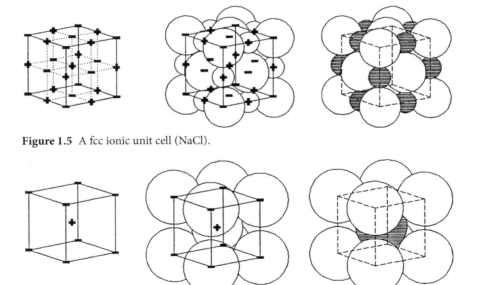

Figure 1.5 A fcc ionic unit cell (NaCl).

Figure 1.6 A bcc ionic unit cell (CsCl).

1.2.3   Packing Factors

As explained earlier, the *packing factor* is defined as the fraction of space in the crystal occupied by atoms, where once again, it is assumed the atom to be an incompressible sphere. The packing factor is conveniently calculated by first determining the number of atoms in the unit cell of the crystal, calculating the spherical volumes of those atoms and dividing by the total volume of the unit cell. For example, a simple cubic structure would have a packing factor of 0.52. This number is obtained by first finding the volume of *one* sphere of radius, r. (Figure 1.2(a) shows that there is one single atom per unit cell.) This volume is then divided by the unit cell volume which is determined by cubing the edge length, 2r, as is shown by the following equation.

$$\frac{\left(\frac{4}{3}\right)\pi r^3}{(2r)^3} = 0.52 \tag{1.4}$$

For metals, the calculation of the packing factor depends only on the metal structure and not on the atom size. This is not the case for ionic solids when at least two different atom sizes must be taken into account.

Properties of materials fall into one of two categories: *extensive* and *intensive*. An *extensive* property is one that depends on the size of the material; volume and mass are examples of such properties. An *intensive* property is independent of size and the measurement of that property always yields the same value whether a small or large amount of the material is being used to make the measurement. *Density* is one such property.

The *unit cell* was described earlier as a representative portion of the crystal lattice. If the cell is truly representative, it must possess the same qualities as the crystal that it represents. It is therefore theoretically possible to use a single unit cell as a basis for calculating intensive properties. (Note: the term *basis* is used to indicate that amount of materials on which a calculation is being performed. In many problems, amounts are not specified and as long as the problem involves intensive properties, the choice of a basis is arbitrary and left to the individual performing the calculation.)

1.2.4   Directions and Planes

Many crystal phenomena are directional. For example, when a metal crystal is plastically deformed (i.e., permanently distorted), parts of the crystal move relative to other parts. This slippage occurs in certain predictable directions and along certain predictable atom planes. Many crystal properties are also directional, e.g., the elastic modulus, ductility and conductivity. It is important, therefore, that directions and planes of a crystal be identifiable.

The right-handed three-dimensional coordinate system that will be used in this discussion is given on the left side of Figure 1.7. In the two cubes shown in Figure 1.7, five directions (indicated by vectors) are also shown. Directions are represented symbolically by [hkl], where h, k, and l are three small whole numbers called *indices* that correspond to the cartesian coordinates x, y, and z. Note that there are no commas between the numbers. The indices for a given direction are determined as follows. The base of the vector is chosen as the origin of the coordinate system. The coordinates of the tip of the arrowhead are then determined. Using direction (a) as an example, the coordinates of the arrowhead are (0,0,1). (Note: the physical coordinates of the arrowhead would have length dimensions and be given in terms of a, the cell constant. In this example, those coordinates are (0,0,a). For purposes of determining the indices, however, the cell constant is considered to be one mathematical unit and the coordinates are given as pure numbers.) The coordinates of the arrowhead tip yield the indices. The directions (a), (b), and (c) are represented by [001],[111] and [110], respectively. The tip of the arrowhead for direction (d) has the coordinates (1,0,-1) and is represented by [10$\bar{1}$]. Note that a negative direction is indicated by a bar over the index. For (e), the arrowhead has the coordinates (1,0,1/2), which, following the rules for finding the indices given so far, would yield [1 0 1/2]. However, the indices should be small whole numbers. Since the direction indicated by vector (e) also passes through the points (2,0,1) and (4,0,2), that direction could also be feasibly represented by [201] and [402]. In other words, multiplying the coordinates and indices by a positive number does not alter the direction. The convention used here is to multiply the indices by the smallest positive integer that will convert all three to small whole numbers. Direction (e) should be therefore represented by [201].

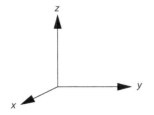 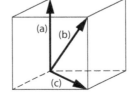 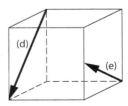

Figure 1.7 Crystal directions.

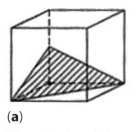 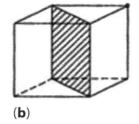 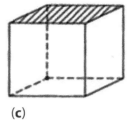

(a) **(b)** **(c)**

Figure 1.8 Crystal planes.

The atom planes of a crystal are represented by the symbol (hkl), where once again h, k, and l are small whole numbers. (It is important that the reader distinguish between the symbols used for coordinates and planes. Both employ parentheses, but the coordinates of a point are separated by commas and need not be small whole numbers. No commas appear in the symbol for a plane.) It takes three points to fix the position of a plane in space (see Figure 1.8), unless the points are along the same straight line. The indices of a plane are sometimes referred to as *Miller indices*, and will not be described here. The reader is referred to the literature for additional details.

1.3   Crystallography of Real Crystals (CRC)

Besides structural defects, real crystals also contain impurities (as long as the temperature is above absolute zero). The word *impurity* implies a mixture of at least two substances. In crystals, the mixture; can be one of two types: a single-phase or homogeneous mixture, and a multiphase or heterogeneous mixture. A single-phase mixture is a *true* solution in which the solute is dispersed on an atomic level. This means that the atoms, ions, or molecules of the solute are surrounded mainly by particles of the solvent. A multiphase mixture is not *mixed* on an atomic level and is therefore not the same throughout its volume; it has different parts with different detectable properties. A multiphase metal alloy such as steel at room temperature may appear to be of uniform composition to the naked eye, but through a microscope (and with proper sample preparation), two different phase or parts can be detected. It should be pointed out that in a multiphase mixture, if the phases or parts could be conveniently separated into single phase, these single phases would still be mixtures. A 50-50 w/o (weight percent) alloy of silver and aluminum, for example, forms two solid phases, which are assigned the symbols, α and β. (Note: A 93 w/o silver- 7 w/o aluminum alloy is called *sterling silver*). Both phases are easily seen through a microscope. The α phase is composed of fcc silver crystals with aluminum atoms occupying about 5-10% of the lattice sites; the β phase is made up of fcc aluminum crystals with silver atoms occupying about 1-3% of the lattice sites. This type of crystal impurity where the impurity atom replaces one of the solvent atoms (the aluminum impurities in the α phase, for example) is called a *substitution* impurity, and the mixture is often referred to as a *substitutional alloy*. There is another type of impurity, the *interstitial* impurity, where the impurity atoms does not occupy a lattice site, but instead is wedged in among the solvent atoms. This type of impurity is the subject of the next subsection.

1.3.1   Interstitial Impurities

The second type of impurity that occurs in crystals is the *interstitial* impurity. When carbon dissolves in iron, it forms such an impurity---the carbon atoms do not replace any iron atoms but wedge into the open spaces among the iron atoms. Iron is a *polymorph*. A *polymorph* is a material that can form more than one crystal structure. At room temperature, and up to 912 °C (1673 °F), the stable crystal structure of iron is bcc (an iron phase given the symbol α). From 912°C to around 1400°C (2550°F), the stable structure is fcc (assigned the symbol γ), and from 1400°C to the melting point at 1540 °C (2800°F), bcc given the symbol, δ.

Carbon has a much higher solubility in fcc iron (about 2.1 w/o at 1150 °C) than in bcc iron (about 0.02 w/o maximum in the α phase and slightly higher in the δ phase). This fact is surprising, at first glance, because the bcc structure is not as well packed as the fcc structure. Metals with a bbc structure have a packing factor of 0.68 and a coordination number of 8; fcc metals have a packing factor 0.72 and a coordination number of 12. This indicates that there is more empty space in the bcc form of iron (which is true), and therefore more carbon atoms should fit into that space (which is false). This apparent discrepancy can be explained by the fact that the greater amount of empty space in the bcc crystal is distributed over openings that are very non-spherical in shape. These non-spherical spaces can comfortably accommodate spherical atoms only if the atoms are extremely small, about one-eighth the size of the carbon atom.

The interstitial holes in fcc iron and bcc iron is pictorially represented in Figure 1.9.

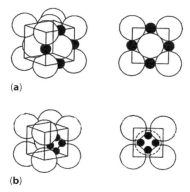

(a)

(b)

Figure 1.9 Interstitial spherical holes in (a) fcc iron, (b) bcc iron.

1.4 Materials of Construction

The selection of the materials with which to build a chemical or petro-chemical process plant is a very important part of process design. It is desirable to select the least expensive material that can be used for any particular piece of equipment in the plant. (Note that "equipment" here refers to piping, values and fittings as well as to pumps, columns, heat exchangers, etc.) In order to be acceptable, however, a material must have adequate mechanical strength and adequate resistance to chemical attack (corrosion) under plant operating conditions.

Mechanical strength requirements depend on the process conditions (pressure and temperature), the size of the equipment, and the equipment's own support requirements. In general, the larger the piece of equipment and the higher the pressure, the greater strength is needed. This can be provided either by using thicker materials or by selecting materials with inherently higher strength. For most materials, strength decreases with increasing temperature; so high temperature processes require thicker or different materials.

Corrosion resistance depends on the chemicals being processed and the conditions under which they are handled. Very often, special alloy metals or other exotic materials are required to avoid rapid corrosion of the equipment. Designers usually try to select materials that will have predicted lifetimes of 20 to 50 years or more actual under process conditions. Sometimes this is impossible, and the equipment must be frequently replaced.

Carbon steel and common plastics, such as polyvinyl chloride and polyethylene, are the least expensive materials used in process plants. The selection process starts with them, then considers successively more expensive materials until one is found that has the required mechanical and corrosion-resistance properties.

1.5   Resistivity

The electrical behavior of certain solids, particularly semiconductors, can be explained (and, to some extent, quantified) using atomic-level phenomena. This was not done to any appreciable degree earlier where the connection between the atomic and macroscopic levels was not as well established, at least in a quantitative sense.

Ohm's law states that, when a potential or voltage difference is placed across a conductor, the resulting current is proportional to the voltage difference. More succinctly stated

$$E = IR \qquad (1.5)$$

where E = voltage difference, volt
I = current, amp
R = resistance, ohm

The resistance term R is a proportionality constant that depends on the nature of the conductor, including its size and shape. Although resistance is a strong function of the material that the conductor is made of, it cannot be called a property of that material. For a quantity to be properly called a property, it must be a function only of the material of which it is made.

If the conductor is a wire of length, l, and cross-sectional area, A, it can be shown by experimentation that the wire's resistance depends almost entirely on four variables: the material, the temperature, the cross-sectional area, and the length. For convenience, the temperature may be lumped with the material to reduce the number of independent variables. In other words, copper at 1000 °F is considered a different "material" than copper at 70°F, which it is. This results in the functional relationship

$$R = f(material, T, A) \qquad (1.6)$$

Laboratory experimentation also shows that resistance is directly proportional to wire length and inversely proportional to cross-sectional area. Removing the length and area dependence from R should leave a variable that depends only on the material (at a given temperature), or, in other words, should leave a variable that is, by definition, a property. This is shown in the equation.

$$R = \rho\left(\frac{1}{A}\right) \tag{1.7}$$

Where ρ is the resistivity of the material. The resistivities for many materials at specified temperatures can be found in the literature. Typical units for are ohm·in, etc.

1.6 Semiconductors

Semiconductors fall into one of two categories: *intrinsic* and *extrinsic*. A semiconductor that conducts in the pure state is an *intrinsic semiconductor*. One that has had to have material added to it to get it to conduct is an *extrinsic semiconductor*. Extrinsic semiconductors are commonly used in most electronic applications. The *matrix* or *solvent* material is usually a single crystal of silicon or germanium. Both these elements belong to the carbon family and, like carbon (diamond), form crystals in which there are three-dimensional networks of covalent bonds. In diamond, silicon, and germanium crystals, each atom is covalently bonded to four neighboring atoms arranged in tetrahedral fashion. The angle between any two of the four bonds is the tetrahedral angle of 109.5°.

There are various classes of materials, metals, and alloys. Perry and Green [1] provided the following categorization:

1. Ferrous metals
2. Ferrous alloys
3. Organic non-metallics
4. Thermoplastics

These are also divided into low-temperature metals and high-temperature materials.

A detailed and expanded treatment of materials science and engineering is available in the following three references:

1. W.F. Smith, *Formulations of Materials Science and Engineering*, McGraw-Hill, New York City, NY, 2004. (ref(2))
2. L. VanVlack, *Materials Science for Engineers*, Addison Wesley, Reading, MA, 1970. (ref(3))

3. W. Callister, *Materials Science and Engineering*, 3rd edition, John Wiley & Sons, Hoboken, NJ, 1985. (ref(4))

1.7   Illustrative Open-Ended Problems

This and the last section provide open-ended problems. Solutions *are* provided for the three problems in this Section in order for the reader to obtain a better understanding of these problems which differ from the traditional problems/illustrative examples. The first problem is relatively straightforward while the third (and last problem) is somewhat more difficult and/or more complex. Note that solutions are not provided for the 35 open-ended problems provided in the next Section.

Problem 1: Discuss the periodic table in layman terms.

Solution: At the time of publication, a total of 118 elements were known. These elements vary widely in location, concentration, and abundance on planet Earth. For example, over 75 percent of the Earth's crust consists of oxygen and silicon. Interestingly, approximately 65 percent by mass of the human body is oxygen. As the number and information on elements increased, chemists attempted to find similarities in elemental as well as chemical behavior. These efforts ultimately resulted in the development of the *Periodic Table*. This has gone through significant changes over the years, with the latest "form" arranging elements in order of increasing atomic number and with elements having similar properties placed in vertical columns known as *groups*.

Problem 2: Discuss structural defects in crystals.
(Comment: There are many types of structural defects in crystals, but only a few will be cited. For a more thorough discussion of these defects, the reader is referred to the literature.)

Solution: There is no such thing as a *perfect* crystal. The reason why imperfections exist in crystals is explained below. Just as the organization of atoms in a solid has a critical role in determining material properties, so to does the occasional *breakdown* of this organization have an important effect. No study of crystallography would be complete without a companion study of crystal imperfections.

Unless a system is at the temperature of absolute zero atomic motion must be occurring inside the crystal. *Temperature* is described as a measure of the *average* kinetic energy of the particles in a system. The use of the word *average* implies that these particle kinetic energies are *distributed*—most particles have energies somewhere around the average, but a few have either very low or excessively high energies. (This distribution of particle energies is called the *Maxwell-Boltzmann distribution*.) Below the melting point, most atoms, ions or molecules of a solid exhibit their kinetic energies by vibrating about fixed positions. In a crystal, the fixed positions are the lattice sites. A relatively small number of particles, however, have kinetic energies that are greater than the bonding energies that hold the other particles close to their assigned sites. These particles are capable of moving away from these sites and causing structural defects in crystals.

Structural defects fall into three categories: point, line and area defects. Two of the more common point defects are the *vacancy* and the *interstitialcy*. A vacancy is defined as a lattice site or position that is unoccupied. If the central atom in a unit cell became one of the *high energy* atoms mentioned earlier, it would move out of the center position, squeeze its way past the vertex atoms (the atoms are not really *hard* spheres and are capable of compression), and end up somewhere else in the lattice. The vertex atoms would tend to collapse in toward the center, but a gaping hole, the *vacancy*, would remain. The center of the cell is now referred to as a vacant site. The movement of the surrounding atoms out of their positions is generally slight and is referred to as *lattice strain*. If the high energy atom discussed above ended up wedged somewhere in the lattice where it does not belong, it becomes an *interstitial defect*. The *interstitial defect* is defined as an atom or ion occupying a non-lattice position. Obviously, the presence of the interstitial defect also produces lattice strain by displacing the neighboring atoms from their positions.

Problem 3: Describe Bragg's Law in technical terms.

Solution: When a pencil-thin beam of monochromatic x-rays (i.e., x-rays of fixed wavelength) is directed at a single crystal, each atom in the crystal, acting individually, scatters a tiny fraction of the radiation in all directions. The amount scattered by one atom is far too small to be measured; the combined scattering by all atoms of the crystal, however, while still only a small fraction of the incident radiation, can be detected fairly easily. The combined scattered, or *diffracted*, radiation does not occur in all directions as does that scattered by the individual atom. When a number of x-ray beams are added together, the combined intensity will be the sum of the individual intensities only if all the individual beams are completely in

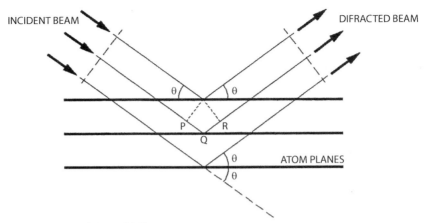

Figure 1.10 Bragg's law of diffraction.

phase. If the individual beams are out of phase with each other, the result will be mutual destructive interference and the total intensity will be from the whole crystal is intense enough to be detected only when all the rays scattered from the individual atoms are completely in phase. This can happen only when a set of atomic planes in the crystal is aligned at a definite angle to the incident x-ray beam.

The above situation is illustrated in Figure 1.10. Note that the angle, θ, must be such that the *difference in path length* traveled between adjacent segments of the x-ray beam (PQR) is exactly one full wavelength (or an integral number of wavelengths). This insures that the diffracted segments are in phase. If this path length difference is any other than an integral number of wavelength, the result is destructive interference, and in effect, mutual annihilation of the diffracted beam.

The relationship existing among the wavelength, λ, the angle of incidence, θ_{hkl}, and the interplanar distance, d_{hkl}, is given by *Bragg's Law of Diffraction*:

$$n\lambda = 2d_{hkl} \sin\theta_{hkl} \tag{1.8}$$

where h, k, and l represent the aforementioned Miller indices of the diffracting plane and n is the number (integer) of wavelengths that make up the path length difference (PQR in Figure 1.10). The value of n is referred to as the *order of diffraction* and may usually be assumed to be unity. If this relationship is not satisfied, diffraction cannot occur. This means that

SPECIMEN

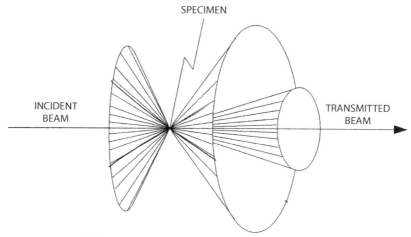

INCIDENT
BEAM

TRANSMITTED
BEAM

Figure 1.11 X-ray diffraction of finely divided powder.

directing a monochromatic x-ray beam at a single crystal will, in all probability, not result in diffraction unless care is taken to align a set of planes in the crystal at the proper angle of incidence. Taking the (200) planes as an example, the proper angle of incidence would be, according to Bragg's law:

$$\theta_{hkl} = \sin^{-1}(\lambda/2d_{200}) \tag{1.9}$$

If, instead of a single crystal, however, a finely divided powder of the crystalline sample is introduced into the x-ray beam, diffraction is guaranteed. The reason for this is that, in the finely divided powder, all possible crystal orientations are represented and, hence, at least a few of the crystalline grains are bound to have their (hkl) planes aligned at the Bragg angle, θ_{hkl}, to the incident beam. Each beam diffracted from one of the properly aligned crystallites contributes to a cone of radiation which is concentric with the incident beam; the semi-apex angle of this cone is twice the Bragg angle, or $2\theta_{hkl}$. Note that, since the angle formed by the diffracting plane and the incident beam is θ, and the angle formed by the diffracting plane and the diffracted beam is also θ, the angle formed by the transmitted beam and the diffracted beam is 2θ (see Figure 1.11).

The diffraction or Bragg angle, θ, cannot be *directly* measured because there is no instrumentation available for detecting the orientations of atom planes. (These plane orientations can be *calculated*, but not *measured*.) The angle, 2θ, however, can be measured because it is the angle between two x-ray beams (the transmitted beam and the diffracted beam) and the positions of x-ray beams *can* be detected. In fact, the only measurements that need to be taken in powder x-ray diffraction are those of the angle 2θ. All information such as d-spacings, cell constants, and atomic/ionic radii comes from those measurements.

1.8   Open-Ended Problems

This last Section of the chapter contains open-ended problems as they relate to Materials Science and Engineering. No detailed and/or specific solution is provided; that task is left to the reader, noting that each problem has either a unique solution or a number of solutions or (in some cases) no solution at all. These are characteristics of open-ended problems described earlier.

There are comments associated with some, but not all, of the problems. The comments are included to assist the reader while attempting to solve the problems. However, it is recommended that the solution to each problem should initially be attempted *without* the assistance of the comments.

There are 35 open-ended problems in this Section. As stated above, if difficulty is encountered in solving any particular problem, the reader should next refer to the comment, if any is provided with the problem. The reader should also note that the more difficult problems are generally located at or near the end of the Section.

1. Describe the early history associated with material science and engineering.
2. Discuss the recent advances in materials science and engineering.
3. Select a refereed, published article on materials science and engineering from the literature and provide a review.
4. Provide some normal everyday domestic application involving the general topic of materials science and engineering.
5. Develop an original problem in materials science and engineering that would be suitable as an illustrative example in a book.
6. Prepare a list of the various books that have been written on materials science and engineering. Select the three

best and justify your answer. Also select the three weakest books and justify your answer.

7. Define and discuss materials science and engineering in layman terms.
8. Why is the general subject of materials science and engineering important to the chemical engineer?
9. Attempt to improve on the present state of chemical nomenclature.
10. Define atomic number and mass number and explain their difference(s).   Also discuss the basis/reasoning of representing the atomic mass unit as a Dalton.
11. Define an isotope in layman terms.
12. In your own words, provide a description of nuclear chemistry.
13. Discuss the impact nanotechnology has on the teaching of materials science and engineering.
14. Explain in technical detail the wave nature of light.
15. Explain the electronegative spectrum in both technical and layman terms.
16. Describe the history of the periodic table. (Comment: See Problem 1 in the previous Section.)
17. Attempt to improve the present format of the periodic table. (Comment: See Problem 1 in the previous Section.)
18. Discuss the role the Boltzmann constant plays in materials science and engineering.
19. Describe Bragg's Law in layman terms. (Comment: See problem 3 in the previous Section)
20. Provide a brief description of ferrous metals and alloys.
21. List some of the various organic non-metals.
22. Describe each of the following materials:

    a. High silicon cast irons
    b. Stainless steel
    c. Hastalloy

23. Describe the various alloys of aluminum.
24. What is the difference between brass and bronze?
25. List and describe the various thermoplastics.
26. Provide examples of low-temperature materials and high-temperature materials.
27. List some of the various inorganic nonmetallics

28. Provide your own thoughts on the future of ceramic materials.
29. Describe the various corrosion-testing methods.
30. Provide some key properties of the following materials:

    - Wood
    - Natural rubber
    - Carbon and graphite
    - Asphalt

31. Obtain the thermal conductivities of a dozen common insulating materials. (Comment: This will require reviewing the materials literature.)
32. Explain why most pierced ears are allowed to heal with a gold post in the opening.
33. Describe the relationship between the modulus of elasticity of an amorphous polymer and the glass transition temperature. Also provide the explanation in graphical form.
34. You have been hired as a consultant to provide a recommendation to replace gold-plated pieces in computer chips with either copper or steel. Pieces prepare an abstract of a report you would submit in this recommendation.
35. A company's research laboratory determines that the density of platinum is 21.51 g/cc. However, the value listed in the literature is 21.45 g/cc. Assuming that each value is accurate to 4 significant digits, that the discrepancy is due to atom vacancies alone, and that the vacancies have no effect on the crystal volume, determine what fraction of the unit cells with measured density contains vacancies, or, equivalently, the number of vacancies per unit cell. (Comment: One should first note that the number of vacancies per unit cell is equivalent to the fraction of unit cells. To demonstrate this fact, first assume that a unit cell cannot contain more than one vacancy. This is a valid assumption; a miniscule number of cells may, in fact, have more than once vacancy, but because of the extremely small number of vacancies (one out of 369 sites, in this case), one can expect the vacancies to be widely distributed throughout the crystal, and the chances of two vacancies occupying the same cell to be almost non-existent. If 10,000 unit cells contain

11 vacancies (as in this problem), one may conclude that 11 of those cells contain on the average, one vacancy each. This is the same as saying that 11/10,000 or 0.011 of the unit cells contain vacancies.

References

1. D. Green and R. Perry (editors), *Perry's Chemical Engineers' Handbook*, 8th edition, McGraw-Hill, New York City, NY, 2008.
2. W.F. Smith, *Foundations of Materials Science and Engineering*, McGraw-Hill, New York City, NY, 2004.
3. L. VanVlack, *Materials Science for Engineers*, Addison Wesley, Reading, MA, 1970.
4. W. Callister, *Materials Science and Engineering*, 3rd edition, John Wiley & Sons, Hoboken, NJ, 1985.

2

Applied Mathematics

This chapter is concerned with applied mathematics. As with all the chapters in Part II, there are several sections: overview several specific technical topics, illustrative open-ended problems, and, open-ended problems. The purpose of the first section is to introduce the reader to the subject of applied mathematics. As one might suppose, a comprehensive treatment is not provided, although several sections addressing additional specific technical topics are included. The section contains three open-ended problems; the authors' solutions (there may be other solutions) are also provided. The last section contains 45 problems; *no* solutions are provided here.

2.1 Overview

This overview section is concerned—as can be noted from its title—with applied mathematics. As one might suppose, it was not possible to address all topics directly or indirectly related to applied mathematics. However, additional details may be obtained from either the references provided at the end of this Overview section and/or at the end of the chapter.

Note: Those readers already familiar with the details associated with this subject may choose to bypass this overview.

The chemical engineer learns early in one's career how to use equations and mathematical methods to obtain exact answers to a large range of relatively simple problems. However, these techniques are often not adequate for solving real-world problems, although the reader should note that one rarely needs exact answers in technical practice. Most real-world engineering and, to a lesser degree, science applications are inexact because they have been generated from data or parameters that are measured, and thus represent only approximations. What one is likely to require and/or desire in a realistic situation is either an approximate answer or one having reasonable accuracy from an engineering point of view.

As noted above, the solution to a chemical engineering (or scientific) problem usually requires an answer to an equation or equations, and the answer(s) may be approximate or exact. Obviously an exact answer is preferred, but because of the complexity of some equations, exact solutions may not be attainable. Furthermore, an answer that is precise may not be necessary; for this condition, one may resort to another approach – a solution that has come to be defined as a *numerical method*. Unlike the exact solution, which is continuous and in closed form, numerical methods provide an inexact (but often reasonably accurate) solution.

Today's computers have had a tremendous impact on the chemical engineering profession, including engineering design, computation, and data processing. The ability of computers to handle large quantities of data and to perform mathematical operations described above at tremendous speeds permits the analysis of many more applications and more engineering variables than could possibly be handled on the slide rule – the trademark of chemical engineers (including one of the authors) of yesteryear. Scientific calculations previously estimated in lifetimes of computation time are currently generated in seconds and, in many occasions, microseconds, and in some rare instances, nanoseconds [1].

Although the chapter is titled "Applied Mathematics" the material presented is primarily concerned with numerical methods. This subject was taught in the past as a means of providing chemical engineers with ways to solve complicated mathematical expressions that they could not otherwise solve. A brief overview of the numerical methods below is given to provide the chemical engineer (as well as other engineers) with some insight into what many of the currently used software packages are actually doing. The authors have not attempted to cover all the topics of numerical methods.

Topics that traditionally fall in the domain of this subject and receive brief treatment include:

1. Differentiation and integration;
2. Simultaneous linear algebraic equations;
3. Nonlinear algebraic equations;
4. Ordinary and partial differential equations; and
5. Optimization.

Since a detailed treatment of each of the above topics is beyond the scope of this presentation, the reader is referred to the literature [2–4] for a more extensive analysis and additional information. The remainder of this section examines the five topics listed above.

2.2 Differentiation and Integration

Several differentiation methods are available to generate expressions for a derivative. One of the authors has provided information in an earlier work. Some useful *analytical* derivatives in engineering calculations are also provided [4].

Numerous chemical engineering and science problems require the solution of integral equations. In a general sense, the problem is to evaluate the function on the right hand side (RHS) of Equation 2.1:

$$I = \int_a^b f(x)\,dx \tag{2.1}$$

where I is the value of the integral. There are two key methods employed in their solution: analytical and numerical. If $f(x)$ is a simple function, it may be integrated analytically. For example if $f(x) = x^2$

$$I = \int_a^b x^2\,dx = \frac{1}{3}(b^3 - a^3) \tag{2.2}$$

If, however, $f(x)$ is a function too complex to integrate analytically; (e.g., $\log[\tanh(e^{x^3-2})]$, one may resort to any of the numerical methods available. Two simple numerical integrations methods that are commonly employed in engineering practice are the *trapezoidal rule* and *Simpson's rule* [4].

2.3   Simultaneous Linear Algebraic Equations

The chemical engineer often encounters problems that not only contain more than two or three simultaneous algebraic equations but also those that can be nonlinear as well. There is, therefore, an obvious need for systematic methods of solving simultaneous linear and simultaneous nonlinear equations [2,5]. This section will address the linear sets of equations; information on nonlinear sets is available in the literature [6].

Consider the following set of n equations:

$$a_{11}x_1 + a_{12}x_2 + \ldots + a_{1n}x_n = y_1$$
$$a_{21}x_1 + a_{22}x_2 + \ldots + a_{2n}x_n = y_2$$

$$\ldots$$

$$\ldots$$

$$a_{n1}x_1 + a_{n2}x_2 + \ldots + a_{nn}x_n = y_n \tag{2.3}$$

where a is the coefficient of the variable x and y is a constant. The above set is considered to be linear as long as none of the x-terms are nonlinear, e.g., x_2^2 or $\ln x_1$. Thus, a linear system requires that all terms in x be linear. The system of linear algebraic equations may be set in *matrix* form:

$$
\begin{bmatrix}
a_{11} & a_{12} & \cdots & a_{1n} \\
a_{12} & a_{22} & \cdots & a_{2n} \\
\cdots & \cdots & \cdots & \cdots \\
a_{n1} & a_{n2} & \cdots & a_{nn}
\end{bmatrix}
\begin{bmatrix}
x_1 \\
x_2 \\
\cdots \\
x_n
\end{bmatrix}
=
\begin{bmatrix}
y_1 \\
y_2 \\
\cdots \\
y_n
\end{bmatrix}
\tag{2.4}
$$

Methods of solution available for solving these linear sets of equations include:

1. Gauss-Jordan reduction;
2. Gauss elimination;
3. Gauss-Seidel;
4. Cramer's rule; and
5. Cholesky's methods.

Ketter and Prawler [3] provide several excellent illustrative examples.

2.4 Nonlinear Algebraic Equations

The subject of the solution to a nonlinear algebraic equation is considered in this section. Although several algorithms are available in the literature, the presentation will key on the Newton-Raphson (NR) method of evaluating the root(s) of a nonlinear algebraic equation.

The solution to the equation

$$f(x) = 0 \qquad (2.5)$$

is obtained by guessing a value for x, i.e., x_{old}, that will satisfy the above equation. This value is continuously updated to x_{new} using the equation (the prime represents a derivative)

$$x_{new} = x_{old} - \frac{f(x_{old})}{f'(x_{old})}$$

$$\frac{x_{new} - x_{old}}{0 - f(x_{old})} = \frac{1}{f(x_{old})} \qquad (2.6)$$

$$x_{new} = x_{old} - \frac{f(x_{old})}{f'(x_{old})}$$

until either little or no change in $(x_{new} - x_{old})$ or $(x_{new} - x_{old})/x_{old}$ is obtained. One can also express this operation graphically (see Figure 2.1). Noting that

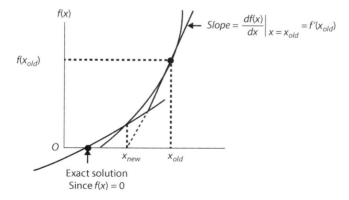

Figure 2.1 Newton-Raphson method for nonlinear equations.

$$\frac{df(x)}{dx} \approx \frac{\Delta f(x)}{\Delta x} = \frac{f(x_{old})-0}{x_{old}-x_{new}} = f'(x_{old}) \tag{2.7}$$

one may rearrange Equation 2.7 to yield Equation 2.8 below. The x_{new} then becomes the x_{old} in the next calculation.

$$x_{new} = x_{old} - \frac{f(x_{old})}{f'(x_{old})} = f(x)/dx \tag{2.8}$$

This method is also referred to as Newton's Method of Tangents (NMT), and is a widely used method for improving a first approximation of a root via the aforementioned equation of the form $f(x)=0$.

2.5   Ordinary and Partial Differential Equation

The Runge-Kutta (RK) method is one of the most widely used techniques in chemical engineering practice for solving first-order differential equations. For the equation

$$\frac{dy}{dx} = f(x,y) \tag{2.9}$$

the solution takes the form

$$y_{n+1} = y_n + \frac{h}{6}(D_1 + 2D_2 + 2D_3 + D_4) \tag{2.10}$$

where
$$D_1 = hf(x,y)$$
$$D_2 = hf\left(x_n + \frac{h}{2}, y_n + \frac{D_1}{2}\right)$$
$$D_3 = hf\left(x_n + \frac{h}{2}, y_n + \frac{D_2}{2}\right)$$
$$D_4 = hf\left(x_n + h, y_n + D_3\right) \tag{2.11}$$

The term h represents the increment in x. The term y_n is the solution to the equation x_n, and y_{n+1} is the solution to the equation at x_{n+1} where $x_{n+1} = x_n + h$. Thus, the RK method provides a straightforward means for developing expressions for Δy; i.e., $y_{n+1} - y_n$, in terms of the function $f(x,y)$ at various "locations" along the interval in question.

The RK method can also be applied if the function in question also contains the independent variable or more than one differential equation or to treat higher-order differential equations.

Many practical problems in chemical engineering applications involve at least two independent variables; i.e., the dependent variable is defined in terms of (or is a function of) more than one independent variable. The derivatives describing these independent variables are defined at *partial* derivatives. Differential equations containing partial derivatives are referred to as partial differential equations (PDEs).

The three main PDEs encountered in chemical engineering practice are briefly introduced below employing T (e.g., the temperature as the dependent variable), t (time) and x,y,z (position) as the independent variables. Note that any dependent variable; e.g., pressure, concentration, etc., could also have been selected in Equations 2.12 to 2.14 below.

The parabolic equation:

$$\frac{\partial T}{\partial t} = a \frac{\partial^2 T}{\partial z^2} \tag{2.12}$$

The elliptical equation:

$$\frac{\partial^2 T}{\partial x^2} + \frac{\partial^2 T}{\partial y^2} = 0 \tag{2.13}$$

The hyperbolic equation:

$$\frac{\partial^2 T}{\partial t^2} = a \frac{\partial^2 T}{\partial x^2} \tag{2.14}$$

The preferred numerical method of solutions of PDEs involve finite differences [3].

2.6   Optimization

Optimization has come to mean different things to different people. However one might offer the following generic definition for many chemical engineers: "Optimization is concerned with determining the 'best' solution to a given problem". Alternatively, a dictionary would offer something to the effect: "to make the most of… develop or realize to the utmost extent… often the most efficient or optimum use of…". This process or operation in chemical engineering practice is required in the solution of problems that involve the maximization or minimization of a given function.

A significant number of optimization problems challenge the practicing chemical engineer. The optimal design of process equipment as well as industrial processes has been an ongoing concern to the practicing chemical engineer, and for some, might define the function and goal of applied engineering. The practical attainment of an optimum design is generally a result of factors that include mathematical analysis, empirical information, and both the subjective and objective experiences of the chemical engineer.

In a general sense, optimization applications can be divided into four categories:

1. The number of independent variables involved;
2. Whether the optimization is "constrained";
3. Time-independent systems; and
4. Time-dependent systems.

In addition, if no unknown factors are present, the system is defined as *deterministic*, while a system containing experimental errors and/or other random factors is defined as *stochastic*.

Formal optimization techniques have as their goal the development of procedures for the attainment of an optimum in a system that can be characterized mathematically. The mathematical characterization may be:

1. partial or complete;
2. approximate or exact, and/or;
3. empirical or theoretical.

The resulting optimum may be a final implementable design or a guide to a practical design and a criterion by which practical designs are to be judged. In either case, the optimization techniques should serve as an important part of the total effort in the design of the units, structure, and control of not only equipment but also industrial process systems.

Optimization is qualitatively viewed by many as a tool in decision-making. It often aids in the selection of values that allow the chemical engineer to better solve a problem. In its most elementary and basic form, one may say—as noted above—that optimization is concerned with the determination of the "best" solution to a given problem. As noted, optimization is required in the solution of many general problems in chemical engineering and applied science—in the maximization (or minimization) of a given function(s); in the selection of a control variable to facilitate the realization of a desired condition; in the scheduling of a series of operations or events to control completion dates of a given project; and, in the development of optimal layouts of organizational units within a given design space, etc. [4]

The optimization problem has been described succinctly by Aris [7] as "getting the best you can out of a given situation." Problems amenable to solution by mathematical optimization techniques have these general characteristics:

1. One or more independent variables whose values must be chosen to yield a viable solution; and
2. Some measure of "success" is available to distinguish between the viable solutions generated by different choices of these variables.

Mathematical optimization techniques are used for guiding the problem solver to that choice of variables that maximizes the goodness measure (profit, for example) or that minimizes some badness measure (cost, for example).

One of the most important areas for the application of mathematical optimization techniques is in engineering design. Applications include [8]:

1. The generation of the "best" functional representations (e.g. curve fitting);
2. The design of optimal control systems;
3. Determining the optimal height (or length) of a mass transfer unit;
4. Determining the optimal diameter of a unit;
5. Finding the best equipment materials of construction;
6. Generating operating schedules; and
7. Selecting operating conditions.

A detailed and expanded treatment of applied mathematics is available in the following two references:

1. R. Ketter and S. Prawler, *Modern Methods of Engineering Computations*, McGraw-Hill, New York City, NY, 1969 [3].
2. L. Theodore, *Chemical Engineering: The Essential Reference*, McGraw-Hill, New York City, NY, 2014 [9].

2.7   Illustrative Open-Ended Problems

This section provides open-ended problems. However, solutions *are* provided for the three problems in this section in order for the reader to hopefully obtain a better understanding of these problems, which differ from the traditional problems/illustrative examples. The first problem is relatively straightforward, while the third (and last problem) is somewhat more difficult and/or complex. Note that solutions are not provided for the 45 open-ended problems in the next Section.

Problem 1: Provide a general description in layman terms of optimization methods.

Solution: The optimization problem has been described by some (see pervious section) as "getting the best you can out of a given situation." Problems amenable to solution by mathematical optimization techniques are those that generally have one or more independent variables whose values must be chosen to yield a viable solution; and possess some measure of *success* to distinguish between the many viable solutions generated by different choices of these variables. Mathematical optimization techniques are used for guiding the problem-solver to that choice of variables that *maximizes* the *approximation measure* (profit, for example) or that *minimizes* some *approximation measure* (cost, for example). In addition, one of the most important areas for the application of mathematical optimization techniques is in engineering design; and, these methods have wide applicability for large classes of problems involving the search for extreme functional values. Applications vary from the generation of "best" functional representations (curve fitting, for example) to the design of optimal operating conditions.

Problem 2: Define the Laplace Transform and provide several transforms of some elementary functions.

Solution: Assume $F(t)$ is a function of t specified for $t > 0$. The Laplace transform of $F(t)$ is usually denoted by $L[F(t)]$ and is defined by

$$L\left[F\left(t\right)\right] = f\left(s\right) = \int_0^{\infty} e^{-st} F\left(t\right) dt \qquad (2.15)$$

where the parameter t is usually considered to be real; however, it can also be complex. In addition, $L[F(t)]$ exists if it *converges* for some value of s; otherwise it does not exist.

The Laplace transforms of some simple functions are provided in Table 2.1. The reader may choose to refer to Chapter 9, Illustrative open-ended Problem 1.

Problem 3: Discuss the subject of regression analysis as it applies to *scatter diagrams*.

Solution [10]: It is no secret that many statistical calculations are now performed with the help of spreadsheets or packaged programs. This statement is particularly true for regression analysis. Often, the use of packaged programs reduces or eliminates one's fundamental understanding of regression analysis.

Chemical engineers encounter applications that require the need to develop a mathematical relationship between data for two or more variables. For example, if y (a dependent variable) is a function of or depends on x (an independent variable) i.e., $y = f(x)$, one may be required to express this (x, y) data in equation form. This process is referred to as *regression analysis*, and

Table 2.1 Laplace Transforms of Simple Functions

F(t)	L[F(t)]
1	$\dfrac{1}{s}$; $s > 0$
t	$\dfrac{1}{s^2}$; $s > 0$
e^{at}	$\dfrac{1}{s-a}$; $s > a$
sin *at*	$\dfrac{a}{s^2 + a^2}$; $s > 0$
cos *at*	$\dfrac{s}{s^2 + a^2}$; $s > 0$

the regression method most often employed is the method of *least squares*. An important step in this procedure—which is often omitted—is to prepare a plot of *y* vs. *x*. The result, referred to as a *scatter diagram*, could take on any form. Three such plots are provided in Figure 2.2 (a to c).

The first plot (a) suggest a linear relationship between *x* and *y*; i.e.,

$$Y = a_o + a_1 X \tag{2.16}$$

The second graph (b) appears to be represented by a second order (or parabolic) relationship:

$$Y = a_o + a_1 X + a_2 X^2 \tag{2.17}$$

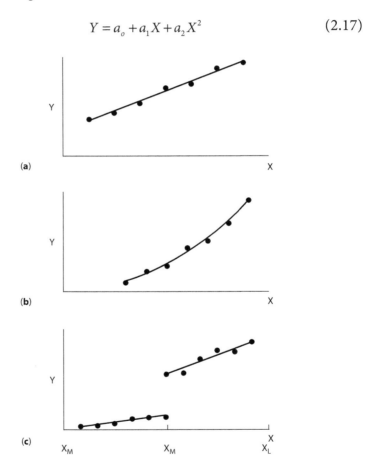

Figure 2.2 Scatter diagrams: (a) linear relationship, (b) parabolic relationship, and (c) dual-linear relationship.

The third plot (c) suggests a linear model that applies over two different ranges; i.e., it could represent the data:

$$Y = a_o + a_1 X; X_o < X < X_M \tag{2.18}$$

and

$$Y = a_o' + a_1' X; X_M < X < X_L \tag{2.19}$$

This multiequation model finds application in representing adsorption equilibria, multiparticle size distributions, quantum energy relationships, etc. In any event, a scatter diagram and individual judgment can suggest an appropriate model at an early stage in the analysis.

2.8   Open-Ended Problems

This last Section of the chapter contains open-ended problems as they relate to applied mathematics. No detailed and/or specific solution is provided; that task is left to the reader, noting that each problem has either a unique solution or a number of solutions or (in some cases) no solution at all. These are characteristics of open-ended problems described earlier.

There are comments associated with some, but not all, of the problems. The comments are included to assist the reader while attempting to solve the problems. However, it is recommended that the solution to each problem should initially be attempted *without* the assistance of the comments.

There are 45 open-ended problems in this Section. As stated above, if difficulty is encountered in solving any particular problem, the reader should next refer to the comments if any are provided with the problem. The reader should also note that the more difficult problems are generally located at or near the end of the Section.

1. Describe the early history associated with the general subject of mathematics.
2. Discuss the recent advances in applied mathematics.
3. Select a refereed published article on applied mathematics from the literature and provide a review.
4. Provide some normal everyday domestic applications involving the general topic of mathematics.
5. Develop an original problem in applied mathematics that would be suitable as an illustrative example in a book.

6. Prepare a list of the various books that have been written on applied mathematics. Select the three best and justify your answer. Also select the three weakest books and justify your answer.

7. Describe the term "machine language program" in layman terms.

8. Chemical engineers are aware that most numerical calculations are, by their very nature, inexact. The errors are primarily due to one of three sources:
   • inaccuracies in the original data;
   • lack of precision in carrying out elementary operations; and
   • inaccuracies introduced by approximate method(s) of solution.

   Of particular significance are the errors due to the *round-off* and the inability to carry more than a certain number of significant figures in a given calculation. Terms such as *absolute error*, *relative error*, and *truncation error* have a very real meaning. And frequently, an analysis parallel to this question must be carried out to establish the reliability of a given answer. Describe the above four italicized terms in layman terms.

9. Error due to roundoff was not considered to be too difficult when calculations were carried out by hand or by desk calculators. Added places and/or error terms could be introduced with little additional work. However, there is evidence that available elementary "error theories" do not seem to be adequate for estimation, with any real degree of certainty, of the roundoff and truncation errors that result when modern high-speed digital computers are used. Develop an improved method to quantify errors that can arise in those numerical calculations.

10. There are a host of topics that reside under the applied mathematics umbrella. List these topics in order of importance and justify your answer.

11. Discuss the difference(s) between analytical mathematics and numerical methods. Which is most important? Explain your choice.

12. Discuss the difference between analog and digital computers.

13. Provide, in technical detail, the various methods for solving simultaneous linear algebraic equations.

14. Provide, in technical detail, the various methods for solving nonlinear equations.

15. Provide, in technical detail, the various methods for solving ordinary differential equations.

16. Provide, in technical detail, the various methods for solving partial differentiation equations.

17. Provide, in technical detail, the various methods for solving optimization problems.

18. Refer to Problem 2 in the previous section. Provide a short paragraph describing how differential equations can be solved by intuition.

19. Refer to Problem 2 in the previous section. Provide a short paragraph describing how differential equations can be solved by analytical methods.

20. Refer to Problem 2 in the previous section. Provide a short paragraph describing how differential equations can be solved by analytical numerical methods.

21. Refer to Problem 2 in the previous Section. Discuss how separation of variables is employed in the analytical solution of differential equations.

22. Refer to Problem 2 in the previous Section. Discuss how the Fourier Series is employed in the analytical solution of differential equations.

23. Refer to Problem 2 in the previous Section. Discuss how Bessel functions are employed in the analytical solution of differential equations.

24. Refer to Problem 2 in the previous Section. Discuss how the Error function is employed in the analytical solution of differential equations.

25. A $10,000 penalty is imposed on an oil service company every time the sulfur (S) content of a 20,000-gallon delivery of oil to an industrial facility is in excess of *one half percent*. Comment on whether the oil company should be penalized if the percent sulfur content is:
    - 0.5
    - 0.55
    - 0.546
    - 0.545
    - 0.51
    - 0.505
    - 0.50001

26. Use any suitable method to linearize the following equations

$$ae^y = \ln x + bx \qquad (2.20)$$

$$\ln y = a + \frac{b}{x^2} + \frac{c}{x} \qquad (2.21)$$

$$cy = \frac{1}{\ln(ax - b)} \qquad (2.22)$$

27. Fluid is flowing from a storage tank 10 ft in diameter. The drop in the tank level was observed at various times as follows (see Table 2.1). Develop an equation describing the instantaneous flow rate in gal/min as a function of time, by any graphical method.

28. Refer to the previous problem. Select an equation to describe the displacement as a function of time, and use this equation to solve the problem.

29. Develop another (and hopefully improved) method of numerically evaluating an integral.

30. Develop another (and hopefully improved) method of numerically evaluating a derivative.

31. Develop another (and hopefully improved) method of solving an ordinary differential equation.

32. Develop another (and hopefully improved) method of solving a nonlinear algebraic equation.

Table 2.1 Storage Tank Problem

Displacement from top, ft	Time, min
0.0	0
3.9	30
5.9	60
7.5	90
8.9	120

33. Develop another (and hopefully improved) method of solving a set of linear algebraic equations.

34. Develop another (and hopefully improved) method of solving an optimization problem.

35. Monte Carlo simulation is a procedure for mimicking observations of a random variable that permits verification of results that would ordinarily require difficult mathematical calculations or extensive experimentation. The method normally uses computer programs called *random number generators*. A random number is a number selected from the interval (0,1) in such a way that the probabilities that the number comes from any two subintervals of equal "length". For example, the probability that the number is in the subinterval (0.1, 0.3) is the same as the probability that the number is in the subinterval (0.5, 0.7). Thus, random numbers are observations on a random variable x having a uniform distribution in the interval (0,1). This means that the probability distribution function (PDF)—defined in the Probability and Statistics Chapter in Part II—of x is specified by

$$f(x) = 1; \quad 0 < x < 1 \tag{2.23}$$

$$f(x) = 0; \quad \text{elsewhere} \tag{2.24}$$

The above PDF assigns equal probability to subintervals of equal length in the interval (0,1). Using random number generators, Monte Carlo simulation can generate observed values of a random variable having any specified PDF. For example, to generate observed values of t, the time to failure, when t is assumed to have a pdf specified by $f(t)$, one first uses the random number generator to generate a value x between 0 and 1. The solution is an observed value of the random variable t having a PDF specified by $f(t)$ [8].

The above provides a technical definition of Monte Carlo simulation. Your task is to describe the Monte Carlo simulation in layman terms.

36. Is Monte Carlo simulation a topic that should be addressed as a mathematics operation?

37. Describe the various searching schemes that are employed in optimization.
38. Develop a general procedure to employ to determine the optimum operating conditions in a plant.
39. Develop an optimum design for a new plant.
40. Develop the optimum design of a plant retrofit.
41. Describe advances in numerical methods this century.
42. Discuss the role the weighted-sum method of analysis in optimization studies.
43. In an attempt to mathematically describe the behavior of an operating system, a young engineer discovers that he/she has generated six equations that contain five variables. Suggest some *simple* approaches that the youngster can employ to best describe the system of concern.
44. In an attempt to mathematically describe the behavior of an operating system, the same young engineer discovers that he/she has generated five equations that contain six variables. Suggest some *simple* approaches that the youngster can employ to best describe the system of concern.
45. Develop a simplified manual procedure (not employing a computer) to generate, for any number, its:
    - square root
    - cube root
    - n^{th} root

References

1. Adapted from: M. Moyle, *Introduction to Computers for Engineers*, John Wiley & Sons, Hoboken, NJ, 1967.
2. B. Carnahan and J. Wilkes, *Digital Computing and Numerical Methods,* John Wiley & Sons, Hoboken, NJ, 1973.
3. R. Ketter and S. Prawler, *Modern Methods of Engineering Computation*, McGraw-Hill, New York City, NY, 1969.
4. J. Reynolds, J. Jeris, and L. Theodore, *Handbook of Chemical and Environments Engineering Calculations,* John Wiley & Sons, Hoboken, NJ, 2004.
5. Personal notes: L. Theodore, East Williston, NY, 1969.
6. L. Theodore, Personal notes, East Williston, NY, 1994.
7. R. Aris, *Discrete Dynamic Programming,* Blaisdell, New York City, NY, 1964.
8. D. Green and R. Perry, *Perry's Chemical Engineers' Handbook,* 8th edition, McGraw-Hill, New York City, NY, 2008.

9. L. Theodore, *Chemical Engineering: The Essential Reference*, McGraw-Hill, New York City, NY, 2014.

10. S. shaefer and L. Theodore, *Probability and Statistics Applications in Environmental Science,* CRC Press/ Taylor & Francis Group, Boca Raton, FL, 2007.

3

Stoichiometry

This chapter is concerned with stoichiometry. As with all the chapters in Part II, there are several sections: overview, several specific technical topics, illustrative open-ended problems, and open-ended problems. The purpose of the first section is to introduce the reader to the subject of stoichiometry. As one might suppose, a comprehensive treatment is not provided although several sections addressing additional specific technical topics are included. The next section contains three open-ended problems; the authors' solution (there may be other solutions) is also provided. The last section contains 44 problems; *no* solutions are provided here.

3.1   Overview

This overview section is concerned with stoichiometry. As one might suppose, it was not possible to address all topics directly or indirectly related to stoichiometry. However, additional details may be obtained from either the references provided at the end of this Overview and/or at the end of the chapter.

Note: Readers already familiar with the details associated with this subject may choose to bypass this Overview.

In order to better understand the design as well as the operation and performance of equipment in the chemical industry, it is necessary for chemical engineers (as well as applied scientists) to understand the fundamentals and principles underlying stoichiometry. How can one predict what products will be emitted from effluent streams? At what temperature must a unit be operated to ensure the desired performance? How much energy in the form of heat is given off? Is it economically feasible to recover this heat? Is the design appropriate? The answers to these questions are rooted not only in stoichiometry but also in the various theories of chemistry, physics, and applied economics.

The remaining topics covered in this section include:

1. The Conservation Law
2. The Conservation Laws for Mass, Energy and Momentum
3. Stoichiometry

Note: the bulk of the material in this chapter has been drawn from the original work of Reynolds [1].

3.2   The Conservation Law

Mass, energy and momentum are all conserved. As such, each quantity obeys the general conservation law below, as applied within a system.

$$
\begin{Bmatrix} quantity \\ into \\ system \end{Bmatrix} - \begin{Bmatrix} quantity \\ out\ of \\ system \end{Bmatrix} + \begin{Bmatrix} quantity \\ generated \\ in\ system \end{Bmatrix} = \begin{Bmatrix} quantity \\ accumulated \\ in\ system \end{Bmatrix} \quad (3.1)
$$

Equation (3.1) may also be written on a *time* basis:

$$
\begin{Bmatrix} rate\ of \\ quantity \\ into \\ system \end{Bmatrix} - \begin{Bmatrix} rate\ of \\ quantity \\ out\ of \\ system \end{Bmatrix} + \begin{Bmatrix} rate\ of \\ quantity \\ generated \\ in\ system \end{Bmatrix} = \begin{Bmatrix} rate\ of \\ quantity \\ accumulated \\ in\ system \end{Bmatrix} \quad (3.2)
$$

The conservation law may be applied by the practitioner at the *macroscopic, microscopic,* or *molecular* level. One can best illustrate the differences in these methods with an example. Consider a system in which a fluid is flowing through a cylindrical tube (see Figure 3.1). One can define the system as the fluid contained within the tube between points 1 and 2 at any time.

If one is interested in determining changes occurring at the inlet and outlet of the system, the conservation law is applied on a macroscopic level to the entire system. The resultant equation describes the *overall* changes occurring *to* the system without regard for internal variations *within* the system. This approach is usually applied by the practicing chemical engineer.

The microscopic approach is employed when detailed information concerning the behavior *within* the system is required, and this is often requested of and by the chemical engineer or scientist. The conservation law is then applied to a *differential* element within the system which is large compared to an individual molecule, but small compared to the entire system. The resultant equation is then expanded, via integration, to describe the behavior of the entire system. This is defined by some as the *transport phenomena* approach [2,3].

The molecular approach involves the application of the conservation law to individual molecules. This leads to a study of statistical and quantum mechanics—both of which are beyond the scope of this text. In any case, the description of individual molecules at the molecular level is of little value to the practicing chemical engineer. However, the statistical averaging of molecular quantities in either a differential or finite element within a system leads to a more meaningful description of the behavior of the system.

The macroscopic approach is primarily adopted and applied in this text, and little to no further reference to microscopic or molecular analyses will be made. This chapter's aim, then, is to express the laws of conservation for mass, energy, and momentum in algebraic or finite difference form.

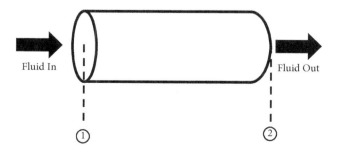

Figure 3.1 Conservation law application.

3.3   Conservation of Mass, Energy, and Momentum [1]

The *conservation law* for mass can be applied to any process, equipment, or system. The general form of this law is given by Equations (3.3) and (3.4).

$$
\left\{\begin{array}{c} mass \\ in \end{array}\right\} - \left\{\begin{array}{c} mass \\ out \end{array}\right\} + \left\{\begin{array}{c} mass \\ generated \end{array}\right\} = \left\{\begin{array}{c} mass \\ accumulated \end{array}\right\} \qquad (3.3)
$$

$$
I \quad - \quad O \quad + \quad G \quad = \quad A
$$

or on a time rate basis by

$$
\left\{\begin{array}{c} rate\ of \\ mass \\ in \end{array}\right\} - \left\{\begin{array}{c} rate\ of \\ mass \\ out \end{array}\right\} + \left\{\begin{array}{c} rate\ of \\ mass \\ generated \end{array}\right\} = \left\{\begin{array}{c} rate\ of \\ mass \\ accumulated \end{array}\right\} \qquad (3.4)
$$

$$
I \quad - \quad O \quad + \quad G \quad = \quad A
$$

The *law of conservation of mass* states that mass can neither be created nor destroyed. Nuclear reactions, in which interchanges between mass and energy are known to occur, provide a notable exception to this law. Even in chemical reactions, a certain amount of mass-energy interchange takes place. However, in normal chemical engineering applications, nuclear reactions do not occur and the mass-energy exchange in chemical reactions is so minuscule that it is not worth taking into account.

The *law of conservation of energy*, which like the law of conservation of mass, applies for all processes that do not involve nuclear reactions, states that energy can neither be created nor destroyed. As a result, the energy level of the system can change only when energy crosses the system boundary, i.e.,

$$
\Delta\left(\text{Energy level of system}\right) = \text{Energy crossing boundary} \qquad (3.5)
$$

(Note: The symbol "Δ" means "change in".) Energy crossing the boundary can be classified in one of two different ways: *heat*, Q, or *work*, W. *Heat* is energy moving between the system and the surroundings by virtue of a temperature driving force and heat flows from high temperature to low temperature. The entire system is not necessarily at the same temperature; neither are the surroundings. If a portion of the system is at a higher temperature than a portion of the surroundings and as a result, energy is transferred from the system to the surroundings, that energy is classified as heat. If part of the system is at a higher temperature than another part of the system and energy is transferred between the two parts, that energy is *not* classified as heat because it is *not* crossing the boundary. Work is also energy moving between the system and surroundings, but the driving force here is something other than temperature difference, e.g., a mechanical force, a pressure difference, gravity, a voltage difference, a magnetic field, etc. Note that the definition of work is a *force acting through a distance*. All of the examples of driving forces just cited can be shown to provide a force capable of acting through a distance [4].

The energy level of a system has three contributions: kinetic energy, potential energy, and internal energy. Any body in motion possesses kinetic energy. If the system is moving as a *whole*, its kinetic energy, E_k, is proportional to the mass of the system and the square of the velocity of its center of gravity. The phrase "as a whole" indicates that motion inside the system relative to the system's center of gravity does not contribute to the E_k term, but rather to the internal energy term. The terms *external kinetic energy* and *internal kinetic energy* are sometimes used here. An example would be a moving railroad tank car carrying propane gas. (The propane gas is the system.) The center of gravity of the propane gas is moving at the velocity of the train, and this constitutes the system's external kinetic energy. The gas molecules are also moving in random directions relative to the center of gravity and this constitutes the system's internal energy due to motion inside the system, i.e., internal kinetic energy. The potential energy, E_p, involves any energy the system as a whole possesses by virtue of its position (more precisely, the position of its center of gravity) in some force field, e.g., gravity, centrifugal, electrical, etc., that provides the system with the potential for accomplishing work. Again, the phase "as a whole" is used to differentiate between *external* potential energy, E_p, and *internal* potential energy. *Internal potential energy refers* to potential energy due to force fields inside the system. For example, the electrostatic force fields (bonding) between atoms and molecules provide these particles with the potential for work. The *internal energy*, U, is the sum of all internal kinetic and internal potential energy contributions [4].

The *law of conservation of energy*, which is also called the *first law of thermodynamics*, may now be written as

$$\Delta(U + E_k + E_p) = Q + W \tag{3.6}$$

or equivalently as

$$\Delta U + \Delta E_k + \Delta E_p = Q + W \tag{3.7}$$

It is important to note the sign convention for Q and W adapted for the above equation. Since any term is always defined as the final minus the initial state, both the heat and work terms must be positive when they cause the system to gain energy, i.e., when they represent energy flowing from the surroundings to the system. Conversely, when the heat and work terms cause the system to lose energy, i.e., when energy flows from the system to the surroundings, they are negative in sign. This sign convention is not universal and the reader must take care to check what sign convention is being used by a particular author when referring to the literature. For example, work is often defined in some texts as positive when the system does work on the surroundings [4,5].

The conservation law for momentum is treated in Chapter 5.

3.4 Stoichiometry [1]

When chemicals react, they do so according to a strict proportion. When oxygen and hydrogen combine to form water, the ratio of the amount of oxygen to the amount of hydrogen consumed is always 7.94 by mass and 0.500 by moles. The term *stoichiometry* refers to this phenomenon, which is sometimes called the *chemical law of combining weights*. The reaction equation for the combining of hydrogen and oxygen is

$$2H_2 + O_2 = 2H_2O \tag{3.8}$$

In chemical reactions, atoms are neither generated nor consumed, merely rearranged with different bonding partners. The manipulation of the coefficients of a reaction equation so that the number of atoms each element on the left of the equation is equal to that on the right is referred to as *balancing* the equation. Once the equation is balanced, the whole number molar ratio that must exist between any two components of the reaction can be determined simply by observation; these are known as *stoichiometric ratios*. There are three such ratios (not counting the reciprocals) in the above reaction. These are:

2 mol H_2 consumed/mol O_2 consumed
1 mol H_2O generated/mol H_2 consumed
2 mol H_2O generated/mol O_2 consumed

The unit mole represents either the *gmol* or the *lbmol*. Using molecular weights, these stoichiometric ratios (which are molar ratios) may easily be converted to mass ratios. For example, the first ratio above may be converted to a mass ratio by using the molecular weights of H_2 (2.016) and O_2 (31.999) as follows:

$$(2 \text{ gmol } H_2 \text{ consumed})(2.016 \text{ g/gmol}) = 4.032 \text{ g } H_2 \text{ consumed}$$
$$(1 \text{ gmol } O_2 \text{ consumed})(31.999 \text{ g/gmol}) = 31.999 \text{ g } H_2 \text{ consumed}$$

The mass ratio between the hydrogen and oxygen consumed is therefore

$$4.032/31.999 = 0.126 \text{ g } H_2 \text{ consumed/g } O_2 \text{ consumed}$$

These molar and mass ratios are used in material balances to determine the amounts or flow rates of components involved in chemical reactions.

Multiplying a balanced reaction equation through by a constant does nothing to alter its meaning. The reaction used as an example above is often written

$$H_2 + \frac{1}{2}O_2 = H_2O \tag{3.9}$$

In effect, the stoichiometric coefficients of Equation (3.8) have been multiplied by 0.5. There are times, however, when care must be exercised because the solution to the problem depends on the manner or form the reaction is written. This is the case with chemical equilibrium problems and problems involving thermochemical reaction equations. These are addressed in the next chapter.

There are three different types of material balances that may be written when a chemical reaction is involved: the *molecular balance*, the *atomic balance*, and the *"extent of reaction" balance*. It is a matter of convenience which of the three types is used. Each is briefly discussed below.

The molecular balance is the same as that described earlier. Assuming a steady-state continuous reaction, the accumulation term, a, is zero for all components involved in the reaction, the balance equation (3.3) becomes

$$I + G = O + C \tag{3.10}$$

where C = consumption

If a total material balance is performed, the above form of the balance equation must be used if the amounts or flow rates are expressed in terms of moles, e.g., lbmol or gmol/h, since the total number of moles can change during a chemical reaction. If, however, the amounts or flow rates are given in terms of mass, e.g., kg or lb/h, the G and C terms may be dropped since mass cannot be gained or lost in a chemical reaction. Thus,

$$I = O \qquad\qquad (3.11)$$

In general, however, when a chemical reaction is involved, it is usually more convenient to express amounts and flow rates using moles rather than mass.

A material balance that is not based on the chemicals (or molecules), but rather on the atoms that make up the molecules, is referred to as an atomic balance. Since atoms are neither created nor destroyed in a chemical reaction, the G and C terms equal zero and the balance once again becomes

$$I = O \qquad\qquad (3.11)$$

As an example, consider once again the combination of hydrogen and oxygen to form water

$$2H_2 + O_2 = 2H_2O \qquad\qquad (3.8)$$

As the reaction progresses, O_2 and H_2 molecules (or moles) are consumed while H_2O molecules (or moles) are generated. On the other hand, the number of oxygen atoms (or moles of oxygen atoms) and the number of hydrogen atoms (or moles of hydrogen atoms) do not change. Care must also be taken to distinguish between molecular oxygen and *atomic* oxygen. If, in the above reaction, one starts out with 1000 lbmol of O_2 (oxygen molecules), one is also starting out with 2000 lbmol of O (oxygen atoms).

A detailed an expanded treatment of stoichiometry is available in the following two references:

1. D. Green and R. Perry, *Perry's Chemical Engineers' Handbook*, 8th edition, McGraw-Hill, New York City, NY, 2008 [5].
2. L. Theodore, *Chemical Engineering: The Essential Reference*, McGraw-Hill, New York City, NY, 2014 [6].

3.5   Illustrative Open-Ended Problems

This and the last Section provide open-ended problems. However, solutions *are* provided for the three problems in this Section in order for the reader to hopefully obtain a better understanding of these problems, which differ from the traditional problems/illustrative examples. The first problem is relatively straightforward while the third (and last problem) is somewhat more difficult and/or complex. Note that solutions are not provided for the 44 open-ended problems in the next Section.

Problem 1: There are four methods that can be used to characterize an emission source. These include the use of:

1. emission factors;
2. mass balance considerations;
3. engineering calculations, and;
4. direct emission measurements.

Describe each of these approaches and indicate what conditions one may be utilized over the other.

Solution: One approach to describing four methods for source characterization is provided below, by highlighting the strengths and weaknesses of each.

1. *Emission factors* are emission rates that have been compiled by EPA and/or other regulatory agencies based on data generated from a given source that are normalized to some unit of production or rate of chemical use, i.e., mass of formaldehyde released from a vehicle or vehicle-driven mile or the mass of butadiene released/gal of product transferred, etc. These factors are compiled for an industry and/or a process on a product-specific basis so that emission rates from similar operations can be estimated without having to repeat direct measurements or detailed calculations for a given emission source. Emission factors are very useful in developing screening-level estimates of emission rates from a large number of sources where time and/or money are limited. They do suffer from significant inaccuracies, however, if the factor utilized is obtained from a process not closely related to the one for which emission rates are being predicted.

2. *Mass balance considerations* involve the evaluation of a chemical process or system as a whole in terms of the mass of reactants added to the

system and the mass of products generated on a constituent by constituent basis. This tracking of constituent mass is termed the aforementioned mass balance approach, and because one knows that mass cannot be created nor destroyed, the mass of any species added to the system that cannot be accounted for in the product must either have been retained within the system or released from the system as a liquid, solid, or gaseous stream. This method requires that the mass flow rate of all influent and effluent streams can be quantified accurately, and that the chemical reactions taking place within the system, if any, are understood.

3. *Engineering calculations* involve the prediction of emission rates from chemical processes and product storage, transport, treatment, or disposal systems based on fundamental science and engineering principles. These calculations generally involve the use of chemical property data (solubility, vapor pressure, viscosity, density, diffusivity, etc.), the physical system (volume, depth, surface area, etc.) and operating conditions (flow rate, mixing rate, temperature, etc.) to estimate specific releases from a given process.

4. *Direct emission measurements* involve the direct determination of emission rates from specific sources. These measurements provide input to emission factors and engineering calculation approaches. The direct measurement approach is the only true way to generate accurate emission rates for a given source operating under a given set of conditions.

Problem 2: Develop an equation describing the adiabatic flame temperature of a hydrocarbon as a function of its net heating value and the excess air employed in the (combustion) process.
Comment: Refer to Theodore, et al [7] and Santoleri, et al [8] for details.

Solution: Some reasonable assumptions can be made to simplify the rigorous approach for calculations involving the adiabatic flame temperature. When compared to the rigorous approach, a simpler (and in many instances, a more informative) set of equations results that are valid for purposes of engineering calculation. One such approach is detailed below [8].

1. The sensible enthalpy change associated with the cooling step can in many instances be neglected compared to the net heating value (combustions step) of the combined waste-fuel mixture. For this condition,

$$\Delta H_{st} = \Delta H_e = \Delta H_w = \Delta H_f = 0 \qquad (3.12)$$

2. Although the products of combustion include many components, the major or primary components are nitrogen, carbon dioxide, and water (vapor). The average heat capacities of these components over the temperature range 60–2000°F (the latter being a typically incinerator operating temperature) are 0.27, 0.27, and 0.52 Btu/lb•°F, respectively. The arithmetic average of these three components is 0.35. However, since this product stream consists primarily of nitrogen, the average heat capacity of the combined mixture (not including the excess air) may be assigned a value of approximately 0.3 Btu/lb•°F.

3. The average heat capacity of the (excess) air is ≈0.27 Btu/lb•°F over the temperature range 60-2000°F. This value may be rounded to 0.30.

$$\Delta H_e = m_e (0.3)(T - T_0) \tag{3.13}$$

where m_e = mass of excess air per unit mass of waste-fuel mixture (lb/lb mixture)

4. There is no heat loss under truly adiabatic conditions.

5. Since most of the heating value data are available at 60°F, the reference or standard temperature should be arbitrarily set to this condition, i.e., 60°F, and NHVs should be obtained at approximately this temperature.

6. Perhaps the key assumption in this development is that associated with the stoichiometric air requirements for the combined waste-fuel mixture. Interestingly, the stoichiometric air requirement, v_{se} (ft³ air/lb mixture), divided by the NHV for many hydrocarbons is approximately 0.01 ft³ air/Btu (or 100 Btu/ft³air). Using the density of air at 60°F, this ratio can be converted to approximately 750 lb air/10⁶ Btu or 7.5×10^{-4} lb air/Btu. Thus, for this condition the stoichiometric air requirement (m_{st}) is given by:

$$m_{st} = 7.5 \times 10^{-4} \; NHV \tag{3.14}$$

7. Applying the six assumptions listed above results in the following equation for the adiabatic flame temperature, T, in terms of the excess air, EA, and the net heating value, NHV.

$$T = 60 + \frac{NHV}{(0.3)\left[1 + (1 + EA)(7.5 \times 10^{-4})(NHV)\right]} \qquad (3.15)$$

Note: The units of T and NHV are °F and Btu/lb, respectively; EA is a dimensionless fraction.

Santoleri, et al [8] provides additional details. This topic is also revisited in Part III, Chapter 25—Thermodynamics Term Projects.

Problem 3: The heat-generating unit in a coal-fired power plant may be simply described as a continuous-flow reactor, into which fuel (mass flow rate F) and air (mass flow rate A) are fed, and from which effluents ("flue gas", mass flow rate E) are discharged.

1. Draw a flow diagram representing this process. Show all flows into and out of the unit. Write a mass balance equation for this process.
2. Suppose the fuel contains a mass fraction y of incombustible component C (for example, ash). Assume that all of the ash is carried out of the reactor with the flue gas (note that in reality, a fraction of the ash generated will remain within the heat-generating unit as bottom ash and must be removed periodically). Write a mass balance equation for component C. What is the mass fraction (z) of C in the exit stream E? Quantitatively discuss the effect of increasing the combustion air flow upon z?
3. Suppose a fraction x of the flue gas is recycled to the inlet of the "reactor". (This is commonly done to help suppress the formation of pollutants, primarily NO_x). Redraw the flow diagram, including the recycle stream (R). Write a mass balance equation for the overall process, and a mass balance equation around the reactor only. Quantitatively discuss how these equations are different from that in [1].
4. An air pollution control device is added to the exhaust stream. It is able to remove x% of the incombustible pollutant C by scrubbing with water. Let S be the mass flow rate of scrubbed material in the water stream. Add this air pollution control unit to the flow diagram, and include all process

streams into and out of the unit. Express S in terms of F, A, and E for various % values of x.

Solution

1. The mass balance is simply F+A=E as indicated in the simple flow diagram in Figure 3.2
2. Since C is not combustible, $C_{in} = C_{out}$, so a component mass balance on C gives

$$yF +(0) A = xE \tag{3.16}$$

Using $E = F + A$, yields $yF = z(F + A)$, and $z = y\,(F/(F + A))$. Therefore, increasing the flow rate A decreases the mass fraction of C in the exhaust (a dilution effect). (This is the reason that *effluent* concentrations are often given at a specified excess air concentrations).

3. The new flow diagram can be drawn as indicated below in Figure 3.3. The overall mass balance is unchanged at:

$$F+A=E \text{ (unchanged)} \tag{3.17}$$

The mass balance around the reactor becomes:

$$F + A + xE = (1 + x)\,E \tag{3.18}$$

$$F + A = E \text{ (also unchanged)} \tag{3.19}$$

Note that one must specify the recycle amount (x) or some other information regarding concentrations to determine the recycle rate since it cannot be determined from these equations alone.

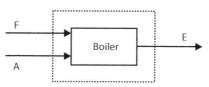

Figure 3.2 Boiler flow diagram.

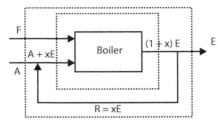

Figure 3.3 Revised boiler flow diagram.

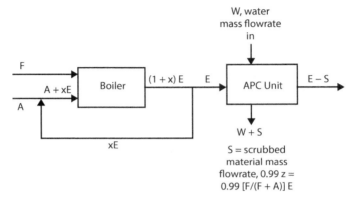

Figure 3.4 Boiler flow diagram with air pollution control unit.

4. The final revised flow diagram with the air pollution control unit takes the form provided in Figure 3.4.

3.6   Open-Ended Problems

This last Section of the chapter contains open-ended problems as they relate to stoichiometry. No detailed and/or specific solution is provided; that task is left to the reader, noting that each problem has either a unique solution or a number of solutions or (in some cases) no solution at all. These are characteristics of open-ended problems described earlier.

There are comments associated with some, but not all, of the problems. The comments are included to assist the reader while attempting to solve the problems. However, it is recommended that the solution to each problem should initially be attempted without the assistance of the comments.

There are 44 open-ended problems in this Section. As stated above, if difficulty is encountered in solving any particular problem, the reader should next refer to the comment, if any is provided with the problem.

The reader should also note that the more difficult problems are generally located at or near the end of the Section.

1. Select a refereed, published article on stoichiometry from the literature and provide a review.
2. Provide some normal everyday domestic applications involving the general topic of stoichiometry.
3. Develop an original problem in stoichiometry that would be suitable as an illustrative example in a book.
4. Prepare a list of the various books which have been written on stoichiometry. Select the three best and justify your answer. Also select the three weakest books and justify your answer.
5. Provide a layman's definition of stoichiometry.
6. Provide a technical definition of stoichiometry.
7. The relationship between the mass of reactants and products involved in a chemical reaction can be derived from a consideration of the equation for the reaction in question and the molecular weights of the materials involved. Should this type of analysis be performed on a mass, mole, or molecular (atomic) basis? Explain your response.
8. Interview a practicing chemical engineer and request his description of the chemical engineering profession and what a chemical engineer does.
9. List the various sources of information available in literature for the chemical engineer.
10. Search the internet (or library) and report on the careers of at least three major players in the chemical engineering field.
11. Develop another method of scientific notation of numbers. Comment: Should a log scale be considered?
12. Develop another set of notations that can be employed in chemical engineering.
13. Develop an equation that relates the atomic number of an element with its atomic weight.
    Comment: A table of (international) atomic weights is obviously required.
14. List the various systems of units employed by chemical engineers. Also, develop another system of units.
15. Develop another set of units for pressure.
16. There are presently four temperature scales that the practicing engineer employs. Develop another temperature scale that one might consider using in certain applications.

Comment: Carefully review the °C, °F, K, and °R temperature scales and the relationship between each of the scales. In addition, should consideration be given to basing the new scale on the melting and/or boiling point of a particular element or compound?

17. List the quantities that are conserved. How many conservation laws are there?

18. The composition of a gas, liquid, or solid mixture can be expressed in many ways. List and describe at least three of them.

19. Describe the concept of a "basis" in stoichiometric calculations to a layman.

20. Select common examples from everyday life that can be describe as *purge*, *bypass*, and *recycle*.

21. Obtain the density or specific gravity of a solid from several sources in the literature. Explain the variation of values.
    Comment: Review the various definitions of density for solids.

22. Generate a series of equations that could be used to describe the vapor pressure of water as a function of temperature.

23. Many years ago, one of the authors as part of a consulting job was asked to convert the Mollier Chart into equation form. You have been assigned that task.
    Comment: This will require regressing Mollier Chart information data. Some probability and statistics background is required.

24. Provide a layman's definition of the *wet-bulb* temperature and *dry-bulb* temperature. Also describe the difference between the two.

25. Convert a standard humidity chart into equation form. Include as many properties as possible in the correlation.

26. Describe what, if any, advantages are derived from basing humidity data on a mole basis.

27. Prepare another type of humidity chart that one could potentially obtain a copyright on.

28. Describe how to best evaluate the constants in an equation of state.

29. Develop another real or ideal gas law.
    Comment: Carefully review all the laws that have been developed previously.

30. Describe the Buckingham π approach to dimensional analysis. Include a procedure to follow in employing this method of analysis.

31. List all possible terms that are required for an energy balance on a chemical reactor. Also list the quantitative information required in an energy analysis of the system.

32. All forms of energy must be included in an energy balance. Describe the various forms of energy, including those that are not easily recognized in everyday life. Be sure to consider all forms of energy.

33. It is difficult to distinguish between energy and mass at the atomic and subatomic levels. Discuss this relationship and its relevance to Einstein's energy equation.

34. There are various equations (linear, parabolic, etc.) in the literature that describe heat capacity variations with temperature. Discuss the advantages and disadvantages of the various forms of these describing equations.

35. Select several chemicals from the literature that provide heat capacity variation with temperature. Develop equations of your choice that describe the temperature variation.

36. Describe what, if any, advantages are derived from basing enthalpy of reaction data on a mole basis.

37. Generate a list of sources of thermodynamic data. The sources should provide information on heat capacity, enthalpy, entropy, internal energy, enthalpy of reaction, free energy of reaction, etc.
    Comment: The list could be enormous.

38. A gaseous mixture of 5% methane in air is contained in a heavy-walled steel vessel at 1 atm and 80°F. A voltage was accidentally applied to the vessel resulting in a spark, causing the methane-air mixture to explode.
    • Assuming that the vessel remains intact, calculate the resulting pressure in the vessel if the gases equilibrate at 1,200°F.
    • Repeat the calculations for various mixtures of methane in air.

39. An auditorium has a volume of 100,000 ft^3. The CO_2 concentration in air before a meeting is 0.04 percent. One hour after the meeting room is in session, the CO_2 concentration is 0.14 per cent. The room is ventilated by 50,000 ft^3/min of

fresh air (0.04 percent CO_2). What is the CO_2 concentration 6 hours after the session has ended? How would the presence of an adsorbent in the auditorium affect the results?

40. There is wide scatter between bond dissociation energies of various molecules, particularly hydrocarbons. For example, the bond dissociation energy of H_2 (H-H) and N_2 (N-N) are 104 and 171 kcal/gmol (at 25°C), respectively, while it is 102 kcal/gmol for CH_4 (CH_3-H). Attempt to correlate these dissociation energies at 25°C with the molecules.
    Comment: Bond dissociation energies for a host of molecules are required.

41. The approximate composition of air is provided in Table 3.1. Attempt to correlate the above composition data with atomic weight.

42. Attempt to improve on Raoult's equation by developing a generalized equation (or equations) that can be employed to describe *non-ideal* vapor-liquid equilibrium.

43. Describe Einstein's equation pertaining to the relationship between mass and energy in an essay (1000 words).

44. Refer to the previous problem. Discuss if Einstein's equation would still apply on the surface of the moon... or another planet.

Table 3.1 Air Concentration

Gas	Percent by Volume
Nitrogen	78
Oxygen	21
Argon	0.94
Carbon dioxide	3.1×10^{-2}
Neon	1.5×10^{-3}
Helium	5.0×10^{-3}
Krypton	1.0×10^{-4}
Xenon	9.0×10^{-6}

References

1. J. Reynolds, *Material and Energy Balances*, A Theodore Tutorial, East Williston, NY, originally published by the USEPA/APTI, RTP, NC, 1992.
2. L. Theodore, *Transport Phenomena for Engineers*, Theodore Tutorials, East Williston, NY, originally published by International Textbook Co., Scranton, PA, 1971.
3. R. Byrd, W. Stewart, and Lightfoot, *Transport Phenomena*, 2nd edition, John Wiley & Sons, Hoboken, NJ, 2002.
4. L. Theodore and J. Reynolds, *Thermodynamics*, A Theodore Tutorial, East Williston, NY, originally published by the USEPA/APTI, RTP, NC, 1991.
5. D. Green and R. Perry (editors), *Perry's Chemical Engineers' Handbook,* 8th edition, McGraw-Hill, New York City, NY, 2008.
6. L. Theodore, *Chemical Engineering: The Essential Reference*, McGraw-Hill, New York City, NY, 2014.
7. L. Theodore, F. Ricci, and T. VanVliet, *Thermodynamics for the Practicing Engineer*, John Wiley & Sons, Hoboken, NJ, 2009.
8. J. Santoleri, J. Reynolds, and L. Theodore, *Introduction to Hazardous Waste Incineration*, 2nd edition, John Wiley & Sons, Hoboken, NJ, 2000.

4

Thermodynamics

This chapter is concerned with thermodynamics. As with all the chapters in Part II, there are several sections: overview, several specific technical topics, illustrative open-ended prolems, and open-ended problems. The purpose of the first section is to introduce the reader to the subject of thermodynamics. As one might suppose, a comprehensive treatment is not provided although numerous references are included. The several technical topics followed by three open-ended problems; the authors' solution (there may be other solutions) is also provided. The last section contains 39 problems; *no* solutions are provided here.

4.1 Overview

This overview section is concerned—as can be noted from its title—with thermodynamics. As one might suppose, it was not possible to address all topics directly or indirectly related to thermodynamics. However, additional details may be obtained from either the references provided at the end of this Overview section and/or at the end of the chapter.

Note: Those readers already familiar with the details associated with this subject may choose to bypass this Overview.

Thermodynamics was once defined as "the science that deals with the intertransformation of heat and work". The fundamental principles of thermodynamics are contained in the first, second, and third laws of thermodynamics. These principles have been defined as "pure" or "theoretical" thermodynamics. These laws were developed and extensively tested in the latter half of the 19th century and are essentially based on experience. (The third law was developed later in the 20th century).

Practically all thermodynamics, in the ordinary meaning of the term, is "applied thermodynamics" in that it is essentially the application of these three laws, coupled with certain facts and principles of mathematics, physics, and chemistry, to problems in chemical engineering. The fundamental laws are of such generality that it is not surprising that these laws find application in other disciplines, including physics, chemistry, plus environmental and mechanical engineering.

As described in the previous chapter, the first law of thermodynamics is a conservation law for energy transformations. Regardless of the types of energy involved in processes—thermal, mechanical, electrical, elastic, magnetic, etc.—the change in energy of the system is equal to the difference between energy input and energy output. The first law also allows free "convertibility" from one form of energy to another, as long as the overall energy quantity is conserved. Thus, this law places no restriction on the conversion of work into heat, or on its counterpart—the conversion of heat into work.

The brief discussion of energy-conversion devices above leads to an important second-law consideration—i.e., that energy has quality as well as quantity. Because work is 100% convertible to heat, whereas the reverse situation is not true, work is a more valuable form of energy than heat. Although it is not as obvious, it can also be shown through the second-law principles and arguments that heat has *quality* in terms of its temperature. The higher the temperature at which heat transfer occurs, the greater the potential for energy transformation into work. Thus, thermal energy stored at higher temperatures is generally more useful to society than that available at lower temperatures. While there is an immense quantity of energy stored in the oceans and the earth's core, for example, its present availability to society for performing useful tasks is essentially nonexistent. Theodore et al [1] also provided a qualitative review of the second law.

The choice of topics to be reviewed in this chapter was initially an area of concern. After some deliberation, it was decided to provide an introduction to four areas that many have included (at one time or another) in this broad engineering subject. These are detailed below:

1. Enthalpy Effects
2. Second Law Calculations
3. Phase Equilibrium
4. Chemical Reaction Equilibrium

The reader should note that the bulk of the material in this chapter has been drawn from L. Theodore and J. Reynolds, *Thermodynamics*, A Theodore Tutorial, originally published by the USEPA/APTI, RTP, NC in 1991 [2].

4.2   Enthalpy Effects

There are many different types of enthalpy effects; these include:

1. sensible (temperature);
2. latent (phase);
3. dilution (with water), e.g., HCl with H_2O;
4. solution (nonaqueous), e.g., HCl with a solvent other than H_2O; and
5. reaction (chemical).

This section is only concerned with effects (1) and (5). Details on effects (2–4) are available in the literature [1–3].

4.2.1   Sensible Enthalpy Effects

Sensible enthalpy effects are associated with temperature. There are methods that can be employed to calculate these changes. These methods include the use of:

1. enthalpy values;
2. average heat capacity values; and
3. heat capacity as a function of temperature.

Detailed calculations on methods (1–3) are provided by Theodore, et al [1,2]. Two expressions for heat capacity are considered in topic (3) employing a, b, c constants and α, β, γ constants.

If enthalpy values are available, the enthalpy change is given by

$$\Delta h = h_2 - h_1; \text{ mass basis} \tag{4.1}$$

$$\Delta H = H_2 - H_1; \text{ mole basis} \tag{4.2}$$

If average molar heat capacity data are available,

$$\Delta H = \overline{C_p} \Delta T \tag{4.3}$$

where $\overline{C_p}$ = average molar value of C_p in the temperature range ΔT. Average molar heat capacity data are provided in the literature [1–5].

A more rigorous approach to enthalpy calculations can be provided if heat capacity variation with temperature is available. If the heat capacity is a function of the temperature, the enthalpy change is written in differential form:

$$dH = C_p dT \tag{4.4}$$

If the temperature variation of the heat capacity is given by

$$C_p = a + \beta T + \gamma T^2 \tag{4.5}$$

Equation (4.4) may be integrated directly between some reference or standard temperature (T_0) and the final temperature (T_1).

$$\Delta H = H_1 - H_0 \tag{4.6}$$

$$\Delta H = \alpha(T_1 - T_0) + (\beta/2)(T_1^2 - T_0^2) + (\gamma/3)(T_1^3 - T_0^3) \tag{4.7}$$

Equation (4.4) may also be integrated if the heat capacity is a function of temperature of the form:

$$C_p = a + bT + cT^{-2} \tag{4.8}$$

The enthalpy change is then given by

$$\Delta H = a\left(T_1 - T_0\right) + \left(b/2\right)\left(T_1^2 - T_0^2\right) + c\left(T_1^{-1} - T_0^{-1}\right) \qquad (4.9)$$

Tabulated values of α, β, γ and a, b, c for a host of compounds (including some chlorinated organics) are available in the literature [2].

4.2.2   Chemical Reaction Enthalpy Effects

The equivalence of mass and energy needs to be addressed qualitatively. This relationship is only important in nuclear reactions, the details of which are beyond the scope of both this chapter and this text. The energy-related effects discussed here arise because of the rearrangement of electrons outside the nucleus of the atom. However, it is the nucleus of the atom that undergoes rearrangement in a nuclear reaction, releasing a significant quantity of energy; this process occurs with a minuscule loss of mass. The classic Einstein equation relates energy to mass, as provided in Equation (4.10).

$$\Delta E = (\Delta m)c^2 \qquad (4.10)$$

where ΔE = change in energy
Δm = decrease in mass
c = velocity of light

The standard enthalpy (heat) of reduction can be calculated from standard enthalpy of formation data. To simplify the presentation that follows, examine the authors' favorite equation:

$$aA + bB = cC + dD \qquad (4.11)$$

If the above reaction is assumed to occur at a standard (or reference) state, the standard enthalpy of reaction, ΔH^0, is given by

$$\Delta H^0 = c(\Delta H_f^0)_C + d(\Delta H_f^0)_D - a(\Delta H_f^0)_A - b(\Delta H_f^0)_B \qquad (4.12)$$

where $\left(\Delta H_f^0\right)_i$ = standard enthalpy of formation of species i.

Thus, the (standard) enthalpy of a reaction is obtained by taking the difference between the (standard) enthalpy of formation of products and reactants. If the (standard) enthalpy of reaction or formation is negative (exothermic), as is the case with most combustion reactions, then energy is liberated due to the chemical reaction. Energy is absorbed and ΔH^0 is positive (endothermic).

Tables of enthalpies of formation and reaction are available in the literature (particularly thermodynamics text/reference books) for a wide variety of compounds. [1] It is important to note that these are valueless unless the stoichiometric equation and the state of the reactants and products are included.

Theodore, et al [1,2] provide equations to describe the effect of temperature on the enthalpy of reaction. For heat capacity data in α, β, γ form:

$$\Delta H_T^0 = \Delta H_{298}^0 + \Delta\alpha\,(T-298) + \left(\frac{1}{2}\right)\Delta\beta\left(T^2 - 298^2\right) + \left(\frac{1}{3}\right)\Delta\gamma\left(T^3 - 298^3\right)$$

(4.13)

For the reaction presented in Equation (4.11):

$$\Delta a = ca_C + da_D - aa_A - ba_B$$
$$\Delta \beta = c\beta_C + d\beta_D - a\beta_A - b\beta_B$$
$$\Delta \gamma = c\gamma_C + d\gamma_D - a\gamma_A - b\gamma_B$$

(4.14)

For heat capacity in a, b, c form,

$$\Delta H_T^0 = \Delta H_{298}^0 + \Delta a\,(T-298) + \left(\frac{1}{2}\right)\Delta b\left(T^2 - 298^2\right) + \Delta c\left(T^{-1} - 298^{-1}\right)$$ (4.15)

4.3   Second Law Calculations [2]

The law of conservation of energy has already been defined as the *first law of thermodynamics.* Its application allows calculations of energy relationships associated with all kinds of processes. The "limiting" law is called the *second law of thermodynamics* (SLT). Applications involve calculations for maximum power output from a power plant and equilibrium yields in chemical reactions. In principle, this law states that water cannot flow uphill and heat cannot flow from a cold to a hot body of its own accord. Other defining statements for this law that have appeared in the literature are provided below:

1. Any process, the sole net result of which is the transfer of heat from a lower temperature level to a higher one is impossible.

2. No apparatus, equipment, or process can operate in such a way that its only effect (on system and surroundings) is to convert heat absorbed completely into work.
3. It is impossible to convert the heat taken into a system completely into work in a cyclical process.

The second law also serves to define another important thermodynamics function called *entropy*. It is normally designated as S. The change in S for a *reversible adiabatic process* is always zero:

$$\Delta S = 0 \tag{4.16}$$

For liquids and solids, the entropy change for a system undergoing a temperature change from T_1 to T_2 is given by

$$\Delta S = C_p \ln(T_2 / T_1); \ C_p = \text{constant} \tag{4.17}$$

The entropy change of an ideal gas undergoing a physical change of state from P_1 to P_2 at a constant temperature T is given by

$$\Delta S_T = R \ln(P_1 / P_2); \ \text{Btu} / \text{lbmol}°\text{R} \tag{4.18}$$

The entropy change of one mole of an ideal gas undergoing a physical change of state from T_1 to T_2 at a constant pressure is given by

$$\Delta S_p = C_p \ln(T_2 / T_1); \ C_p(\text{gas}) = \text{constant} \tag{4.19}$$

Correspondingly, the entropy change for an ideal gas undergoing a physical change from (P_1, T_1) to (P_2, T_2) is

$$\Delta S = R \ln(P_1 / P_2) + C_p \ln(T_2 / T_1) \tag{4.20}$$

Some fundamental facts relative to the entropy concept are discussed below. The entropy change of a system may be positive (+), negative (-), or zero (0); the entropy change of the surroundings during this process may likewise be positive, negative, or zero. *However*, note that the total entropy change, ΔS_T, must be equal to or greater than zero:

$$\Delta S_T \geq 0 \tag{4.21}$$

The equality sign applies if the change occurs *reversibly* and *adiabatically*.

Second law calculations will be revisited in Chapter 6 and in Part III, Chapter 27, Heat Transfer Term Projects.

The *third law of thermodynamics* is concerned with the absolute values of entropy. By definition; the entropy of all pure crystalline materials at absolute zero temperature is exactly zero. Note however, that chemical engineers are usually concerned with *changes* in thermodynamic properties, including entropy.

4.4   Phase Equilibrium

Relationships governing the equilibrium distribution of a substance between two phases, particularly gas and liquid phases, are the principal subject matter of phase-equilibrium thermodynamics. These relationships form the basis of calculational procedures that are employed in the design and the prediction of the performance of several mass transfer equipment and processes [6].

The most important equilibrium phase relationship is that between a liquid and a vapor. Raoult's and Henry's Laws theoretically describe liquid-vapor behavior and, under certain conditions, are applicable in practice. Raoult's Law is sometimes useful for mixtures of components of similar structure. It states that the partial pressure of any component in the vapor phase is equal to the product of the vapor pressure of the pure component and the mole fraction of that component in the liquid:

$$p_i = p_i' x_i \qquad (4.22)$$

where p_i = partial pressure of component i in the vapor
p_i' = vapor pressure of pure i at the same temperature
x_i = mole fraction of component i in the liquid

This expression may be applied to all components. If the gas phase is ideal, this equation becomes

$$y_i = \left(p_i' / P \right) x_i \qquad (4.23)$$

where y_i = mole fraction of component i in the vapor
P = total system pressure

Unfortunately, relatively few mixtures follow Raoult's law. Henry's law is a more empirical relation used for representing data for many systems:

$$p_i = H_i x_i \qquad (4.24)$$

where H_i = Henry's law constant for component i (in units of pressure) If the gas behaves ideally, the above equation may be written as

$$y_i = m_i x_i \qquad (4.25)$$

where m_i = constant (dimensionless)

In some engineering applications, mixtures of condensable vapors and noncondensable gases must be handled. A common example is water vapor and air; a mixture of organic vapors and air is another such example that often appears in air pollution applications. Condensers can be used to control organic emissions to the atmosphere by lowering the temperature of the gaseous stream, although an increase in pressure will produce the same result. The calculation for this is often accomplished using the *phase equilibrium constant K*. This constant has been referred to in industry as *a phase componential split factor* since it provides the ratio of the mole fractions of a component in the two equilibrium phases. The defining equation is

$$K_i = y_i / x_i \qquad (4.26)$$

where K_i = phase equilibrium constant for component i (dimensionless).

As a first approximation, K_i is generally treated as a function only of the temperature and pressure. For ideal gas conditions, K_i may be approximated by

$$K_i = p_i' / P \qquad (4.27)$$

where p' is the vapor pressure.

Many of the phase equilibrium calculations involve hydrocarbons. Fortunately, most hydrocarbons approach ideal gas behavior over a fairly wide range of temperatures and pressures. Values for K_i for a large number of hydrocarbons are provided in two DePriester nomographs, which are available in the literature. These two nomographs or charts were originally developed by DePriester in 1953 [8]. Additional details are provided in Problem 3 in the next section.

The aforementioned DePriester charts are a valuable source of vapor-liquid equilibrium data for many hydrocarbons that approach ideal behavior. However, it should be noted that the DePriester chart data are based on experimental data. The fact that these compounds approach ideal behavior allows the data to be presented in a simple form, i.e., as a function solely of temperature and pressure.

Although Raoult's law was included in the early analysis, non-ideal deviations can be accounted for by two theoretical models that have been verified by rather extensive experimental data. The two models are:

1. Wilson's method [9]
2. NRTL model [10]

Additional details are available in the literature [1].

4.5   Chemical Reaction Equilibrium

With regard to chemical reactions, two important questions are of concern to the engineer: (1) how *far* will the reaction go; and (2) how *fast* will the reaction go? Chemical thermodynamics provides the answer to the first question; however, it tells nothing about the second. Reaction rates fall within the domain of chemical reaction kinetics. To illustrate the difference and importance of both questions in an engineering analysis of a chemical reaction, consider the following process: Substance A, which costs 1 cent/ton, can be converted to B, which is worth 1 million dollars/mg, by the reaction A→B. Chemical thermodynamics will provide information on the maximum amount of B that can be formed. If 99.99% of A can be converted to B, the reaction would then *appear* to be economically feasible, from a *thermodynamic* point of view. However, a *kinetic* analysis might indicate that the reaction is so slow that, for all practical purposes, its rate is vanishingly small. For example, it might take 10^6 years to obtain a 10^{-6}% conversion of A. The reaction is then economically unfeasible. Thus, it can be seen that both equilibrium and kinetic effects must be considered in an overall engineering analysis of a chemical reaction [2]. This topic receives treatment in Chapter 8.

A rigorous, detailed presentation of this equilibrium topic is beyond the scope of this chapter and this text. However, superficial treatment is presented to provide at least a qualitative introduction to chemical reaction equilibrium. As will be shown in Chapter 8, if a chemical reaction is conducted in which reactants go to products, the products will be formed at a rate governed (in part) by the concentration of the reactants and conditions such as temperature and pressure. Eventually, as the reactants form products and the products react to form reactants, the *net* rate of reaction must equal zero. At this point, equilibrium will have been achieved.

Chemical reaction equilibrium calculations are structured around a thermodynamic term referred to as *free energy*, G. This so called energy is a thermodynamic property that cannot be easily defined without some basic grounding in thermodynamics. No attempt will be made to define it here, and the interested reader is directed to the literature [2] for further development of this term. Note that free energy has the same units as enthalpy, and may be used on a mole or total mass basis.

Consider the *equilibrium* reaction:

$$aA + bB = cC + dD \tag{4.28}$$

For this reaction

$$\Delta G^0_{298} = c\left(\Delta G^0_f\right)_C + d\left(\Delta G^0_f\right)_D - a\left(\Delta G^0_f\right)_A - b\left(\Delta G^0_f\right)_B \tag{4.29}$$

The standard free energy of reaction ΔG^0 may be calculated from standard free energy of formulation data, in a manner similar to that for the standard enthalpy of reaction. The following equation is used to calculate the *chemical reaction equilibrium constant K* at a temperature *T*:

$$\Delta G^0_T = -RT \ \ln(K) \tag{4.30}$$

The effect of temperature on the standard free energy of reaction, $\Delta G_T^{\,0}$, and the chemical reaction equilibrium constant, K, is available in the literature [1–4] and has been developed in a manner similar to that presented earlier for the effect of temperature on the enthalpy of reaction. Once the chemical reaction equilibrium constant (for a particular reaction) has been determined, one can proceed to estimate the quantities of the participating species at equilibrium [1–4].

A detailed and expanded treatment of thermodynamics is available in the following four references.

1. L. Theodore, F. Ricci, and T. VanVliet, *Thermodynamics for the Practicing Engineer*, John Wiley & Sons, Hoboken, NJ, 2009 [1].
2. J. Smith, H VanNess, and M. Abbott, *Chemical Engineering Thermodynamics*, 6th edition, McGraw-Hill, New York City, NY, 2001 [2].

3. D. Green and R. Perry (editiors), *Perry's Chemical Engineers' Handbook*, 8th edition, McGraw-Hill, New York City, NY, 2008 [5].

4. L. Theodore, *Chemical Engineering: The Essential Reference*, McGraw-Hill, New York City, NY, 2014 [11].

4.6   Illustrative Open-Ended Problems

This and the last Section provide open-ended problems. However, solutions *are* provided for the three problems in this Section in order for the reader to hopefully obtain a better understanding of these problems, which differ from the traditional problems/illustrative examples. The first problem is relatively straightforward while the third (and last problem) is somewhat more complex. Note that solutions are not provided for the 39 open-ended problems in the next Section.

Problem 1: Define vaporization and discuss its importance to the chemical engineer.

Solution: At any one temperature, all molecules move with an average velocity that is independent of state or structure. The temperature at which a phase change occurs can be determined by an energy balance. At the temperature where the energy of any molecule is sufficient to overcome the attractive forces of its neighbors, it will change from a condensed phase to a vapor phase. The phenomenon is the same whether the condensed phase is liquid or solid. When a solid vaporizes, the term *sublimation* (rather than vaporization) is used to describe the phenomenon.

There is a wide distribution of energies among molecules at any one temperature. Hence, if one considers the external surface of a condensed phase as a location from which molecules can escape, some molecules will always have sufficient energy to overcome the attractive forces of their neighbors. If the movement is not constrained, these molecules will move (or escape) into the vapor; this process is defined as *evaporation*. Since the molecules on the surface of a condensed phase are attracted inward on only three sides, it is reasonable to postulate that all the escaping molecules originate at the interphase.

Problem 2: This open-ended problem is concerned with a discussion of vapor-solid equilibria. The relation at constant temperature between the amount of substance adsorbed by an adsorbent (solid) and the equilibrium partial pressure or concentration is called the *adsorption isotherm*. The adsorption isotherm is the most important and by far the most often used of the various equilibria data.

Most available data on adsorption systems are determined at equilib-rium conditions. *Adsorption equilibrium* is the set of conditions at which the number of molecules arriving on the surface of the adsorbent equals the number of molecules that are leaving. An adsorbent is then said to be "saturated with vapors" and can adsorb no more vapors. Equilibrium determines the maximum amount of vapor that may be adsorbed on the solid at a given set of operating conditions. Although a number of vari-ables affect adsorption, the two most important ones in determining equi-librium for a given system are temperature and pressure. Three types of equilibrium graphs and/or data are used to describe adsorption systems: *isotherm* at constant temperature, *isobar* at constant pressure, and *isostere* at constant amount of vapors adsorbed.

The most common and useful adsorption equilibrium data is the adsorption isotherm. The isotherm is a plot of the adsorbent capacity vs the partial pressure of the adsorbate at a constant temperature. Adsorbent capacity is usually given in weight percent, usually expressed as grams of adsorbate per 100 g of adsorbent. Figure 4.1 shows a typical example of an adsorption isotherm for carbon tetrachloride on activated carbon. Graphs of this type are used to estimate the size of adsorption systems. Attempts have been made to develop generalized equations that can predict adsorp-tion equilibrium from physical data. This is very difficult because adsorp-tion isotherms take many shapes depending on the atomic forces involved. Isotherms may be concave upward, concave downward, or "S" shaped. To date, most of the theories agree with data only for specific adsorbate-systems and are valid over limited concentration ranges.

Although this problem is concerned with isotherms, two additional adsorption equilibrium relationships are the aforementioned isostere and the isobar. The isostere is usually provided as a plot of the ln p vs

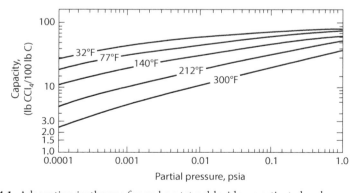

Figure 4.1 Adsorption isotherms for carbon tetrachloride on activated carbon.

$1/T$ at a constant amount of vapor adsorbed. Adsorption isostere lines are usually straight for most adsorbate-adsorbent systems. The isostere is important in that the slope of the isostere (approximately) corresponds to the heat of adsorption. The isobar is a plot of the amount of vapors adsorbed vs. temperature at a constant partial pressure. However, the adsorption isotherm is by far the most commonly used equilibrium relationship.

Relative to the isotherm, develop two equations that have been employed to describe vapor-solid adsorption equilibria data.

Comment: Refer to the work of Theodore [11] and Theodore and Ricci [12].

Solution: Several models have been proposed to describe the afore-mentioned vapor-solid equilibrium phenomena. Two such models are described below to represent the variation of the amount of adsorption per unit area or unit mass with partial pressure.

Freundlich proposed the equation:

$$Y = kp^{1/n} \tag{4.31}$$

where Y is the weight or volume of gas (or vapor) adsorbed per unit area or unit mass of adsorbent and p is the equilibrium partial pressure. The k and n terms are the empirical constants dependent on the nature of solid and adsorbate, and on the temperature. Equation (4.31) may be rewritten as follows. Taking logarithms of both sides,

$$\log Y = \log k + \left(\frac{1}{n}\right)\log p \tag{4.32}$$

If $\log Y$ is now plotted against $\log p$, a straight line should result with slope equal to $1/n$ and an ordinate intercept equal to $\log k$. Although the requirements of the equations are met satisfactorily at lower pressures, the experimental points curve away from the straight line at higher pressures, indicating that this equation does not have general applicability in repro-ducing adsorption of gases (or vapors) by solids.

A much better equation for isotherms was deduced by Langmuir from theoretical considerations. The final form is given as

$$Y = \frac{ap}{1+bp} \tag{4.33}$$

which can be rewritten as

$$\frac{p}{Y} = \frac{1}{a} + \left(\frac{b}{a}\right)p \qquad (4.34)$$

Since a and b are constants, a plot of p/Y vs. p should yield a straight line with slope equal to b/a and an ordinate intercept equal to $1/a$.

Other models of potential interest include those developed by Brunauer, Emmett, and Taylor [13], Polyani [14] and Dubinin and coworkers [15–16]

Problem 3: Discuss the applicability of the DePriester charts and how the information provided by the charts can be converted to a more useful form.
Comment: Refer to literature [1–3] for additional details.

Solution: One possible approach to representing the DePrister charts in equation form for the various components (I) is to evaluate the A, B, C, D, E coefficients in the equation [17].

$$K(I) = A(I) + B(I)*T + C(I)*T^2 + D(I)*T^3 + E(I)*T^3 \qquad (4.35)$$

This equation provides the variation of K with T. The effect of P on K may be approximated by

$$K(I)_p = K(I)_{p_0} (p_0 / p); \; p_0 = 1 \text{ atm} \qquad (4.36)$$

Numerical values for the five coefficients for a variety of compounds have been developed by Theodore [17].
Another approach is to express the relationship as

$$K(T)' = A(I) + B(I)*T + C(I)*T^2 \qquad (4.37)$$

Values of A, B, and C are also provided by Theodore [17] in tabular form.
It is important to note that the DePriester charts discussed earlier are based on the assumption of ideal gas behavior. This assumption is generally not valid for most mixtures containing inorganic components. No simple and reliable method is available for estimating K for both inorganics and organics in an inorganic-organic mixture.

4.7 Open-Ended Problems

This last Section of the chapter contains open-ended problems as they relate to thermodynamics. No detailed and/or specific solution is provided; that task is left to the reader, noting that each problem has either a unique solution or a number of solutions or (in some cases) no solution at all. These are characteristics of open-ended problems described earlier.

There are comments associated with some, but not all, of the problems. The comments are included to assist the reader while attempting to solve the problems. However, it is recommended that the solution to each problem should initially be attempted *without* the assistance of the comments.

There are 39 open-ended problems in this section. As stated above, if difficulty is encountered in solving any particular problem, the reader should next refer to the comment, if any is provided with the problem. The reader should also note that the more difficult problems are generally located at or near the end of the section.

1. Describe the early history associated with thermodynamics.
2. Discuss the recent advances in thermodynamics.
3. Select a refereed, published article on thermodynamics from the literature and provide a review.
4. Develop an original problem in thermodynamics that would be suitable as an illustrative example in a book.
5. Prepare a list of the various books which have been written on thermodynamics. Select the three best and justify your answer. Also select the three weakest books and justify your answer.
6. Provide your interpretation of the principle of corresponding states.
7. Describe the role critical properties play in non-ideal gas behavior calculations.
8. Provide a layman's definition of a reversible process.
9. Is it possible to circumvent the second law?
   Comment: Carefully review the various definitions provided in the literature for the second law.
10. Some individuals have claimed (and several books have been written) that Earth and the humans that inhibit Earth are all doomed. The basis of their argument is that the entropy of the universe is constantly on the rise so that a time will come when there will be no useful energy

available for doing useful work. Discuss the pros and cons of this argument.

Comment: Consider how long will it take for Earth to reach a state when there will be no useful energy available.

11. One of the authors once requested on an examination that the student provide a layman's definition and explanation of the second law. Provide your answer.

12. One of the authors once requested on an examination that the student provide a layman's definition and explanation of entropy. Provide your answer.

13. One of the authors once requested on an examination that the student provide a layman's definition and explanation of exergy. Provide your answer.

14. Describe the difference between entropy and exergy. How are they similar? Which approach better describes the second law?

15. Define the following terms:
    - Ideal work
    - Lost work
    - Minimum work
    - Maximum work

    Also discuss how these terms are related.

16. Define and discuss the third law of thermodynamics in both technical and layman terms.

17. Describe the various methods available in the literature for predicting heat capacity values (as opposed to obtaining values experimentally).

    Comment: Check both the internet and the literature.

18. Describe the various methods available in the literature for predicting standard enthalpy of reaction values (including combustion).

19. Describe the various methods available in the literature for predicting normal boiling point.

20. It has been proposed to use the sensible heat from the flue gas of a combustion device to reduce the energy needs of the combustion process. Discuss the advantages and disadvantages of the proposal.

    Comment: The total energy content of the gas should be taken into consideration based on the mass involved.

21. Refer to saturated steam-ice data. Generate an equation describing the specific volume of ice and steam as a function of temperature and pressure.

22. Refer to saturated steam-ice data. Generate an equation describing the specific volume of ice and steam as a function of enthalpy and pressure.

23. Refer saturated steam-ice data. Generate an equation describing the specific volume of ice and steam as a function of entropy and pressure.

24. Convert enthalpy-concentration data for H_2SO_4-H_2O at various temperatures into equation form.
    Comment: Refer to the literature [1–3] for data.

25. Convert enthalpy-concentration data for $NaOH$-H_2O at various temperatures into equation form.
    Comment: Refer to the literature [1–3] for data.

26. Obtain enthalpy of mixing at infinite dilution data for various gases and solids dissolved in a variety of liquids (not only water).

27. Can the standard enthalpy of reaction ever be positive? Explain.
    Comment: Carefully review the definition of the standard enthalpy of reaction.

28. Describe the effect of pressure on the enthalpy of reaction.
    Comment: Is this effect important in most real-world applications?

29. Define and explain relative volatility in layman terms.

30. Outline the various procedures available to calculate the phase equilibrium constant, and comment on their advantages and disadvantages.

31. Develop another law that combines/integrates the features of both Henry's and Raoult's laws.
    Comment: Carefully review both laws.

32. Develop another method to produce vapor-liquid equilibrium in nonideal solutions.

33. Describe the *Wilson method* employed for vapor-liquid equilibrium in layman terms [9].

34. Describe thr NRTL method employed for vapor-liquid equilibrium in layman terms [10].

35. Discuss the differences between the Wilson and NRTL methods for describing vapor-liquid equilibrium.

36. Develop another method that combines the features of both the NRTL and Wilson methods to describe non-ideal phase equilibria.
    Comment: Refer to the literature [2, 9–11].

37. Define and describe the chemical reaction equilibrium constant and the standard free energy.
38. Discuss the differences and the relationship between the free energy and the standard free energy.
39. Outline the various procedures available to calculate the chemical reaction equilibrium constant, and comment on their advantages and disadvantages.

References

1. L. Theodore, F. Ricci, and T. VanVliet, *Thermodynamics for the Practicing Engineer*, John Wiley & Sons, Hoboken, NJ, 2009.
2. L. Theodore and J. Reynolds, *Thermodynamics*, A Theodore Tutorial, East Williston, NY, originally published by the USEPA/APTI, RTP, NC, 1991.
3. J. Smith, H. Van Ness, and M. Abbott, *Chemical Engineering Thermodynamics*, 6th edition, McGraw-Hill, New York City, NY, 2001.
4. K. Pitzer, *Thermodynamics*, 3rd edition, McGraw-Hill, New York City, NY, 1995.
5. R. Perry and D. Green (editors), *Perry's Chemical Engineers' Handbook*, 8th edition, McGraw-Hill, New York City, NY, 2008.
6. Personal Notes, L. Theodore, East Williston, NY, 1990.
7. L. Theodore and F. Ricci, *Mass Transfer Operations for the Practicing Engineer*, John Wiley & Sons, Hoboken, NJ, 2010.
8. C. DePriester, *Chem Eng. Prog. Symp. Ser.*, 49(7), 42, New York City, NY, 1953.
9. G. Wilson, *J.Am. Chem. Soc.*, 86, 27–130, Washington DC, 1964.
10. H. Renon and J. Prausnitz, *AIChE J.*, 14, 135–144, New York City, NY, 1968.
11. L. Theodore, *Chemical Engineering: The Essential Reference*, McGraw-Hill, New York City, NY, 2014.
12. L. Theodore and F. Ricci, Class Note, Manhattan College, Bronx, NY John Wiley & Sons, Hoboken, NJ, 2010.
13. S. Brunauer, P. H. Emmett, and E. Teller, *J. Am. Chem. Soc.*, 60, 309, location unknown 1938.
14. K. Polanyi, *Tran. Faraday Soc.*, 28, 316, location unknown 1932.
15. J. Dubinin, *Chem Rev.*, 60, 235, location unknown 1960.
16. J. Dubinin, *Chemistry and Physics of Carbon*, Vol. 2, P. L. Walker, ed., Marcel Dekker, New York City, NY, 51, 1966.
17. Personal Notes, L. Theodore, East Williston, NY, 1974.

5

Fluid Flow

This chapter is concerned with fluid flow. As with all the chapters in Part II, there are several sections: overview, several specific technical topics, illustrative open-ended problems, and open-ended problems. The purpose of the first section is to introduce the reader to the subject of fluid flow. As one might suppose, a comprehensive treatment is not provided although several technical sections are included. The next section contains three open-ended problems; the authors' solutions (there may be other solutions) are also provided. The third (and final) section contains 42 problems; *no* solutions are provided here.

5.1 Overview

This overview section is concerned—as can be noted from its title—with fluid flow. As one might suppose, it was not possible to address all topics directly or indirectly related to fluid flow. However, additional details may be obtained from either the references provided at the end of this overview section and/or at the end of the chapter.

Note: Those readers already familiar with the details associated with this subject may choose to bypass this Overview.

This section is introduced by examining the units of some of the common quantities that are encountered in fluid flow. The momentum of a system is defined as the product of the mass and velocity of the system.

$$\text{Momentum} = (\text{Mass})(\text{Velocity}) \tag{5.1}$$

The engineering units (one set) for momentum are therefore, lb·ft/s. The units of time rate of change of momentum (hereafter referred to as the rate of momentum) are simply the units of momentum divided by time:

$$\text{Rate of Momentum} = \frac{\text{lb} \cdot \text{ft}}{\text{s}^2} \tag{5.2}$$

The above units can be converted to lb_f if multiplied by an appropriate conversion constant. The conversion constant in this case is

$$g_c = 32.2 \frac{(\text{lb} \cdot \text{ft})}{(\text{lb}_f \cdot \text{s}^2)} \tag{5.3}$$

This equation serves to define the conversion constant g_c. If the rate of momentum is divided by g_c as 32.2 $(\text{lb} \cdot \text{ft})/(\text{lb}_f \, \text{s}^2)$—the following units result

$$\text{Rate of momentum} = \left(\frac{\text{lb} \cdot \text{ft}}{\text{s}^2} \right)\left(\frac{\text{lb}_f \cdot \text{s}^2}{\text{lb} \cdot \text{ft}} \right) \tag{5.4}$$

$$\equiv \text{lb}_f$$

One may conclude from the above dimensional analysis that a force is equivalent to a rate of momentum [1]. The notation employed in the development that follows are those normally appearing in the chemical engineering literature and will therefore (in many instances) not be redefined.

Fluids are classified based on their rheological (viscous) properties. These are detailed below [1,2]:

1. Newtonian fluids: fluids that obey Newton's law of viscosity, i.e., fluids in which the shear stress is linearly proportional to the velocity gradient [2]. All gases are considered Newtonian fluids and nearly all liquids of a simple chemical formula are considered Newtonian fluids. Newtonian liquid examples are water, benzene, ethyl alcohol, hexane and sugar solutions.
2. Non-Newtonian fluids: fluids that do not obey Newton's law of viscosity; they are generally complex mixtures, e.g., polymer solutions, slurries, etc.

The remainder of this Chapter addresses the following topics:

1. Basic Laws
2. Key Fluid Flow Equations
3. Prime Movers
4. Fluid – Particle Applications

The reader should note that the bulk of the material in this chapter has been drawn from I. Farag, *Fluid Flow*, A Theodore Tutorial, East Williston, NY, originally published by the USEPA/APTI, RTP, NC [2]. In addition, topic (2) – Key Fluid Flow Equations – are highlighted in the presentation to follow.

5.2   Basic Laws

The conservation law for energy finds application in many chemical process units such as heat exchangers, reactors, and distillation columns, where shaft work plus kinetic and potential energy changes are negligible compared with heat flows and either internal energy or enthalpy changes. Energy balances (see two previous chapters) on such units therefore reduce to $Q = \Delta E$ (for a closed non-flow system) or $\dot{Q} = \Delta \dot{H}$ (an open flow system) [3].

Applying the conservation law of energy mandates that all forms of energy entering the system equal that of those leaving. Expressing all terms in consistent units (e.g., energy per unit mass of fluid flowing), results in the total energy balance presented in Equation (5.5).

$$P_1 V_1 + \frac{v_1^2}{2g_c} + \frac{g}{g_c} + E_1 + Q + W_s = P_2 V_2 + \frac{v_2^2}{2g_c} + \frac{g}{g_c} z_2 + E_2 \qquad (5.5)$$

Equation (5.5) may also be written as

$$\frac{v_1^2}{2g_c} + \frac{g}{g_c}z_1 + H_1 + Q + W_s = \frac{v_2^2}{2g_c} + \frac{g}{g_c}z_2 + H_2 \qquad (5.6)$$

or simply

$$\frac{\Delta v^2}{2g_c} + \frac{g}{g_c}\Delta z_1 + \Delta H = Q + W_s \qquad (5.7)$$

Note that Δ refers to a difference between the value at station 2 (the usually designation for the outlet) minus that at stations 1 (the inlet).

5.3   Key Fluid Flow Equations

5.3.1   Reynolds Number

The Reynolds number, Re, is a dimensionless quantity, and is a measure of the relative ratio of inertia to viscous forces in the fluid:

$$Re = \rho VL / \mu$$
$$= VL / v \qquad (5.8)$$

$$\text{where } L = \text{a characteristic length}$$
$$V = \text{average velocity}$$
$$\rho = \text{fluid density}$$
$$\mu = \text{dynamic (or absolute) viscosity}$$
$$v = \text{kinematic viscosity}$$

In flow through round pipes and tubes, L is the length and D is the diameter. The Reynolds number provides information on flow behavior. It is particularly useful in scaling up bench-scale or pilot plant data to full-scale applications.

Laminar flow is usually encountered at a Reynolds number, Re, below approximately 2100 in a circular duct, but it can persist up to higher Reynolds numbers. Under ordinary conditions of flow, the flow (in circular ducts) is turbulent at a Reynolds number above approximately 4000. Between 2100-4000, where the type of flow may be either laminar or turbulent, the predictions are unreliable. The Reynolds numbers at which the

fluid flow changes from laminar to transition or to turbulent are termed
critical numbers. In the case of flow in circular ducts there are two critical
Reynolds numbers, namely the aforementioned 2100 and 4000. Different
Re criteria exist for geometries other than pipes [2].

5.3.2   Conduits

Fluids are usually transported in tubes or pipes. Generally speaking,
pipes are heavy-walled and have a relatively large diameter. Tubes are
thin-walled and often come in coils. Pipes are specified in terms of their
diameter and wall thickness. The nominal diameters range from 1/8 to 30
inches for steel pipes. Standard dimensions of steel pipe are available in
the literature [1,3] and are known as IPS (iron pipe size) or NPS (nominal
pipe size). The pipe wall thickness is indicated by the schedule number.
Tube sizes are indicated by the outside diameter. The wall thickness is usually given by the outside diameter. The wall thickness is usually given by
the Birmingham wire gauge (BWG) number. The smaller the BWG, the
heavier the tube.

5.3.3   Mechanical Energy Equation – Modified Form

Abulencia and Theodore [3] provide the following equation:

$$\frac{\Delta P}{\rho} + \frac{\Delta v^2}{2g_c} + \frac{g}{g_c}\Delta z - \eta W_s + \sum F = 0 \tag{5.9}$$

This was defined as the *mechanical energy equation*. Equation (5.9) was
rewritten without the pump work and friction terms.

$$\frac{\Delta P}{\rho} + \frac{\Delta v^2}{2g_c} + \frac{g}{g_c}\Delta z = 0 \tag{5.10}$$

This equation was defined as the basic form of the *Bernoulli equation*.
Equation (5.9) was also written as

$$\frac{P_1}{\rho}\frac{g_c}{g} + \frac{v_1^2}{2g} + z_1 = \frac{P_2}{\rho}\frac{g_c}{g} + \frac{v_2^2}{2g} + z_2 + h_s\frac{g_c}{g} + h_f\frac{g_c}{g} \tag{5.11}$$

The h terms were included above to represent the loss of energy due to friction in the system. Frictional loss can take several forms. An important chemical engineering problem is the calculation of these losses. It was noted (earlier) that the fluid can flow in either of two modes – laminar or turbulent. For laminar flow, an equation is available from basic theory to calculate friction loss in a pipe. In practice, however, fluids (particularly gases) are rarely moving in laminar flow.

5.3.4   Laminar Flow Through a Circular Tube

Fluid flow in circular tubes (or pipes) is encountered in many applications, and is always accompanied by friction. Consequently, there is energy loss, indicating a pressure drop in the direction of flow. One can theoretically derive the h_s term for laminar flow [1]. The equation can be shown to take the form

$$h_s = \frac{32\mu v L}{\rho g_c D^2} \qquad (5.12)$$

for a fluid flowing only through a straight cylinder of diameter D and length L. A friction factor, f, that is dimensionless, may now be defined (for laminar flow):

$$f = \frac{16}{Re} \qquad (5.13)$$

so that Equation (5.12) takes the form

$$h_s = \frac{4fLv^2}{2g_c D} \qquad (5.14)$$

Although this equation describes friction loss across a conduit of length, L, it can also be used to predict the pressure drop due to friction per unit length of conduit, i.e., $\Delta P/L$, by simply dividing the above equation by L.

It should also be noted that another friction factor term exists, which differs from that presented in Equation (5.13). In this other case, f_D is defined as

$$f_D = \frac{64}{Re} \qquad (5.15)$$

The f_D term is used to distinguish the difference of Equation (5.13) from that of Equation (5.15). In essence:

$$f_D = 4f \qquad (5.16)$$

The term f is defined as the Fanning friction factor while f_D is defined as the Darcy or Moody friction factor [4]. Care should be taken as to which of the friction factors are being used in calculations. This will become more apparent shortly. In general, chemical engineers employ the Fanning friction factor; other engineers prefer the Darcy (or Moody) factor. This Chapter employs the Fanning friction factor.

Employing Equation (5.14), Equation (5.11) may be extended in the absence of pump work and rewritten as

$$\frac{\Delta P}{\rho} + \frac{\Delta v^2}{2g_c} + \Delta z \frac{g}{g_c} + \frac{4fLv^2}{2g_c D} = 0 \qquad (5.17)$$

The symbols Σh_c and Σh_e, representing the sum of the contraction and expansion losses, respectively, may also be added to the equation as provided below in Equation (5.18). (These effects will be discussed later in this Section).

$$\frac{\Delta P}{\rho} + \frac{\Delta v^2}{2g_c} + \Delta z \frac{g}{g_c} + \frac{4fLv^2}{2g_c D} + \Sigma h_c + \Sigma h_e = 0 \qquad (5.18)$$

5.3.5   Turbulent Flow Through a Circular Conduit

It is important to note that almost all the key fluid flow equations presented for laminar flow apply as well to turbulent flow, provided the appropriate friction factor is employed. The effect of the Reynolds number of the Fanning friction factor is provided in Figure 5.1. Note that Equation (5.13) appears on the far left-hand side of Figure 5.1.

In the turbulent regime, the "roughness" of the pipe becomes a consideration. In his original work on the friction factor, Moody [4] defined the

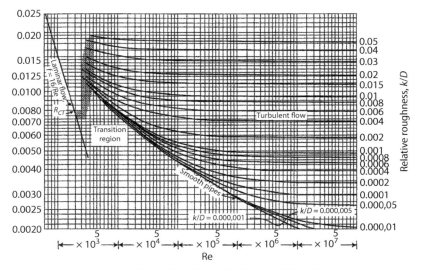

Figure 5.1   Fanning Friction factor (*f*) – Reynolds Number (Re) plot

term k (or ε), as the roughness and the ratio, k/D, as the relative roughness. Thus, for rough pipes/tubes in turbulent flow:

$$f = f(Re, k/D) \tag{5.19}$$

This equation reads that the friction factor is a function of *both* the Re and k/D. However, as can be seen in Figure 5.1, the dependency on the Reynolds number is a weak one in turbulent flow.

5.3.6   Two Phase Flow

The simultaneous flow of two phases in pipes (as well as other conduits) is complicated by the fact that the action of gravity tends to cause settling and "slip" of the heavier phase; the result is that the lighter phase flows at a different velocity in the pipe than does the heavier phase. The results of this phenomena are different depending on the classification of the two phases, the flow regime, and the inclination of the pipe (conduit). As one might suppose, the major industrial application in this area is gas (G) – liquid (L) flow in pipes.

See Abulencia and Theodore [3] for additional details on this class of flow. This topic is revisited in Part III, Chapter 26 – Fluid Flow Term Projects.

5.3.7   Prime Movers

Three devices thay convert electrical energy into the mechanical energy that is applied to various streams are discussed in this section. These devices are: fans, which move low-pressure gases; pumps, which move liquids and liquid-solid mixtures such a slurries, suspensions, and sludges; and compressors, which move (compress) high-pressure gases.

There are three general process classifications of prime movers—centrifugal, rotary, and reciprocating—that can be selected. Except for special applications, centrifugal units are normally employed in industry. These units are usually rated in terms of the four characteristics listed below:

1. Capacity: the quantity of fluid discharge per unit time (the mass flow rate);
2. Increase in pressure, often reported for pumps as *head*: head can be expressed as the energy supplied to the fluid per unit weight and is obtained by dividing the increase in pressure (the pressure change) by the fluid density;
3. Power: the energy consumed by the mover per unit time; and
4. Efficiency: the energy supplied to the fluid divided by the energy supplied to the unit.

Finally, the net effect of almost all prime movers is to increase the pressure of the fluid.

5.3.8   Valves and Fittings

Pipes and tubing (and other conduits) are used for the transportation of gases, liquids, and slurries. These ducts are often connected and may also contain a variety of valves and fittings, including expansion and contraction units. Types of connecting conduits include:

1. threaded;
2. bell-and-spigot;
3. flanged; and
4. welded.

Extensive information on these classes of connections is available in the literature [1,3].

5.4   Fluid-Particle Applications

Fluid-particle studies find wide applications in practice. Consider the following scenario. If a particle is initially at rest in a stationary gas and is then set in motion by the application of a constant external force or forces, the resulting motion occurs in two stages. The first period involves acceleration, during which time the particle velocity increases from zero to some maximum velocity. The second stage occurs when the particle achieves this maximum velocity and remains constant. During the second stage, the particle is not accelerating. This final, constant, and maximum velocity attained is defined as the *terminal settling velocity* of the particle. Most particles reach their terminal settling velocity almost instantaneously. These velocities are given by the following three equations: For the Stokes' law range, i.e., $Re < 2.0$

$$v = \frac{f d_p^2 \rho_p}{18\mu} \tag{5.20}$$

For the intermediate range, i.e. $2.0 < Re < 500$,

$$v = \frac{0.153 f^{0.71} d_p^{1.14} \rho_p^{0.71}}{\mu^{0.43} \rho^{0.29}} \tag{5.21}$$

Finally, for Newton's law range, i.e. $Re > 500$,

$$v = 1.74(f d_p \rho_p \rho)^{0.5} \tag{5.22}$$

Note that Re for these equations is based on the particle diameter, d_p. Keep in mind that f denotes the external force per unit mass of particle. One consistent set of units (English) for the above equations is ft/s^2 for f, ft for d_p, lb/ft^3 for ρ, lb/ft·s for μ, and ft/s for v [5].

When particles approach sizes comparable to the mean free path of other fluid molecules, the medium can no longer be regarded as continuous since particles can fall between the molecules at a faster rate than that predicted by aerodynamic theory. Cunningham's correction factor [6] is introduced to Stokes' law to allow for this *slip*.

$$v = \frac{g d_p^2 \rho_p}{18\mu} C$$

(5.23)

where C is the Cunningham correction factor (CCF), and

$$C = 1 + \frac{2A\lambda}{d_p}$$

(5.24)

The term A is given by $1.257 \times 10^{0.40} \exp(-1.10 d_p/2\lambda)$ and λ is the mean-free path of the fluid molecules (6.53×10^{-6} cm for ambient air). The CCF is usually applied to particles equal to or smaller than 1 micron. Applications include particulate air pollution studies [5] and nanotechnology [7].

5.4.1 Flow Through Porous Media

The flow of a fluid through porous media and/or a packed bed occurs frequently in chemical process applications and represents an extension of the fluid-particle discussion presented previously. Some chemical engineering examples include flow through a fixed-bed catalytic reactor, flow through an adsorption tower, and flow through a filtration unit. An understanding of this type of flow is also important in the study of some particle dynamics applications and fluidization [8,9].

A porous medium is a continuous (relatively speaking) solid phase with many void spaces called pores. Examples include sponges, paper, sand, and concrete. Packed beds of porous material are used in a number of chemical engineering operations, e.g., distillation, adsorption, filtration, and drying. Porous media are divided into:

1. impermeable media: solid media in which the pores are not interconnected, e.g., foamed polystyrene; and
2. permeable media: solid media in which the pores are interconnected, e.g., packed columns and catalytic reactors.

5.4.2 Filtration

Another fluid-particle application involves filtration. This operation is one of the most common chemical engineering applications that involve the flow of fluids through packed beds. As carried out industrially, it is similar to the filtration carried out in the chemical laboratory using a filter paper in a funnel. The object is still the separation of a solid from the fluid in which it is carried and the separation is accomplished by allowing (usually by force) the fluid through a porous filter. The solids are trapped within the pores of the filter and (primarily) build up as a layer on the surface of this filter. The fluid, which may be either gas or liquid, passes through the bed of the solids and through the retaining filter. Abulencia and Theodore [3] provide additional filtration details plus design and predictive equations.

5.4.3   Fluidization

Fluidization is yet another fluid-particle application. It may be viewed as the operation in which a fluid (gas or liquid) transforms fine solids into a fluid-like state. Excellent particle-fluid contact results. Consequently, fluidized beds are used in many applications, e.g., oil cracking, zinc roasting, coal combustion, gas desulfurization, heat exchanges, plastic coating, and fine powder granulation. Once again, Abulencia and Theodore [3] provide details.

A detailed and expanded treatment of fluid flow principles is available in the following two references.

1. P. Abulencia and L. Theodore, *Fluid Flow for the Practicing Chemical Engineer*, John Wiley & Sons, Hoboken, NJ, 2009.[3]
2. L. Theodore, *Chemical Engineering: The Essential Reference*, McGraw-Hill, New York City, NY 2014 [10].

5.5   Illustrative Open-Ended Problems

This and the last Section provide open-ended problems. However, solutions *are* provided for the three problems in this Section in order for the reader to hopefully obtain a better understanding of these problems which differ from the traditional problems/illustrative examples. The first problem is relatively straightforward while the third (and last problem) is somewhat more difficult and/or complex. Note that solutions are not provided for the 42 open-ended problems in the next Section.

Problem 1: A pump is no longer capable of delivering the required flow rate to a system. Rather than purchase a new pump, you have been asked to list and/or describe what steps can be taken to resolve the problem. Replacing the pump is not an option.

Solution: Since replacing the pump is not an option, some of the many other options can include:

1. Carefully check the pump, including the clogging of screens and/or intakes, impellers.
2. Decrease pipe size(s), i.e., the diameter (if feasible).
3. Decrease the pipe length (if possible).
4. Eliminate unimportant valves, expansion and contraction joints, etc.., in order to reduce the pressure drop.
5. Decrease the viscosity of the water by increasing its temperature.

Problem 2: A pressurized airplane flying at an altitude of 35,000 ft experiences a structural failure resulting in the loss of a portion of the external skin of the fuselage. The airplane may be considered a cylindrical vessel, 6 m in diameter and 50 m in length. The area of the hole created in the side of the plane is initially 4 m² and is located in the center of the plane, immediately above the wing. The pressure inside the plane before the failure is 1.0 bar and the pressure outside the plane is 0.85 bar. It is estimated that a lateral thrust in excess of 100 Newtons will result in structural failure of the fuselage.

1. Estimate the initial thrust (Newtons) normal to the fuselage when the hole opens up.
2. Determine the maximum area of the hole that can develop prior to structural failure of the fuselage.
3. Specify some design limitations for structural components of the fuselage (ribs and cross braces) based on the results of part (2).
4. Comment on the results.

Solution:

1. The force of the failure is calculated as

$$\text{Force} = (\text{Pressure})(\text{Area}) = (1 - 0.85 \text{ bar})(101.325 \text{N/m}^2 \cdot \text{bar})(4\text{m}^2)$$
$$= 60.8 \text{ N}$$

2. The maximum area of a hole that can develop prior to structural failure of the fuselage is based on the maximum thrust that is sustainable without failure:

$$\text{Area}_{\text{max}} = \frac{100\text{N}}{(1 \text{ bar} - 0.85 \text{ bar})(101.325 \text{N}/m^2 \cdot \text{bar})}$$
$$= 6.58 m^2$$

    3. Assuming that rivets connecting skin metal to ribs and cross braces are adequate to withstand forces of a skin failure, one option could include that supports be provided so that unsupported skin sections do not exceed approximately 6 m² in area for safety considerations.

    4. There are several assumptions in the above solution that need to be carefully reviewed and analyzed.

Problem 3: The suggested method of calculating the pressure drop of gas-liquid mixtures flowing in pipes is essentially that was originally proposed by Lockhart and Martinelli [10]. The basis of their correlation is that the two-phase pressure drop is equal to the single-phase pressure drop for either phase (G or L) multiplied by a factor that is a function of the single-phase pressure drops of the two phases. The equations for the total pressure drop per unit length Z, $(\Delta P/Z)_T$, are written as:

$$(\Delta P/Z)_T = Y_G (\Delta P/Z)_G \qquad (5.25)$$

$$(\Delta P/Z)_T = Y_L (\Delta P/Z)_L \qquad (5.26)$$

The terms Y_L and Y_G are functions of the variable X:

$$Y_G = F_G(X) \qquad (5.27)$$

$$Y_L = F_L(X) \qquad (5.28)$$

where

$$X = \left[\frac{(\Delta P / Z)_L}{(\Delta P / Z)_G} \right]^{0.5} \qquad (5.29)$$

Based on the above equation, the relationship between Y_L and Y_G is therefore given by

$$Y_G = X^2 Y_L \qquad (5.30)$$

The single-phase pressure-drops $(\Delta P/Z)_L$ and $(\Delta P/Z)_G$ can be calculated by assuming that each phase is flowing alone in the pipeline, and the phase in question is traveling at its superficial velocity where superficial velocities are based on the full cross-sectional area, S, of the pipe. Thus,

$$v_L = q_L/S \qquad (5.31)$$

and

$$v_G = q_G/S \qquad (5.32)$$

where v_L = liquid-phase superficial velocity
v_G = gas-phase superficial velocity
q_L = liquid-phase volume flow rate
q_G = gas-phase volume flow rate
S = pipe cross-sectional area

Note that either Equation (5.25) or (5.26) can be employed to calculate the pressure drop.

The functional relationships for Y_L and Y_G in Equations (5.27) and (5.28) were also provided by Lockhart and Martinelli [11] in terms of X for phase classification under different flow conditions. For gas-liquid flows, semi-empirical data were provided for the following three flow categories:

1. gas (turbulent flow) – liquid (turbulent flow);
2. gas (turbulent flow) – liquid (viscous flow); and
3. gas (viscous flow) – liquid (viscous flow).

For their correlation [10,11] ϕ_{vv} was expressed as

$$\left(\frac{\Delta P}{Z}\right)_{vv} = \phi^2_{vv}\left(\frac{\Delta P}{Z}\right)_G \qquad (5.33)$$

where ϕ_{vv} is a function of a dimensionless group, X_{vv}. The magnitude of ϕ_{vv} for values of X_{vv} is provided in Table 5.1.

Convert the above information into equation form.

Solution: The results in Table 5.1 were expressed in terms of Y_L and Y_G, both of which are functions of X_{vv}. VanVliet [14] subsequently regressed the data to a model of the form given below:

$$Y_{G(vv)} = 1.1241 + 3.7085X + 6.7318X^2 - 11.542X^3;\ X < 1 \qquad (5.34)$$

$$= 10 - 10.405X + 8.6786X^2 - 0.9167X^3;\ 1 < X < 10 \qquad (5.35)$$

$$= -78.333 + 7.3223X + 1.8957X^2 - 0.0087X^3;\ X > 10 \qquad (5.36)$$

$$Y_{L(vv)} = 3.9794X^{-1.6583}; X < 1 \tag{5.37}$$

$$= 6.4699X^{-0.556}; 1 < X < 10 \tag{5.38}$$

$$= 3.7013X^{-0.2226}; X > 10 \tag{5.39}$$

The above is just one correlation. The reader is encouraged to attempt to develop an improved set of equations.

5.6   Open-Ended Problems

This last Section of the chapter contains open-ended problems as they relate to fluid flow. No detailed and/or specific solution is provided; that task is left to the reader, noting that each problem has either a unique solution or a number of solutions or (in some cases) no solution at all. These are characteristics of open-ended problems described earlier.

Table 5.1 ϕ_{vv} vs $\sqrt{X_{vv}}$

$\sqrt{X_{vv}}$	ϕ_{vv}
0.2	1.40
0.4	1.69
0.6	1.93
0.8	2.16
1	2.44
2	3.81
3	5.15
4	6.4
6	8.7 (limit of experimental data)
.	.
.	.
.	.
	∞

There are comments associated with some, but not all, of the problems. The comments are included to assist the reader while attempting to solve the problems. However, it is recommended that the solution to each problem should initially be attempted *without* the assistance of the comments.

There are 42 open-ended problems in this Section. As stated above, if difficulty is encountered in solving any particular problem, the reader should next refer to the comment, if any is provided with the problem. The reader should also note that the more difficult problems are generally located at or near the end of the Section.

1. Describe the early history associated with fluid flow.
2. Discuss the recent advances in fluid flow technology.
3. Select a refereed, published article on fluid flow from the literature and provide a review.
4. Provide some normal everyday domestic applications involving the general topics of fluid flow.
5. Develop an original problem that would be suitable as an illustrative example in a book on fluid flow.
6. Prepare a list of the various books which have been written on fluid flow and/or fluid mechanics. Select the three best and justify your answer. Also select the three weakest books and justify your answer.
7. Develop a new experimental method to determine the viscosity of a liquid.
   Comment: Carefully review the experimental methods currently employed.
8. Develop a new experimental method to determine the viscosity of a gas.
   Comment: Carefully review the experimental methods currently employed.
9. Develop an equation to describe the viscosity of a slurry.
   Comment: Give due consideration to the concentration, shape and size distribution of the solids.
10. Describe the role surface tension plays in fluid flow.
11. Define and discuss in technical terms, the differences between Newtonian, pseudoplastic, dilatant, and Bingham plastic fluids.
12. Define and discuss in layman terms, the differences between Newtonian, pseudoplastic, dilatant, and Bingham plastic fluids.

13. Develop a new manometer for pressure (drop) measuring purposes.
    Comment: Carefully review the manometer writeup in Abulencia and Theodore's text [3].
14. Develop a new valve for flowing fluids.
15. Develop a new valve for two-phase flow.
16. Develop a new valve for high-viscosity fluids.
17. List some of the decisions that should come into play in the selection of a flow measure device.
18. Describe the various velocity profiles that can develop with flow through a conduit.
19. Discuss the differences between a fan, a pump, and a compressor.
20. Develop a better method of describing the classification and description of pipe standards.
21. Should the Reynolds number be based on the average or maximum velocity in a pipe/conduit? Justify your answer.
22. Describe at least two methods that can be used to determine the gas flow rate in a large underground pipe.
23. One of the authors [12] has argued that adding fine particulates to a gas flowing in a pipe will reduce its pressure drop. Comment on the proposal.
24. Develop/design a new fitting.
25. Develop/design a new steam trap.
    Comment: Review the steam trap literature.
26. Generate an equation describing the effect of both Reynolds number and pipe roughness on the Fanning friction factor.
    Comment: Refer to the literature [3].
27. Develop a general pressure drop equation for conduit flow where the cross-sectional area of the conduit varies.
28. Consider flow through a number of pipes or conduits of varying cross-sectional areas that are arranged in both series and parallel format. Develop a general equation describing the pressure drop across this system.
29. A fan is no longer capable of delivering the required flow rate to a system. Rather than purchase a new fan, you have been asked to list and/or describe what steps can be taken to resolve the problem. See also Problem 1 in the previous Section.
30. Define compressible flow and discuss the role it plays in fluid flow applications.

31. Discuss the advantages and disadvantages of the various methods employed to describe the size of a particle.
32. Define and discuss the differences between sedimentation, centrifugation, and flotation.
33. With reference to porous media, provide an explanation in layman terms of the difference(s) between porosity, void fraction, solid fraction, and permeability.
34. Derive both the Blake -Kozeny and Burke -Plummer equations. Comment on their validity. Also discuss the differences between the two equations.
35. Develop a new equation to describe the Cunningham Correction Factor as a function of particle size.
    Comment: Refer to L. Theodore's "Air Pollution Control Equipment Calculations" text [5].
36. Generate an equation describing the effect of Reynolds number on the (particle) drag coefficient.
    Comment: Refer to L. Theodore's "Air Pollution Control Equipment Calculations" text [5].
37. Discuss the effect of particle shape on the drag coefficient.
38. Develop a new filtration process for slurries.
    Comment: Carefully review the filtration writeup in Abulencia and Theodore's text [3].
39. Describe the procedure you would follow in an attempt to resurrect a pirate's ship loaded with gold that is located approximately 1.5 miles under the Caribbean Ocean. Comment on the problems that will arise in a project of this nature.
40. One option available to a plant manager when a tube within a heat exchanger fails is to simply plug the inlet of the tube. Develop an equation to describe the impact on the pressure drop across the exchanger as a function of both the number of tubes within the exchanger and the number of plugged tubes.
41. Morgano Consultants designed a crushing and grinding unit to operate with a specific discharge particle size distribution. Once the unit was installed and running, it operated with a different and larger particle size distribution. Rather than purchase a new unit, what options are available to bring the unit into compliance with the specified design size and distribution?
42. Doyle Engineers designed a fan to operate with a discharge pressure of 22 psia. Once the fan was installed and

running, it operated with a discharge pressure of 19 psia. Rather than purchase a new unit, what options are available to bring the unit into compliance with the specified design pressure?

References

1. L. Theodore, *Transport Phenomena for Engineers*, Theodore Tutorials, East Williston, NY, originally published by International Textbook CO., Scranton, PA, 1971.
2. I. Farag, *Fluid Flow*, A Theodore Tutorial, Theodore Tutorials, East Williston, NY, originally published by USEPA/ACTI, RTP, NC 1994.
3. P. Abulencia and L. Theodore, *Fluid Flow for the Practicing Chemical Engineer*, John Wiley & Sons, Hoboken, NJ, 2009 [3].
4. L. Moody, *Friction Factors for Dye Flow*, Trans. Am. Soc. Mech. Engrs., 66, 67 1-84, New York City, NY 1944.
5. L. Theodore, *Air Pollution Control Equipment Calculations*, John Wiley & Sons, Hoboken, NJ, 2008.
6. E. Cunningham, *Proc. R. Soc. London Ser.*, 17, 83, 357, location unknown, 1910.
7. L. Theodore, *Nanotechnology: Basic Calculations for Engineers and Scientists*, John Wiley & Sons, Hoboken, NJ, 2007.
8. C. Bennett and J. Myers, *Momentum, Heat, and Mass Transfer*, McGraw-Hill, New York City, NY 1962.
9. S. Ergun, *Fluid Flow Through Packed Columns*, CEP, 48:49, New York City, NY, 1952.
10. D. Green and R. Perry (editors), *Perry's Chemical Engineers' Handbook,* 8th edition, McGraw-Hill, New York City, NY, 2008.
11. L. Theodore, *Chemical Engineering: The Essential Reference*, McGraw-Hill, New York City, NY 2014 [10].
12. R. Lockhart and R. Martinelli, *Generalized Correlation of Two-Phase, Two-Component Flow Data*, CEP, New York City, NY, 45, 39-48, 1949.
13. R. Martinelli et al, *Two-Phase Two-Component Flow in the Viscous Region*, Trans AIChE, New York City, NY, 42, 681-705, 1946.
14. T. VanVliet: Personal correspondence to L. Theodore, Manhattan College, Bronx, NY, 2008.
15. Personal notes: L. Theodore, East Williston, NY 1983.

6

Heat Transfer

This chapter is concerned with heat transfer. As with all the chapters in Part II, there are several sections: overview, several technical topics, illustrative open-ended problems, and open-ended problems. The purpose of the first section is to introduce the reader to the subject of heat transfer. As one might suppose, a comprehensive treatment is not provided although numerous references are included. The second section contains three open-ended problems; the authors' solutions (there may be other solutions) and also provided. The third (and final) section contains 43 problems; *no* solutions are provided here. The reader should also note that four open-ended heat transfer projects can be found in Chapter 27, Part III.

6.1 Overview

This overview section is concerned—as can be noted from its title—with heat transfer. As one might suppose, it was not possible to address all topics directly or indirectly related to heat transfer. However, additional details may be obtained from either the references provided at the end of this Overview section and/or at the end of the chapter.

Note: Those readers already familiar with the details associated with this subject may choose to bypass this Overview.

A difference in temperature between two bodies in close proximity or between two parts of the same body results in heat flow from the higher temperature to the lower temperature. There are three different (and classic) mechanisms by which this heat transfer can occur:

1. conduction,
2. convection, and
3. radiation.

When the heat transfer is the result of molecular motion (e.g., the vibrational energy of molecules in a solid being passed along from molecule to molecule), the mechanism of transfer is *conduction*. When the heat transfer results from macroscopic motion, such as currents in a fluid, the mechanism is that of *convection*. When heat is transferred by electromagnetic waves, the mechanism is defined as *radiation*. In some industrial applications, more than one mechanism; i.e., a combination of mechanisms is usually involved in the transmission of heat. However, since each mechanism is governed by its own set of physical laws, it is beneficial to discuss them independently of each other.

The five topics addressed in the Section include:

1. Conduction
2. Convection
3. Radiation
4. Condensation, Boiling, Refrigeration, and Cryogenics
5. Heat Exchangers

The reader should note that the bulk of the material for this chapter was draw from I. Farag and J. Reynolds, *Heat Transfer*, Theodore Tutorials, East Williston, NY, originally published by the USEPA/APTI, RTP, NC, 1996 [1].

6.2   Conduction

The rate of heat flow by conduction is given by Fourier's law [2].

$$q = -kA\frac{dT}{dx} \qquad (6.1)$$

where q = heat flow rate (Btu/h)
   x = direction of heat flow (ft)
   k = thermal conductivity (Btu/hft°F)
   A = heat transfer area, a plane perpendicular to the x direction (ft^2)
   T = temperature (°F)

The negative sign reflects the fact that heat flow is from a high to low temperature, and therefore the sign of the derivative is opposite that of the direction of the heat flow.

Equation 6.1 may be written in the form of the general transfer rate equation:

$$\text{Transfer rate} = \frac{\text{Driving force}}{\text{Resistance}} \qquad (6.2)$$

Since q in Equation 6.1 is the heat flow rate and ΔT is the driving force, the L/kA (L is the length in the direction of flow) term may be considered to be the resistance to heat flow. This approach is useful when heat is flowing by conduction in sequence through different materials.

Consider, for example, a flat incinerator wall made up of three different layers: an insulating layer, a; steel plate, b; and an outside insulating layer, c. The total resistance to heat flow through the incinerator wall is the sum of the three individual resistances; i.e.,

$$R = R_a + R_b + R_c \qquad (6.3)$$

At steady state, the rate of heat flow through the wall consisting of the three layers is therefore, given by

$$q = \frac{T_1 - T_2}{(L_a / k_a A_a) + (L_b / k_b A_b) + (L_c / k_c A_c)} \qquad (6.4)$$

where k_a, k_b, k_c = thermal conductivity of each section (Btu/h·ft·°F)
A_a, A_b, A_c = area of heat transfer of each section (ft)^2; these are equal for
areas of constant cross section
L_a, L_b, L_c = thickness of each layer (ft)
T_1 = temperature at inside surface of insulating wall a (°F)
T_2 = temperature at outside surface of insulating wall c (°F)

In the above example, the heat is flowing through a slab of constant cross section. In many cases of industrial importance however, this is not the case. For example, in heat flow through the walls of a cylindrical pipe in a heat exchanger or a rotary kiln, the heat transfer area increases with distance displaced from the center of the cylinder. The heat flow in this case is given by

$$q = \frac{kA_{lm}\Delta T}{L} \tag{6.5}$$

The term, A_{lm} in Equation 6.5 represents the average heat transfer area, or more accurately, the log-mean average heat transfer area. This log-mean average can be calculated by

$$A_{lm} = \frac{A_2 - A_1}{\ln(A_2 / A_1)} \tag{6.6}$$

where A_2 = outer surface area of cylinder (ft^2)
A_1 = inner surface area of cylinder (ft^2)

6.3   Convection

The flow of heat from a hot fluid to a cooler fluid through a solid wall is a situation regularly encountered in engineering equipment; examples of such equipment are heat exchangers, condensers, evaporators, boilers, and economizers. The heat absorbed by the cool fluid or given up by the hot fluid may be *sensible* heat, causing a temperature change in the fluid or it may be *latent* heat, cause a phase change such as vaporization or condensation (see also Chapter 3). The rate of heat transfer between the two streams, assuming *no* heat loss to the surroundings, may be calculated by the enthalpy (*h*) change for either fluid:

$$q = \Delta H = \dot{m}_h \left(h_{h1} - h_{h2} \right) = \dot{m}_c (h_{c2} - h_{c1}) \tag{6.7}$$

Equation 6.7 is applicable to the heat exchange between two fluids whether a phase change is involved or not.

In order to design a piece of heat transfer equipment, it is not sufficient to only know the heat transfer rate calculated by the enthalpy balances described above. The rate at which heat can travel from the hot fluid, through the tube walls, and into the cold fluid, must also be considered in the calculation of the contact area. The slower this rate is, for given hot and cold fluid flow rates, the more contact area is required.

The rate of heat transfer through a unit of contact area is referred to as the *heat flux* or *heat flux density* and, at any point along the tube length, is given by

$$\frac{dq}{dA} = U\left(T_h - T_c\right) \tag{6.8}$$

where dq/dA = local heat flux density (Btu/h·ft²)
        U = local overall heat transfer coefficient (Btu/h·ft²·°F)
The use of this overall heat transfer coefficient (U) is a simple, yet powerful, concept. In most applications, it combines both conduction and convection effects, although transfer by radiation can also be included. Methods for calculation the overall heat transfer coefficient are available in the literature [3,4]. In actual practice it is not uncommon for vendors to provide a numerical value for U.

In order to apply Equation 6.8 to the *entire* heat exchanger, the equation must be *integrated*. This cannot be accomplished unless the geometry of the exchanger is first defined. For simplicity, one of the simplest exchangers will be assumed here – the double-pipe heat exchanger (see later section). This device consists of two concentric pipes. The outer surface of the outer pipe is normally well insulated so that no heat exchange with the surroundings may be assumed. One of the fluids flows through the center pipe while the other flows through the annular channel between the pipes. The fluid flows may be either *co-current*, when the two fluids flow in the same direction, or *countercurrent*, where the flows are in opposite directions. However, countercurrent arrangement is more efficient and is more commonly used.

For a heat exchanger, integration of Equation (6.8) along the exchanger length yields (after applying several simplifying assumptions),

$$q = UA\Delta T \tag{6.9}$$

This has come to be defined by some, including the authors of this text, as the *heat transfer equation*. The aforementioned simplifying assumptions are that U and ΔT do not vary along the length. Since this is not actually the case, both U and ΔT must be regarded as averages of some type. A more careful examination of Equation 6.8, assuming that only U is constant, would ultimately indicate that the appropriate average for ΔT is the log-mean average, i.e.,

$$\Delta T_{lm} = \frac{\Delta T_2 - \Delta T_1}{\ln(\Delta T_2 / \Delta T_1)} \tag{6.10}$$

where ΔT_1 and ΔT_2 are the temperature differences between the two fluids at the ends of the exchanger. The area term (A) in Equation 6.9 is the cylindrical contact area between the fluids. However, since a pipe of finite thickness separates the fluids, the cylindrical area (A) must first be defined. Any one of the infinite number of areas between and including the inside and outside surface areas of the pipe may be arbitrarily chosen for this purpose. The usual approach in practice to use either the inside (A_i) our outside (A_0) surface area; the outside area is the more commonly used of the two. Since the value of the overall heat transfer coefficient depends on the area chosen, it should be subscripted to correspond to the area on which it is based. Equation 6.9 based on the outside surface area now becomes

$$q = U_0 A_0 \Delta T_{lm} \tag{6.11}$$

Comparing Equation 6.11 to the general rate equation (6.2), it can be seen that $(U_0, A_0)^{-1}$ may be regarded as the resistance to heat transfer between the two fluids, i.e.,

$$R_o = \frac{1}{U_0 A_0} \tag{6.12}$$

The *total* resistance to heat transfer across the wall therefore may be divided into three contributions: the *inside film*, the *tube wall*, and the *outside film*. This is restated mathematically in Equations 6.13 and 6.14.

$$R_t = R_i + R_w + R_0 \tag{6.13}$$

or

$$\frac{1}{U_0 A_0} = \frac{1}{h_i A_i} + \frac{x}{k A_{lm}} + \frac{1}{h_0 A_0} \tag{6.14}$$

where h_i = inside film coefficient (Btu/h·ft²·°F)
h_o = outside film coefficient (Btu/h·ft²·°F)

These film coefficients are almost always determined experimentally. Many empirical correlations can be found in the literature for a variety of fluids and exchanger geometries. Typical values of film coefficients are available in the literature [3–6].

After a period of service, thin films of foreign materials such as dirt, scale or products of corrosion build up on the tube wall surfaces. As shown in Equations 6.15 and 6.16 these films R_{fi} and R_{fo} introduce additional resistances to heat flow, and reduce the overall heat transfer coefficient:

$$R_t = R_i + R_h + R_w + R_{fo} + R_o \qquad (6.15)$$

For this new condition,

$$\frac{1}{U_0 A_0} = \frac{1}{h_i A_i} + \frac{1}{h_{di} A_i} + \frac{x}{k A_{lm}} + \frac{1}{h_{do} A_o} + \frac{1}{h_0 A_0} \qquad (6.16)$$

where h_{di}=inside fouling factor (Btu/h·ft²·°F)
h_{do}=outside fouling factor (Btu/h·ft²·°F)
Typical values of fouling factors are provided in the literature [3–6].

6.4   Radiation

Radiation becomes important as a heat transfer mechanism only when the temperature of the source or system is very high. The driving force for conduction and convection is the temperature difference between the source and the receptor; the actual temperatures have only minor influence since it is the difference in temperatures that count. For these two mechanisms, it usually does not matter whether the temperatures are 100°F and 50°F or 500°F and 450°F. Radiation on the other hand is strongly influenced by the temperature level; as the temperature level increases, the effectiveness of radiation as a heat transfer mechanism increases rapidly. It follows that, at very low temperatures, conduction and convection are the major contributors to the total heat transfer; however, at very high temperatures, radiation is the controlling or predominant factor. At temperatures in between, the fraction contributed by radiation depends on such factors as the convection film coefficient and the nature of the radiating surface [3]. To cite two extreme examples, for large pipes losing heat by natural convection, the temperature at which radiation accounts for roughly one half of the total heat transmission is around room temperature; for fine wires, this temperature is significantly above that.

A perfect or ideal radiator, referred to as a *black body*, emits energy at a rate proportional to the *fourth* power of the absolute temperature T of the body. When two bodies exchange heat by radiation, the net heat exchange is then proportional to the difference between each T^4. This is shown in Equation 6.17, which is based on the classic Stefan-Boltzmann law of thermal radiation:

$$q = \sigma FA\left(T_h^4 - T_c^4\right) \tag{6.17}$$

where σ = 0.1714 x10^{-8} Btu/hr·ft^2·R^4(Stefan-Boltzmann constant)
   A = area of either surface (chosen arbitrarily)(ft^2)
   F = view factor (dimensionless)
   T_h = absolute temperature of the hotter body (°R)
   T_c = absolute temperature of the colder body (°R)

The view factor (F) above depends on the geometry of the system; i.e., the surface geometries of the two bodies plus the spatial relationship between them, and on the surface chosen for A. Values of F are available in the literature for many geometries [3–5]. It is important to note that Equation 6.17 applies only to black bodies and is valid only for thermal radiation.

Other types of surfaces besides black bodies are less capable of radiating energy, although the T^4 law is generally obeyed. The ratio of energy radiating from one of the *gray* (non-black) bodies to that radiating from the black body under the same conditions is defined as the *emissivity* (ε). For *gray* bodies, Equation (6.17) becomes

$$q = \sigma FF_\varepsilon A\left(T_h^4 - T_c^4\right) \tag{6.18}$$

where F_ε = emissivity function, dependent on the emissivity of each body .

Relationships for F_ε for various geometries are available in the literature [3–5].

6.5   Condensation, Boiling, Refrigeration, and Cryogenics

Phase-change processes involve changes (sometimes significantly) in the density, viscosity, heat capacity, and thermal conductivity of the fluid in question. The heat transfer process and the applicable heat transfer coefficients for boiling and condensation is generally more involved and complicated than that for a single-phase process. It is therefore not surprising that more real-world applications involving boiling and condensation require the use of empirical correlations; many of these are available in the literature [3–6].

One of the main cost considerations when dealing with refrigeration and cryogenics is the cost of building and powering the unit. This is a costly element in the overall economics, so it is important to efficiently transfer heat in the refrigeration and cryogenic processes. Since the cost of equipment can be expensive, there are a number of factors that should be considered when choosing the applicable equipment. Cryogenics plays a major role in the chemical processing industry. Its importance lies in such processes as the recover of valuable feedstocks from natural gas streams, upgrading the heat content of fuel gas, purifying many process and waste streams, producing ethylene, as well as other chemical processes.

6.6   Heat Exchangers

Heat exchangers are defined as equipment that affect the transfer of thermal energy (in the form of heat) from one fluid to another. The simplest exchangers involve the direct mixing of hot and cold fluids. Most industrial exchangers are those in which the fluids are separated by a wall. The latter type, referred to by some as a *recuperator*, can range from a simple plane wall between two flowing fluids to more complex configurations involving multiple passes, fins, or baffles. Conductive and convective heat transfer principles (see earlier subsections) are required to describe and design these units; radiation effects are generally neglected. Kern [3] and Theodore [4] provide an extensive predictive and design calculations, most of which are based on Equation 6.9 – the *heat transfer equation*.

The presentation in this section keys on the description of the various heat exchanger equipment (and their classification), not on the log-mean temperature difference driving force, ΔT_{tm}, the overall heat transfer coefficient, U, etc., developed earlier in the section on convection. However, the design and predictive equation is the *heat transfer equation* provided in Equation 6.9. Three heat exchangers are briefly discussed below.

The *double-pipe* unit consists of two concentric pipes. Each of the two fluids—hot and cold—flow either through the inside of the inner pipe or through the annulus formed between the outside of the inner pipe and the inside of the outer pipe. Generally, it is more economical (from a heat efficiency and design perspective) for the hot fluid to flow through the inner pipe and the cold fluid through the annulus, thereby reducing heat losses to the surroundings. In order to ensure sufficient contacting time, pipes longer than approximately 20 ft are extended by connecting them to *return bends*. The length of pipe is generally kept to a maximum of 20 ft because the weight of the piping may cause the pipes to sag. Sagging may allow the inner pipe to touch the outer pipe, distorting the annulus flow region (as well as the velocity profile/distribution) and disturbing proper operation. When two pipes are connected in a "U" configuration that entails a return bend, the bend is referred to as a *hairpin*. In some instances, several hairpins may be connected in series. Additional information and calculation details are provided by Kern [3] and Theodore [4].

Shell-and-tube (also referred to as tube and bundle) heat exchangers provide a large heat transfer area economically and practically. The tubes are placed in a bundle and the ends of the tubes are mounted in tube sheets. The tube bundle is enclosed in a cylindrical shell, through which the second fluid flows. Most shell-and-tube exchangers used in practice are of welded construction. The shells are built from a piece of pipe with flanged ends and necessary branch connections. The shells consist of seamless pipe up to 24 inches in diameter; they are made of bent and welded steel plates if above 24 inches. Channel sections are usually of built-up construction, with welding-neck forged-steel flanges, rolled-steel barrels and welded-in pass partitions. Shell covers are either welded directly to the shell, or are built-up constructions of flanged and dished heads and welding-neck forged-steel flanges. The tube sheets are usually nonferrous castings in which the holes for inserting the tubes have been drilled and reamed before assembly. Baffles are usually employed to both control the flow of the fluid outside the tubes and provide turbulence.

One method of increasing the heat transfer rate is to increase the surface area of the heat exchanger. This can be accomplished by mounting metal *fins* on a tube in such a way that there is good metallic contact between the base of the fin and the wall of the tube. With this contact, the temperature throughout the fins will approximate that of the temperature of the heating (or cooling) medium and the high thermal conductivity of most metals used in practice reduces the resistance to heat transfer by conduction in the fins. Consequently, the surface will be increased without more tubes [7].

Extended surfaces, or fins, are classified into longitudinal fins, transverse fins, and spine fins. *Longitudinal fins* (also termed straight fins) are attached continuously along the length of the surface. *Transverse or circumferential fins* are positioned approximately perpendicular to the pipe or tube axis and are usually used in the cooling of gases. *Annular fins* are examples of continuous transverse fins. *Spine of peg fins* employ cones or cylinders, which extend from the heat transfer surface, and are used for either longitudinal flow or cross flow.

Other heat exchangers including: evaporators, waste heat boilers, condensers, and quenchers, details of which are provided by Theodore [4]. Material on cooling towers is also available in the literature [8].

A detailed and expanded treatment of heat transfer is available in the following three references [9–11]:

1. D. Green and R. Perry (editors), *Perry's Chemical Engineers' Handbook,* 8th edition, McGraw-Hill, New York City, NY, 2008 [9].
2. L. Theodore, *Chemical Engineering: The Essential Reference,* McGraw-Hill, New York City, NY, 2014 [10].
3. L. Theodore, *Heat Transfer for the Practicing Engineer,* John Wiley & Sons, Hoboken, NJ, 2011 [11].

6.7   Illustrative Open-Ended Problems

This and the last section provide open-ended problems. However, solutions *are* provided for the three problems in this section in order for the reader to hopefully obtain a better understanding of these problems which differ from the traditional problems/illustrative examples. The first problem is relatively straightforward while the third (and last problem) is somewhat more difficult and/or complex. Note that solutions are not provided for the 43 open-ended problems in the next section. Additional case study projects are located in Chapter 27, Part III.

Problem 1: A double-pipe heat exchanger is designed to heat a discharge stream to a required temperature of 105°F. However, the exchanger is currently only partially heating the stream. You have been asked to briefly describe what steps can be taken to get the heat exchanger back to design specifications.

Solution: Depending on the type of heat exchanger, the approaches to bring the heat exchanger back to design specification are different. However, most of them include the following:

1. cleaning of heat exchanger;
2. installation or removal of insulation;
3. correction of leakage;

    4. upgrading or replacement of gaskets;
    5. increasing the mass flow rate of the cooling fluid; and,
    6. change the cooling medium (e.g., have colder water pass through the heat exchanger).

Problem 2: A plant has three streams to be heated (see Table 6.1) and three streams to be cooled (see Table 6.2). Cooling water (90°F supply, 155°F return) and steam (saturated at 250 psia) are available. Calculate the heating and cooling duties and indicate what utility (or utilities) should be employed. Comment on the type of utility that should be employed.

Solution: The sensible heating duties (in units of Btu/h) can now be computed and compared.

    Heating: 7,475,000 + 6,612,000 + 9,984,000 = 24,071,000 Btu/h
    Cooling: 12,600,000 + 4,160,000 + 3,150,000 = 19,910,000 Btu/h
    Heating – Cooling = 24,071,000 – 19,910,000 = 4,161,000 Btu/h

At a minimum, 4,161,000 Btu/h will have to be supplied by steam or another hot medium. The type of heating medium will be dictated by availability, location economics, safety considerations, etc. This problem will be revisited in Part III, Chapter 27 – *Heat Transfer Term Projects*.

Table 6.1 Streams to be Heated in Problem 2

Stream	Flowrate, lb/h	c_p, Btu/lb·F	T_{in}, °F	T_{out}, °F
1	50,000	0.65	70	300
2	60,000	0.58	120	310
3	80,000	0.78	90	250

Table 6.2 Streams to be Cooled in Problem 2

Stream	Flowrate, lb/h	c_p, Btu/lb·F	T_{in}, °F	T_{out}, °F
4	60,000	0.70	420	120
5	40,000	0.52	300	100
6	35,000	0.60	240	90

Table 6.3 Heat Exchanger Duty Requirements for Problem 2.

Stream	Duty, Btu/h
1	7,475,000
2	6,612,000
3	9,984,000
4	12,600,000
5	4,160,000
6	3,150,000

The heat exchanger requirements and the physical layout of the exchangers receive treatment in a later chapter.

Problem 3: It would normally seem that the thicker the insulation on a pipe, the less the heat loss; i.e., increasing the insulation should reduce the heat loss to the surroundings. But, this is not always the case. There is a "critical insulation thickness" above which the system will experience a greater heat loss due to an increase in insulation. This situation arises for "small" diameter pipes when the increase in area increases more rapidly than the resistance of the thicker insulation. Develop an equation describing this phenomena. Provide a graphical analysis of your results.

Comment: Refer to the literature for additional details [3–4].

Solution: Consider the system shown in Figure 6.1. One notes that the area terms for the heat transfer equations in rectangular coordinates are no longer the same in cylindrical coordinates for pipes (e.g., for the inside surface, the heat transfer area is given by $2\pi r_i L$). Applying the general heat transfer equation to a pipe/cylinder system leads to:

$$g = \frac{T_i - T_o}{\dfrac{1}{2\pi r_i L}\left(\dfrac{1}{h_i}\right) + \dfrac{\Delta x_w}{k_w \, 2\pi L r_{lm,w}} + \dfrac{\Delta x_i}{k_i \, 2\pi L r_{lm,i}} + \dfrac{1}{2\pi r L}\left(\dfrac{1}{h_o}\right)}$$

$$= \frac{2\pi L(T_i - T_o)}{\dfrac{1}{r_i h_i} + \dfrac{\ln(r_o / r_i)}{k_w} + \dfrac{\ln(r / r_o)}{k_i} + \dfrac{1}{r h_o}} = \frac{2\pi L\left(T_i - T_o\right)}{f(r)} \qquad (6.19)$$

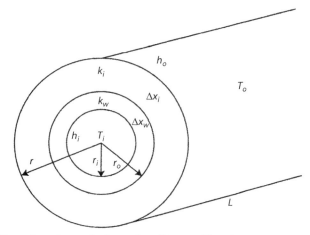

Figure 6.1 Critical insulation thickness for a Pipe; problem 3.

where

$$f(r) = \frac{1}{r_i h_i} + \frac{\ln(r_o / r_i)}{k_w} + \frac{\ln(r / r_o)}{k_i} + \frac{1}{r h_0} \tag{6.20}$$

Assuming that q goes through a maximum or minimum as r is varied, l'Hôpital's rule can be applied to Equation (6.20):

$$\frac{dq}{dr} = 2\pi L (T_i - T_o) \left[\frac{-\dfrac{df(r)}{dr}}{f(r)^2} \right] = 0 \tag{6.21}$$

with

$$-\frac{df(r)}{dr} = -\frac{1}{r k_i} + \frac{1}{r^2 h_0}$$

For $dq/dr = 0$, one may therefore write

$$\frac{df}{dr} = -\frac{df(r)}{dr} = \frac{1}{r k_i} + \frac{1}{r^2 h_0} = 0 \tag{6.22}$$

For this maximum/minimum condition, set $r = r_c$ and solve for r_c.

$$r_c = \frac{k_i}{h_o} \qquad (6.23)$$

The second derivative of dq/dr of Equation 6.21 provides information as to whether g experiences a maximum or minimum at r_c.

$$\frac{d}{dr}\left(\frac{dq_r}{dr}\right) + \frac{1}{r^2 k_i} - \frac{1}{r^3 h_o} = \frac{h_o^2}{k^3} - \frac{2h_o^2}{k^3} - \frac{h_o^2}{k^3}(1-2) \qquad (6.24)$$

Clearly, the second derivative is a negative number, is therefore a *maximum* $r = r_c$. The term q then decreases monotonously as r is increased beyond r_c. However, one should exercise care in interpreting the implications of the above development. This results applies *only* if r_o is *less* than r_c; i.e., it generally applies to "small" diameter pipes/tubes. Thus, r_c represents the outer radius (not the thickness) of the insulation that will maximize the heat loss and at which point any further increase in insulation thickness will result in an increase in heat loss.

A graphical plot of the resistance R versus r is provided in Figure 6.2. (The curve is inverted for the plot of q or r). One notes that the maximum heat loss from a pipe occurs when the critical radium equals the ratio of the thermal conductivity of the insulation to the surface coefficient of heat transfer. This ratio has the dimension of length (e.g., ft). The equation for r_i can be rewritten in terms of a dimensionless number defined as the Biot number Bi.

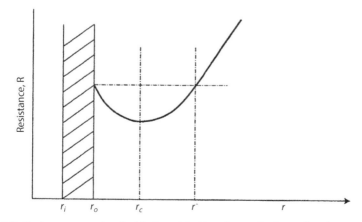

Figure 6.2 Resistance associated with the critical insulation thickness for a bare surface.

$$\frac{k_i}{h_o r_i} = \left(Bi\right)^{-1} \tag{6.25}$$

To reduce q below that for a bare wall ($r = r_o$), r must be greater than r_c; i.e., $r > r_c$. The radius at which this occurs is denoted as r^*. The term r^* may be obtained by solving the equation

$$q_{bare} = q_{1,r^*}. \tag{6.26}$$

so that

$$\frac{2\pi L(T_i - T_o)}{\dfrac{1}{r_i h_i} + \dfrac{\ln(r_o / r_i)}{k_w} + \dfrac{1}{r_o h_o}} = \frac{2\pi L(T_i - T_o)}{\dfrac{1}{r_i h_i} + \dfrac{\ln(r_o / r_i)}{k_w} + \dfrac{\ln(r^* / r_o)}{k_i} + \dfrac{1}{r^* h_o}} \tag{6.27}$$

One may now solve for r^* using a suitable trial-and-error procedure. See Figure 6.2 once again. This topic will be revisited in Chapter 16, Illustrative Open-ended Problem #3.

6.8   Open-Ended Problems

This last Section of the chapter contains open-ended problems as they relate to heat transfer. No detailed and/or specific solution is provided; that task is left to the reader, noting that each problem has either a unique solution or a number of solutions or (in some cases) no solution at all. These are characteristics of open-ended problems described earlier.

There are comments associated with some, but not all, of the problems. The comments are included to assist the reader while attempting to solve the problems. However, it is recommended that the solution to each problem should initially be attempted without the assistance of the comments.

There are 43 open-ended problems in this section. As stated above, if difficulty is encountered in solving any particular problem, the reader should next refer to the comment, if any is provided with the problem. The reader should also note that the more difficult problems are generally located at or near the end of the section.

1. Describe the early history associated with heat transfer.
2. Discuss the recent advances in heat transfer technology.

3. Select a refereed, published article on heat transfer from the literature and provide a review.

4. Provide some normal everyday domestic applications involving the general topic of heat transfer.

5. Develop an original problem that would be suitable as an illustrative example in a book on heat transfer.

6. Prepare a list of the various books which have been written on heat transfer. Select the three best and justify your answer. Also select the three weakest books and, once again, justify your answer.

7. Develop a new method to experimentally determine the heat conductivity of a
   - solid;
   - liquid;
   - gas; and
   - slurry

8. Define tube/pipe *pitch* and *clearance* in layman terms.

9. Discuss the role *bypass* plays in the operation and performance of a heat exchanger.

10. Discuss the role flow arrangements in a heat exchanger can play in increasing heat transfer/recovery.

11. Discuss the effect that noncircular ducts have on the general heat transfer equations and the various heat transfer coefficient correlations.

12. Discuss the effect of conduits with varying cross-sectional areas have on the general heat transfer equation.

13. Consider the impact on the heat transfer performance of a heat exchanger as a function of the heat capacity of various cooling mediums flowing through the tubes of a shell-and-tube unit. Qualitatively discuss this effect as the heat capacity of the medium increases. Also, qualitatively discuss the effect of other physical properties on the exchanger's performance.

14. Most home heating systems employing either natural gas or oil are (relative to industrial systems) inefficient. Outline various procedures that can be applied to increase the thermal efficiency of these units. Include economic considerations in the analysis.

15. Describe the similarities and differences between the atmospheric *lapse rate* and natural convection heat transfer [10].

16. Global warming has become a reality and sea levels around the world have reportedly increased approximately 12

inches. As an authority on heat transfer, the President of the United States has invited you to participate in an emergency meeting whose agenda is concerned with providing the technical community with recommendations to reduce and/or eliminate rising sea levels. Suggest possible solutions.

17. Obtain order of magnitude values of industrial film coefficients for various boiling liquids, various condensing vapors, dropwise condensation for various vapors, film-type condensation for various vapors, boiling liquids, heating of various liquids and vapors, cooling of various liquids and vapors, and superheated steam.
    Comment: This will require a rather extensive investigation.

18. Discuss the differences and problems that can arise when condensing mixed vapors (as opposed to pure vapors).

19. Develop improved operation, maintenance and inspection (OM&I) procedures for heat exchangers.

20. Is there sufficient justification to employ the use of the log-mean temperature difference (LMTD) as the actual/true driving force in a heat exchanger.

21. Develop equations describing the "F" factor employed in shell-and-tube heat exchanger calculations for the following geometries:
    1. One shell pass: 2 or more tube passes
    2. Two shell pass: 4 or more tube passes
    3. Four shell pass: 8 or more tube passes
    4. Six shell pass: 12 or more tube passes
    5. Cross flow with both fluids unmixed
    6. Cross flow with one fluid unmixed
    Comment: Refer to the literature [4].

22. Consider a finned heat exchanger. Qualitatively discuss the effect that the thermal conductivity of the fins have on the heat transfer rate of the exchanger. Also discuss the effect of the heat capacity of the fin material.

23. Ganapathy, in his article titled "Size or Check Waste Heat Borders Quickly," *Hydrocarbon Processing*, New York City, NY, 169-170, September, 1984, provides a chart to design and predict the performance of a boiler. Convert that chart to equation form [12].

24. One option available to a plant manager when a tube within a heat exchanger fails is to simply plug the inlet of the tube. Develop an equation to describe the impact on the heat

transfer performance of the exchanger as a function of both the number of tubes within the exchanger and the number of plugged tubes.

25. Attempt to develop a general all-purpose equation that can be employed to describe the overall heat transfer coefficient for most of the heat exchangers in use today. Clearly define and quantify the terms/ factors that affect the coefficient.

26. Describe the hazards, risks, and safety associated with cryogenic operation and detail procedures that can be implemented to reduce those problems.
    Comment: Refer to L. Theodore and R. Dupont's *Environmental Health and Hazard Risk Assessment: Principles and Calculations text*, CRC Press/Taylor & Francis Group, Boca Raton, FL, 2012 [13].

27. Describe the hazards, risks, and safety associated with boiling operations and detail procedures that can be implemented to reduce those problems.
    Comment: Refer to L. Theodore and R. Dupont's *Environmental Health and Hazard Risk Assessment: Principles and Calculations text*, CRC Press/Taylor & Francis Group, Boca Raton, FL, 2012 [13].

28. Develop a generalized equation to describe the radiation view factor as a function of the geometry for aligned parallel rectangles.
    Comment: Refer to the literature [4].

29. Develop a generalized equation to describe the radiation view factor as a function of the geometry for perpendicular rectangles with a common edge.
    Comment: Refer to the literature [4].

30. Develop a generalized equation to describe the radiation view factor as a function of the geometry for coaxial parallel disks.
    Comment: Refer to the literature [4]

31. Outline Wilson's method for evaluating the *inside* film coefficient for a double pipe heat exchanger unit.
    Comment: Refer to E. Wilson, *Trans. AMSE*, 37, 47, ASME, New York City, NY, 1915 [14] This problem will be revisited in Part III, Chapter 27 – *Heat Transfer Term Projects*.

32. Devise a new method of reducing heat loss from a hot surface.
    Comment: Is insulation the only option?

33. Explain why entropy calculations should be included in some heat exchanger design calculations.

    Comment: Refer to L. Theodore et al, *Thermodynamics for the Practicing Engineer*, John Wiley & Sons, Hoboken, NJ, 2010 [15]

34. Design a new type of heat exchanger.

35. Design a new fin.

36. Develop an optimum design procedure that could be employed with the various classes of heat exchangers.

    Comment: This will require a rather extensive investigation.

37. Develop a comprehensive procedure for designing and predicting the performance of a cooling tower.

38. You have been hired by Theodore Consultants to estimate the energy requirements to maintain a small 10 ft high, 20 ft x 20 ft laboratory facility located along the Equator. Outline how to calculate the energy requirements if the temperature in the lab is to be maintained at 72°F. Clearly specify all information required in performing the analysis e.g., the properties of walls, ceiling and floor.

39. Refer to Problem 3 in the previous Section. Develop an expression for the optimum thickness of insulation from an economic perspective to cover a flat surface if the annual cost per unit area of insulation is directly proportional to the thickness. Neglect the air film (resistance) coefficient.

40. Refer to the previous problem. Resolve the problem but include the air film resistance. The air-film coefficient may be assumed constant and independent of insulation thicknesses.

41. CORENZA Partners designed a shell-and-tube heat exchanger to operate with a maximum discharge temperature of a hot stream to be cooled to 90°F. Once the unit was installed and running, the exchanger operated with a discharge temperature of 105°F. Rather than purchase a new exchanger, what options are available to bring the unit into compliance with the specified design temperature? Include how your answers would be affected if the unit is a shell-and-tube exchanger.

42. Environmentalists are now concerned that Earth's surface (not the atmosphere) will increase in temperature this century. Comment on the validity of this concern by estimating any temperature increases due to thermal activity within the Earth's surface.

43. Lou Theodore, a supposed authority on heat transfer and energy conservation, has indicated that there is a near limited amount of energy in the water covering the Earth's surface. Outline how this energy could be extracted from the water and discuss some of the economic considerations.

References

1. I. Farag and J. Reynolds, *Heat Transfer*, A Theodore Tutorial, Theodore Tutorials, East Williston, NY, originally published by the USEPA/APTI, RTP, NC, 1996.

2. J.B. Fourier, *Théorie Analytique de la Chaleur*, Gauthier-Villars, Paris, 1822; German translation by Weinstein, Springer, Berlin, 1884; *Ann. Chim. Et Phys.*, 37(2), 291 (1828); *Pogg. Ann.*, 13,327 (1828).

3. D. Kern, *Process Heat Transfer*, McGraw-Hill, New York City, NY, 1950.

4. L. Theodore, *Personal Notes,* East Williston, NY, 2000

5. C. Bennett and J. Meyers, *Momentum, Heat and Mass Transfer*, McGraw-Hill, New York City, NY, 1962.

6. D. Kirk and D. Othmer, *Encyclopedia of Chemical Technology*, 4th edition, Vol. 7, "Cryogenics," John Wiley & Sons, Hoboken, NJ, 2001.

7. W. Badger and J. Banchero, *Introduction to Chemical Engineering*, McGraw-Hill, New York City, NY, 1955.

8. L. Theodore and F. Ricci, *Mass Transfer Operations for the Practicing Engineer*, John Wiley & Sons, Hoboken, NJ, 2010.

9. D. Green and R. Perry (editors), *Perry's Chemical Engineers' Handbook,* 8th edition, McGraw-Hill, New York City, NY, 2008.

10. L. Theodore, *Chemical Engineering: The Essential Reference*, McGraw-Hill, New York City, NY, 2014.

11. L. Theodore, *Heat Transfer for the Practicing Engineer*, John Wiley & Sons, Hoboken, NJ, 2011.

12. V. Ganapathy, Size and Check Waste Heat Boilers Quickly, *Hydrocarbon Processing*, New York City, NY, 169-170, September, 1984.

13. L. Theodore and R. Dupont, *Environmental Health and Hazard Risk Assessment: Principles and Calculations*, CRC Press/Taylor & Francis Group, Boca Raton, FL, 2012.

14. E. Wilson, *Trans. AMSE*, 37, 47, ASME, New York City, NY, 1915.

15. L. Theodore, F. Ricci, and T. VanVliet, *Thermodynamics for the Practicing Engineer*, John Wiley & Sons, Hoboken, NJ, 2009.

7

Mass Transfer Operations

This chapter is concerned with mass transfer operations (MTOs). As with all the chapters in Part II, there are sereval sections: overview, sereval technical sections, illustrative open-ended problems, and open-ended. The purpose of the first section is to introduce the reader to the subject of MTO. As one might suppose, a comprehensive treatment is not provided although sereval technical section are included. The next section contains three open-ended problems; the authors' solutions (there may be other solutions) are also provided. The last section contains 43 problems; *no* solutions are provided here.

7.1   Overview

This overview section is concerned—as can be noted from its title—with mass transfer operations (MTO). As one might suppose, it was not possible to address all topics directly or indirectly related to heat transfer. However, additional details may be obtained from either the references provided at the end of this Overview section and/or at the end of the chapter.

Note: Those readers already familiar with the details associated with MTO may choose to bypass this Overview.

There are host of mass transfer operations. The three that are most encountered in practice are absorption, adsorption, and distillation. As such they receive the bulk of the treatment in this chapter. Operations briefly reviewed include liquid-liquid extraction, leaching, humidification, drying, and membrane processes. Other novel separation processes (not reviewed) include:

1. Freeze crystallization
2. Ion exchange
3. Liquid ion exchange
4. Resin adsorption
5. Evaporation foam fractionation
6. Dissociation extraction
7. Crystallization
8. Electrophoresis
9. Vibrating screens

Details on these operations are available in the literature [1].

The topic of stagewise vs. continuous operation needs to be addressed before leaving this Overview section. Stagewise operation is considered first. If two insoluble phases are allowed to come into contact so that the various diffusing components of the mixture distribute themselves between the phases, and if the phases are then mechanically separated, the entire operation is said to constitute one *stage*. Thus, a stage is the unit in which contacting occurs and where the phases are separated; and, a single-stage process is naturally one where this operation is conducted once. If a series of stages are arranged so that the phases are contacted and separated once in each succeeding stage, the entire *multistage* assemblage is called a *cascade* and the phases may move through the cascade in parallel, counter-current, or cross-flow mode [1–3].

In order to establish a standard for the measure of performance, the *ideal*, or *theoretical*, or *equilibrium* stage is employed and it is defined as one where the effluent phases are in equilibrium, so that (any) longer time of contact will bring about no additional change of composition in either phase. Thus, at equilibrium, no further net change of composition state of the phases is possible for a given set of operating conditions. (In actual equipment in the chemical process industries, it is usually not practical to allow sufficient time, even with thorough mixing, to attain equilibrium.)

An *actual* stage does not accomplish as large a change in composition as an equilibrium stage. It is for this reason, that the *fraction stage efficiency* is defined by many as the ratio of a composition change in an actual stage to than in an equilibrium stage. Stage efficiencies for equipment in the chemical process industry range between a few percent to that approaching 100 percent. The approach to equilibrium realized in any stage is then defined as the aforementioned fractional stage efficiency [1,2].

In the case of *continuous-contact* operation, the phases in question flow through the equipment in continuous intimate contact throughout the unit *without* repeated physical separation and contacting. The nature of the method requires the operation to be *either* semi batch or steady-state, and the resulting change in compositions may be equivalent to that given by a fraction of an ideal stage or by more than one stage. Thus, and as noted above, equilibrium between two phases at any position in actual equipment is generally never completely established.

The remainder of this chapter addresses the following four topics.

1. Absorption
2. Adsorption
3. Distillation
4. Other mass transfer proccesses

The reader should note that the bulk of the material in this chapter is drawn from J. Barden and L. Theodore, *Mass Transfer Operations*, A Theodore Tutorial, Theodore Tutorials, East Williston, NY, 1997, originally published by the USEPA/APTI, RTP, NC, 1997 [3].

7.2 Absorption

The removal of one or more selected components from a gas mixture by absorption is one of the most important operations in the field of mass transfer. The process of absorption conventionally refers to the intimate contacting of a mixture of gases with a liquid so that part of one or more of the constituents of the gas will dissolve in the liquid. The contact usually occurs in some type of packed column (addressed later in this section), although plate and spray towers are also used. In gas absorption operations the equilibrium of interest is that between a relatively nonvolatile absorbing liquid (solvent) and a soluble gas (solute). The usual operating data to be determine or estimated for isothermal packed tower systems are the liquid rate(s) and the terminal concentrations or mole fractions. An operating line, that describes operating conditions in the column, is obtained by a mass balance around the column [1–3]. (Details on packing height and diameter follow.)

7.2.1   Packing Height

The height of a packed column is calculated by determining the required number of theoretical separation units and multiplying this number by the packing height. In continuous contact countercurrent operations, the theoretical separation unit is called the transfer unit, and the packing height producing one transfer unit is referred to as the height of a transfer unit. The transfer unit is essentially a measure of the degree of the separation (of the solute out of the gas stream and into the liquid stream) and the average driving force producing this transfer. In actual gas absorption design practice, the number of transfer units N_{OG} can be estimated from Equation 7.2 below One can also show that if Henry's law applies.

The number of transfer units is given by Colburn's equation

$$y = mx \tag{7.1}$$

$$N_{OG} = \ln \frac{\left[\left(\dfrac{y_1 - mx_2}{y_2 - mx_2} \right) \left(1 - \dfrac{1}{A} \right) + \dfrac{1}{A} \right]}{1 - \dfrac{1}{A}} \tag{7.2}$$

where $A = \dfrac{L_m}{mG_m}$

The term A is defined as the *adsorption factor* and is related to the slope of the equilibrium curve where L_m and G_m are the molar liquid and gas flow rates, respectively. If the gas is highly soluble in the liquid and/or reacts with the liquid, Theodore [4,5] has shown that, $N_{OG} = \ln\left(\dfrac{y_1}{y_2} \right)$[4,5] where the y_1 and y_2 refer to the inlet and outlet mole fraction, respectively.

The height of a transfer unit (H_{OG}) is usually determined experimentally for the system under consideration. Information on many different systems using various types of packings has been compiled by the manufacturers of gas absorption equipment and should be consulted prior to design. The data may be in the form of graphs depicting, for the specific system and packing, the H_{OG} versus the gas rate (in lb/h·ft²) with the liquid rate (in lb/h·ft²) as a parameter.

The packing height (Z) is then simply the product of H_{OG} and N_{OG}. Although there are many different approached to determine the column height, the $H_{OG} - N_{OG}$ approach is the most commonly use at the present time for approximate calculations, with the H_{OG} usually being obtained from the manufacturer.

7.2.2   Tower Diameter

Regarding tower diameter, consider a packed column operating at a given liquid rate in which the gas rate is gradually increased. After a certain point, the gas rate is so high that the drag on the liquid is sufficient to keep the liquid from flowing freely down the column. Liquid begins to accumulate and tends to block the entire cross – section for flow. This, of course, both increases the pressure drop and prevents the packing from mixing the gas and liquid effectively, and ultimate some liquid is even carried back up the column. This undesirable condition, know as *flooding*, occurs fairly abruptly and the superficial gas velocity at which it occurs is called the flooding velocity. The calculation of column diameter is based on flooding considerations; the usual operating range is 60% to 70% of the flooding rate.

The most commonly used correlation for pressure drop and diameter in the U.S. Stoneware's generalized correlation (see Figure 7.1). This procedure to determine the tower diameter is as follows:

1. Calculate the abscissa, $(L/G)(\rho/\rho_L)^{0.5}$; L, G = mass divided by time·area
2. Proceed to the flood line and read the ordinate.
3. Solve the ordinate equation for G at flooding.
4. Calculate the tower cross – sectional area, S, for the *fraction* of flooding velocity chosen for operation, f, by the equation

$$S = m/fG = (\text{total lb/s})/(\text{lb/s·ft}^2) = \text{ft}^2 \qquad (7.3)$$

The diameter of the tower is then determined by

$$D = 1.13 S^{0.5} = \text{ft} \qquad (7.4)$$

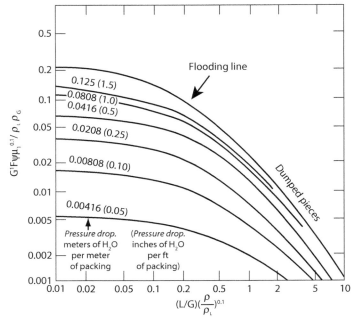

Figure 7.1 Generalized correlation for flooding and pressure drop.

7.2.3   Plate Columns

Plate columns are essentially vertical cylinders in which the liquid and gas are contacted in stepwise fashion (stage operation) on plated or trays [1–3]. The liquid enters at the top and flows downward via gravity. On the way down, it flows across each plate and through a downspout to the plate below. The gas passes upward through openings in the plate, e.g., a sieve tray [1–3], then bubbles through the liquid to form a froth, and passes on to the next plate above. The overall effect is the multiple (repeated) countercurrent contact of a gas and liquid. Each plate of the column is a stage since the fluids on the plate are brought into intimate contact, interphase diffusion occurs, and the fluids are separated. The number of theoretical plates (or stages) is dependent on the difficulty of the separation to be carried out and is determined solely from material balances and equilibrium considerations. Details on plate columns are available in the literature [1-3].

7.2.4   Stripping

Quite often, an absorption column is followed by a liquid absorption process in which the gas solute is removed from the absorbing medium by contact with an insoluble gas. This reverse operation is called *stripping* and is utilized to generate the solute *rich* solvent so that if may (in many cases) be recycled back to the absorption unit. The rich solution enters the stripping unit and the volatile solute is stripped from solution by either reducing the pressure, increasing the temperature, using a stripping gas to remove the vapor solute dissolved in the solvent, or any combination of these process changes. While the concept of stripping is opposite to that of absorption, it is treated in the same manner [6].

7.2.5   Summary of Key Equations [1,2]

The key equations for absorption and stripping calculations for tower height, including a summary of earlier material, are presented below.
    For packed tower absorption:

$$N_{OG} = \frac{\log\left\{(y_1 - mx_2)/(y_2 - mx_2)\right\}\left\{1 - \left(\frac{1}{A}\right)\right\} + \{1/A\}}{1 - \frac{1}{A}} \qquad (7.5)$$

For stripping

$$N_{OG} = \frac{\log\left\{((x_2 - y_1)/m)/((x_1 - y_1)/m)\right\}\{1 - A\} + A}{1 - A} \qquad (7.6)$$

where the subscripts 1 and 2 refer to bottom and top conditions, respectively. In addition, $A = L_m/mG_m$ and $S = 1.0/A$.
    For plate tower absorption

$$N = \frac{\log\left\{(y_{N+1} - mx_0)/(y_1 - mx_0)\right\}\left\{1 - \left(\frac{1}{A}\right)\right\} + \{1/A\}}{\log A} \qquad (7.7)$$

Note: the term ln, rather than log, may also be employed in both the numerator and denominator.

If A approaches unity, Equation (7.7) becomes

$$N = \frac{y_{N+1} - y_1}{y_1 - mx_0}$$

or

$$\frac{y_{N+1} - y_1}{y_{N+1} - mx_0} = \frac{N}{N+1} \tag{7.8}$$

Note that the subscripts 1 and N referred to the top and bottom of the column respectively.

For stripping in plate towers

$$N = \frac{\log\left\{ ((x_0 - y)_{N+1} / m) / ((x_N - y_{N+1}) / m) \right\} \left\{ 1 - \left(\frac{1}{S} \right) \right\} + S}{\log S} \tag{7.9}$$

or

$$\frac{x_0 - x_N}{x_0 - (y_{N+1} / m)} = \frac{S^{N+1} - S}{S^{N+1} - 1} \tag{7.10}$$

If S is approximately 1.0, one may use either of the following equations.

$$N = \frac{x_0 - x_N}{x_0 - (y_{N+1} / m)} \tag{7.11}$$

$$\frac{x_0 - x_N}{x_0 - (y_{N+1} / m)} = \frac{N}{N+1} \tag{7.12}$$

7.3   Adsorption [3]

The material to follow will primarily address *gas* adsorption. However, the bulk of the material presented may also be applied to liquid adsorption.

Adsorption is a mass transfer process in which a solute is removed from a gas stream because it adheres to the surface of a solid. In an adsorption system, the gas stream is passed through a layer of solid particles referred to as the adsorbent bed. As the gas stream passes through the adsorbent bed, the solute absorbs or "sticks" to the surface of the solid adsorbent particles. Eventually, the adsorbent bed becomes "filled" or saturated with the solute. The adsorbent bed must then be desorbed before the adsorbent bed can be reused.

The process of adsorption is analogous to using a sponge to mop up water. Just as a sponge soaks up water, a porous solid (the adsorbent) is capable of capturing gaseous or liquid solute molecules. The stream carrying the solute must then diffuse into the pores of the adsorbent (internal surface) where they are adsorbed. The majority of the molecules are adsorbed on the internal pore surfaces.

The relationship between the amount of substance adsorbed by the absorbent at constant temperature and the equilibrium pressure or concentration is called the adsorbent *isotherm*. The adsorption isotherm is the most important and by far the most often used of the various equilibria relationships which can be employed. Information on adsorption equilibria was addressed in Chapter 4.

Fixed bed adsorbers are the usual industrial choice when adsorption is the desired method of recovery/control. Consider a binary solution containing a solute at concentration C_0 (see Figure 7.2). The gas stream containing the solute (or adsorbent) is passed continuously down through a relatively deep bed of adsorbent which is initially free of adsorbate. The top layer of adsorbent, in contact with the inlet gas entering, at first absorbs the adsorbate rapidly and effectively. The remaining adsorbate is removed in the lower part of the bed. At this *initial* point in time the effluent concentration from the bottom of the bed is essentially zero. The bulk of the adsorption takes place over a relatively narrow adsorption zone (defined as the mass transfer zone, *MTZ*) in which there is a rapid change in concentration. At some later time, roughly half of the bed is saturated with the adsorbate, but the effluent concentration, C_2, is still substantially zero. Finally, at C_3, the lower portion of the adsorption zone has reached the bottom of the bed, and the concentration of adsorbate in the effluent has steadily risen to an appreciable value for the first time. The system is then said to have reached the *breakpoint*. The adsorbate concentration in the effluent gas stream now rises rapidly as the adsorption zone passes through the bottom of the bed, and at C_4 has essentially reached (approached) the initial value, C_0. At this point the bed is essentially fully saturated with adsorbate. The portion of the curve between C_3 and C_4 is termed the *breakthrough* curve.

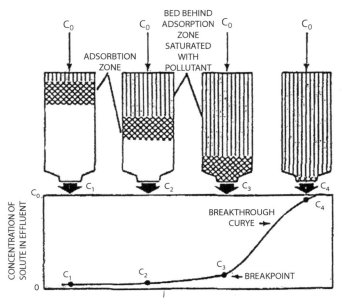

Figure 7.2 Operation Adsorption System.

The *breakthrough capacity* is the bed capacity when traces of adsorbate first begin to appear in the exit gas stream from the adsorption bed. It may be estimated from the following equation.

$$BC = [(0.5)(CAP)(MTZ) + (CAP)(Z\text{-}MTZ)]/Z \qquad (7.13)$$

where BC = breakthrough capacity
  CAP = saturation capacity
  MTZ = mass transfer zone height
  Z = adsorption bed depth.
The usual procedure in practice is to work with a term defined as the *working charge* (or working capacity). It provides a numerical value for the actual adsorbing capacity of the bed of height Z under operating conditions. If experimental data are available, the working charge, WC, may be estimated from

$$WC = CAP\,[(Z - MTZ)/Z] + 0.5\,[MTZ/Z] - HEEL \qquad (7.14)$$

where *HEEL* is the residual adsorbate present following regeneration. Since much of the data requried is rarely available, or just simply ignored, the working charge may be taken to be some fraction, f, of the saturated (equilibrium) capacity of the adsorbent; i.e., $WC = (f)(CAP)$; $0 \leq f \leq 1.0$.

7.3.1   Adsorption Design

A rather simplified overall design procedure for a system adsorbing an organic with activated carbon that consists of two horizontal units (one on/one off) that are regenerated with steam was developed by one of the authors and is provided below[7]:

1. Select adsorbent type and size.
2. Select cycle time; estimate regeneration time; set adsorption time equal to regeneration time; set cycle time equal to twice the regeneration time; generally, try to minimize regeneration time.
3. Set gas throughput velocity v is it usually in the 80 ft/min range but can increase to 100 ft/min.
4. Set the steam/solvent ratio.
5. Calculate (or obtain) WC for the above.
6. Calculate the amount of solvent adsorbed (M_s) during ½ the cycle time (t_{ads}) using the equation

$$M_s = qC_i t_{ads} \; ; \; C_i = \text{inlet solvent concentration} \qquad (7.15)$$

7. Calculate the adsorbent required, M_{AC}:

$$M_{AC} = M_s/WC \qquad (7.16)$$

8. Calculate the adsorbent volume requirement.

$$V_{AC} = M_{AC}/\rho_B; \; \rho_B = \text{carbon bulk density} \qquad (7.17)$$

9. Calculate the face area of the bed.

$$A_{AC} = q/v \qquad (7.18)$$

10. Calculate the bed height, Z.

$$Z = V_{AC}/A_{AC} \qquad (7.19)$$

11. Estimate the pressure drop using any convenient method.

12. Set the L/D (length to diameter) ratio.
13. Calculate L and D, noting that

$$A_{AC} = LD \qquad (7.20)$$

14. Design (structurally) to handle if filled with water.
15. Design vertically if $q < 2500$ actual cubic feet per minute (acfm).

This topic is revisited in Part III, Chapter 28 – *Mass Transfer Operation Term Projects*.

7.3.2   Regeneration

Adsorption processes in practice use various techniques to accomplish regeneration or desorption. The adsorption-desorption cycles are usually classified into four types, used separately or in combination.[8]

1. Thermal swing cycles;
2. Pressure swing cycles use either a low pressure or vacuum to desorb the bed;
3. Purge gas stripping cycles; and,
4. Displacement cycles using an adsorbable purge to displace the previously adsorbed material on the bed.

7.4   Distillation

Distillation is no doubt the most widely used separation process in the chemical and allied industries. Applications range from the rectification of alcohol to the fractionation of crude oil. The separation of liquid mixtures by distillation is based on differences in volatility between the components. The first component to vaporize is considered to be more volatile than the other components in the system. The greater the *relative* volatilities, the easier the separation.

Information on batch and flash distillation is available in the literature. [1–3] This section will primarily review the McCabe-Thiele method [9] and has been primarily adapted from the literature [1–3].

In a distillation column (see Figure 7.3), vapor flows up the column and liquid flows countercurrently down the column. The vapor and liquid are brought into contact on plates as shown in Figure 7.3, or packing. The vapor

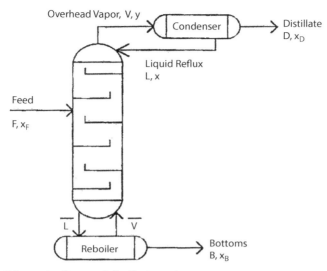

Figure 7.3 Schematic of a trayed distillation column.

from the top of column is sent to a condenser. Part of the condensate from the condenser is returned to the top of the column as reflux to descend counter to the rising vapors. The remainder of the condensed liquid is drawn as product. The ratio of the amount of reflux returned to the column to the distillate product collected is known as the *reflux ratio*. As the liquid stream descends the column it is progressively enriched with low-boiling constituents. The column internals are then used as an apparatus for bringing these streams into intimate contact so that the vapor stream tends to vaporize the low-boiling constituent from the liquid and the liquid stream tends to condense the high-boiling constituent from the vapor. Column-wise distillation involves two sections: an *enriching* (top) and *stripping* (bottom) section, which lies below the feed where the more volatile components are stripped from the liquid; in the enriching or rectifying section above the feed.

A column may consist of one or more feeds and may produce two or more product streams. The product recovered at the top of a column is referred to as the *tops*, while the product at the bottom of the column is referred to as the *bottoms*. Any product(s) drawn at various stages between the top and bottom are referred to as *side streams*. Multiple feeds and product streams do not alter the basic operation of a column, but they do complicate the calculations and analysis of the process to some extent. If the process requirement is to strip a volatile component from a relatively nonvolatile solvent, the rectifying (bottom) section may be omitted, and the unit is then called a stripping column. Virtually pure top and bottom products can be achieved by using many stages or (occasionally) additional columns; however, this is not usually economically feasible.

The top of the column is cooler than the bottom, so that the liquid stream becomes progressively hotter as it descends and the vapor stream becomes progressively cooler as it rises. This heat transfer is accomplished by actual contact of liquid and vapor; and for this purpvose, effective contacting is desirable. Each plate in the column is assumed to approach equilibrium conditions. This type of plate is defined as a *theoretical plate*, i.e., a plate on which the contact between the vapor and liquid is sufficiently good so that the vapor leaving the plate has the same composition as the vapor in equilibrium with the liquid overflow from the plate. The vapor and liquid leaving are related by the aforementioned equilibrium curve. Distillation columns designed on this basis serve as a standard for comparison to actual columns. By such comparisons, it is possible to determine the number of actual plates that are equivalent to a theoretical plate and then to reapply this factor when designing other columns for similar service.

In some operations where the top product is required as a vapor, the liquid condensed is sufficient only to provide reflux to the column, and the condenser is referred to as a *partial condenser*. In a partial condenser, the reflux will be in equilibrium with the vapor leaving the condenser, and is considered to be an equilibrium stage in the development of the operating line when estimating the column height. When the liquid is totally condensed, the liquid returned to the column will have the same compositions as the top product and is not considered to be in an equilibrium state. A *partial reboiler* is utilized to generate vapor to operate the column and to produce a liquid product if necessary. Since both liquid and vapor are in equilibrium, a partial reboiler is considered to be an equilibrium stage as well.

As noted earlier, an operating line can be developed to describe the equilibrium relation between the liquid and vapor components. However, in staged column design, it is necessary to develop an operating line which relates the passing streams (liquid entering and vapor leaving) on each stage in the column. The following analysis will develop operating lines for the top, or enriching (rectifying) section and the bottom, or stripping, section of a column.

To accomplish the above analysis, an overall material balance is written for the condenser as $V = L + D$, which represents the vapor (V) leaving the top stage, the liquid reflux returning (L) to the column from the condenser (reflux), and the distillate (D) collected. See above Figure 7.3 A material balance for a component is written as $Vy = Lx + Dx_D$. This can be rearranged in the form of an equation for a straight line straight, $y = mx + b$ as,

$$y = \frac{L}{V}x + \frac{D}{V}x_D \qquad (7.21)$$

From the overall mass balance, $D = V - L$ so that Equation 7.21 can be written as

$$y = \frac{L}{V}x + \left(1 - \frac{L}{V}\right)x_D \qquad (7.22)$$

where L/V = slope. This is the internal reflux ratio (liquid reflux returned to the column/vapor from the top of the column). Even though Equation 7.22 has been developed around the condenser, it represents the equilibrium relationship of passing liquid/vapor streams and can be applied to the top of the column.

The corresponding operating line in the bottom or stripping section can be developed in a similar manner. The overall material balance around the reboiler and the component material balance reduce to Equation 7.23. Note that the terms with the bars over them represent the flow at the bottom of the column. Again, this material balance has been rearranged in the form for a straight line as

$$y = \frac{B}{\overline{V}}x_B + \frac{\overline{L}}{\overline{V}}x \qquad (7.23)$$

But since $B = \overline{L} + \overline{V}$, Equation 7.23 can be rewritten as

$$y = \left(\frac{\overline{L}}{\overline{V}} - 1\right)x_B - \frac{\overline{L}}{\overline{V}}x \qquad (7.24)$$

Equation 7.24 can be rearranged into a form similar to Equation 7.22.

$$y = -\frac{\overline{L}}{\overline{V}}x + \left(\frac{\overline{L}}{\overline{V}} - 1\right)x_B \qquad (7.25)$$

And, as before, the term $\overline{L}/\overline{V}$ is the slope and $\left(\frac{\overline{L}}{\overline{V}} - 1\right)$ represents the y-intercept. The slope can be calculated from the external reflux ratio by Equation 7.26:

$$\frac{\overline{L}}{\overline{V}} = \frac{\frac{L}{D}\left(x_f - x_B\right) + q(x_D - x_B)}{\frac{L}{D}\left(x_f - x_B\right) + q\left(x_D - x_B\right) - \left(x_D - x_f\right)} \tag{7.26}$$

A procedure for designing a staged distillation column is provided below.

1. Plot the equilibrium data, the top and bottom operating lines and the $y = x$ line on the same graph. This plot is called a McCabe-Thiele diagram as portrayed in Figure 7.4 [9].
2. Step off the number of stages required beginning at the top of the column. The point at the top of the curve represents the composition of the liquid entering (reflux) and gas leaving (distillate) the top of the column. The point at the bottom of the curve represents the composition of liquid leaving (bottoms) and the gas entering (vapor from reboiler) the bottom of the column. All stages are stepped off by drawing alternate vertical and horizontal lines in a *stepwise* manner between the top and bottom operating lines and the equilibrium curve. The number of steps is the number of theoretical stages requires. When a partial reboiler or partial condenser is employed, the number of equilibrium stages stepped off is

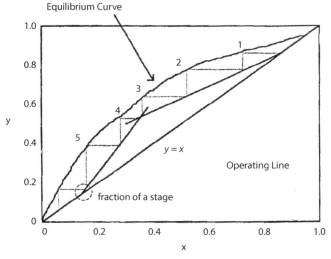

Figure 7.4 McCabe-Thiele Diagram.

$N+1$ and the number of theoretical plates would be N. Thus, if both are to be included, the number of equilibrium stages stepped off would be $N+2$.

3. To determine the actual number of plates required, divide the result in step (2) by the *overall* plate fractional efficiency, typically denoted by E_o. Values can range from 0.4 to 0.8. The actual number of plates can be calculated from

$$N_{act} = N / E_0 \qquad (7.27)$$

4. The column height can be calculated by multiplying the result in step (3) by the tray spacing (9,12,15,18 and 24 inches are typical tray spacings).

5. To determine the *optimum* feed plate location, draw a line from the feed composition on the $y = x$ line, through the intersection of the top/bottom operating lines, to the equilibrium curve. The step straddling the feed line is the correct feed-plate location.

6. Due to the differences in vapor traffic throughout the column, the diameter will vary from the top to the bottom. Typical designs are *swaged* columns, where the diameter is larger at the bottom than the top. A "cone-like" vessel links the bottom of the column with the top of the column. The diameter at the top of a column can be roughly sized by assuming a superficial (column only contains vapor) vapor velocity of 5 ft/s and then using the following equation to calculate the diameter.

$$D = \left(\frac{4q}{\pi v} \right)^{0.5} \qquad (7.28)$$

A somewhat similar procedure is employed to calculate the diameter at the bottom of a column. The interested reader is referred to a mass transfer text which can provide a more detailed analysis [1–3].

As indicated earlier, distillation can be carried out in both trayed and packed columns. The selection of which method to be used depends on the specific application. Packing is frequently comparable in cost with trays and becomes more attractive where a low pressure drop and small liquid hold-up are required. It is generally recommended that packed towers should be used if the column diameter is less than 1.5 ft. Packed columns also provide continuous contacting between the liquid and gas. This imitate contacting occurs on the packing which promotes the heat transfer required to separate the more volatile component from the process stream.

The Fenske equation calculates the minimum number of equilibrium stages required for *multicomponent* separation at total reflux (100% of the liquid reflux is returned to the column). On the other hand, Underwood developed an equation which estimates the maximum number of equilibrium stages required at minimum reflux, and the actual reflux ratio for a multicomponent system. But in order to estimate the number of actual stages, the Gilliland equation correlates the minimum number of stages at total reflux, the minimum reflux ratio, and the actual reflux rate for a multicomponent system. The application of all three of these equations in the design of a distillation column is referred to as the Fenske-Underwood-Gilliland (FUG) method. Details on each of the above are provided in the literature [1,2,10]

7.5   Other Mass Transfer Processes

7.5.1   Liquid-Liquid Extraction [1,2,11]

Liquid-liquid extraction (or liquid extraction) is a process for separating a solute from a solution employing the concentration driving force between two immiscible (non-dissolving) liquid phases. Thus, liquid extraction involves the transfer of solute from one liquid phase into a second immiscible liquid phase. The simplest example involves the transfer of one component from a binary mixture into a second immiscible liquid phase; such is the case with extraction of an impurity from waste water into an organic solvent. Liquid extraction is usually selected when distillation or stripping is impractical or too costly; this situation occurs when the relative volatility for two components falls between 1.0 and 1.2. Treybal [11] provides extensive details. This topic is also revisited in Part III, Chapter 28 – *Mass Transfer Operation Term Projects*.

7.5.2   Leaching

Leaching is the preferential removal of one or more components from a solid by contact with a liquid solvent. The soluble constituent may be solid or liquid, and it may be chemically or mechanically held in the pore structure of the insoluble solids material. The insoluble solid material is often particulate in nature, porous, cellulite with selectively permeable cell walls, or surface activated. In chemical engineering practice, leaching is also referred to by several other names such as extraction, liquid-solid extraction, lixiviation, percolation, infusion, washing, and decantation-settling. The simplest example of a leaching process is in the preparation of a cup of tea. Water is the solvent used to "extract", or leach, tannins and other solids from the tea leaf.

7.5.3 Humidification and Drying

In many unit operations, it is necessary to perform calculations involving the properties of mixtures of air and water vapor. Such calculations often require knowledge of: the amount of water vapor carried by air under various conditions, the thermal properties of such mixtures, and the changes in enthalpy content and moisture content as air containing some moisture is brought into contact with water or wet solids and other similar processes. Some mass transfer operations in the chemical process industry involve simultaneous heat and mass transfer. In the *humidification* process, water or another liquid is vaporized, and the required heat of vaporization must be transferred to the liquid. *Dehumidification* is the condensation of water vapor from air, or, in general, the condensation of any vapor from a gas.

Drying involves the *removal* of relatively small amounts of water from solids. In many applications, such as in corn processing, drying equipment follows an evaporation step to provide an ultra-high solids content product stream. Drying, in either a batch or continuous process, removes liquid as a vapor by passing warm gas (usually air) over, or indirectly heating, the solid phase. The drying process is carried out in one of the three basic dryer types. The first is a continuous tunnel dryer. In a continuous dryer, supporting trays with wet solids are move through an enclosed system while warm air blows over the trays. Similar in concept to the continuous tunnel dryer, rotary dryers consist of an inclined rotating hollow cylinder. The wet solids are fed in one side and hot air is usually passed countercurrently over the wet solids. The dried solids then pass out the opposite side of the dryer unit. The final type of dryer is a spray dryer. In spray dryers, a liquid or slurry is sprayed through a nozzle, and fine droplets are dried by a hot gas, passed either concurrently countercurrently, past the falling droplets. This unit has found wide application in air pollution control [8].

7.5.4 Membrane Processes [1]

Membrane separation processes are one of the newer (relatively speaking) technologies being applied in practice. The subject matter is and has been introduced into the chemical engineering curriculum. There are four major membrane processes of interest to the chemical engineer.

1. Reverse osmosis (hyperfiltration);
2. Ultrafiltration;
3. Microfiltration; and
4. Gas permeation.

The four processes have their differences. The main difference between reverse osmosis (RO) and ultrafiltration (UF) is that the size/diameter of the particles or molecules in solution to be separated is smaller in RO. In microfiltration (MF), the particles to be separated/concentrated are generally solids or colloids rather than molecules in solution. Gas permeation (GP) is another membrane process that employs a non-porous semipermeable membrane to "fractionate" a gaseous stream [12–14].

A detailed and expanded treatment of mass transfer operations is available in the following three references.

1. L. Theodore and F. Ricci, *Mass Transfer Operations for the Practicing Engineer*, John Wiley & Sons, Hoboken, NJ, 2010 [1].
2. R. Treybal, *Mass Transfer Operations*, McGraw-Hill, New York City, NY, 1955 [2].
3. Adapted from, L. Theodore, *Chemical Engineering: The Essential Reference*, McGraw-Hill, New York City, NY, 2014.

7.6   Illustrative Open-Ended Problems

This and the last Section provide open-ended problems. However, solutions *are* provided for the three problems in this Section in order for the reader to hopefully obtain a better understanding of these problems which differ from the traditional problems/illustrative examples. The first problem is relatively straightforward while the third (and last problem) is somewhat more difficult and/or complex. Note that solutions are not provided for the 43 open-ended problems in the next Section.

Problem 1: Provide a technical explanation of the essential difference between stagewise and continuous operation.
Solution: The essential difference between stagewise and continuous-contact operation may be summarized in the following manner. In the case of the stagewise operation, the flow of matter between the phases is allowed to

reduce the concentration difference. If allowed to contact for long enough, equilibrium can be established after which no further transfer occurs. The rate of transfer and the time (of contact) then determine the stage efficiency realized in any particular application. On the other hand, in the case of the continuous-contact operation, the displacement from equilibrium is deliberately maintained and the transfer between the phases may continue without interruption. Economics plays a significant role in determining the most suitable method. [1–3].

Problem 2: One of the more commonly used procedure for pressure drop and column diameter calculations is the U.S. Stonewaring [3] generalized pressure drop correlation as presented in Figure 7.3. Convert the generalized U.S. Stoneware pressure drop-flooding correlation into equation form. Comment: Refer to the literature [1–3].

Solution: There are several correlations available. Chen [16] developed the following equation from which the tower diameter can easily be obtained:

$$D = 16.28 \left(\frac{W}{\phi L} \right)^{0.5} \left(\frac{\rho_L}{\rho_G} \right)^{0.25} \tag{7.28}$$

with

$$\log_{10}\phi = 32.5496 - 4.1288\log_{10}\left(\frac{L^2 A_v \mu_L^{0.2}}{\rho L^2 \varepsilon^3} \right)^{0.5} \tag{7.30}$$

where (employing Chen's notation) A_v is the specific surface area of dry packing (ft^2/ft^3 packed column), L is the liquid flux (gal/min·ft^2 of superficial tower cross section), W is the mass flow rate of gas (lb/h), ε is the void fraction, μ_L is the liquid viscosity (cP), and the density terms are in lb/ft^3.

Problem 3: Consider the absorber system shown in Figure 7.5. Ricci Engineers designed the unit to operate with maximum discharge concentration of 50 ppm. Once the unit was installed and running, the unit operated with a higher discharge concentration. Rather than purchase a new unit, what options are available to get the unit operating at the specified design concentration?

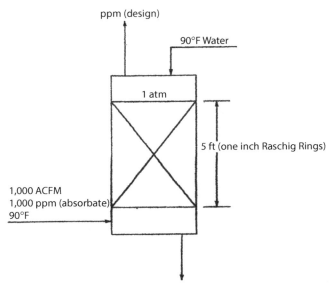

Figure 7.5 Absorber failure to meet design performance.

Solution: This is obviously an open-ended question. One may employ any one or a combination of the following suggestions [1]:

1. Increase the temperature.
2. Decrease the pressure.
3. Increase the height of the packing.
4. Place sprays at the inlet before the packing.
5. Increase the liquid flow rate, but check the pressure drop increase and any potential effect on the fan.
6. Change the liquid, but check the pressure drop increase and any potential effect on the fan.
7. Change packing size and/or type, but check the pressure drop increase and any potential effect on the fan.
8. Process modification: if possible reduce inlet absorbate concentration.
9. Process modification: reduce gas flow rate.
10. Design a new system.
11. Add more packing on top of existing packing (if possible), but check the pressure drop increase and any potential effect on the fan.
12. Fire the design engineer.
13. Shut down the plant.

7.7   Open-Ended Problems

This last section of the chapter contains open-ended problems as they relate to mass transfer operations. No detailed and/or specific solution is provided; that task is left to the reader, noting that each problem has either a unique solution or a number of solutions or (in some cases) no solution at all. These are characteristics of open-ended problems described earlier.

There are comments associated with some, but not all, of the problems. The comments are included to assist the reader while attempting to solve the problems. However, it is recommended that the solution to each problem should initially be attempted *without* the assistance of the comments.

There are 43 open-ended problems in this section. As stated above, if difficulty is encountered in solving any particular problem, the reader should next refer to the comment, if any is provided with the problem. The reader should also note that the more difficult problems are generally located at or near the end of the section.

1. Describe the early history associated with traditional mass transfer operations.
2. Describe the early history associated with traditional membrane technology.
3. Discuss the recent advances in traditional mass transfer technology.
4. Discuss the recent advances in traditional membrane technology.
5. Select a refereed, published article on mass transfer from the literature and provide a review.
6. Develop an original problem that would be suitable as an illustrative example in a book on mass transfer operations.
7. Prepare a list of the various books which have been written on mass transfer operations. Select the three best (hopefully including a book written by one of the authors) and justify your answer. Also select the three weakest books and, once again, justify your answer.
8. Discuss the advantages and disadvantages of batch versus continuous operation in mass transfer operations.
9. Describe the options that are available to represent a multicomponent diffusion operation via an average diffusion coefficient.
   Comment: Refer to Treybal's "Mass Transfer Operations" text [2].

10. Describe the advantages and limitations of employing the concept of an ideal stage in all the applicable mass transfer operations.

11. Discuss the merits of employing Henry's law in gas absorption calculations.

12. Discuss the merits of employing Raoult's law in distillations calculations.

13. Construct a vapor-liquid equilibrium diagram of x vs y for various relative volatility (γ) values. Include $\propto$ values of 0.0, 1.0, and ∞.

14. Discuss the merits of including the activity coefficient in liquid-vapor calculations.

15. Describe the various calculational problems that arise in many of the mass transfer processes that are not isothermal. Also extend the description to adiabatic operations.

16. Design a new packing that could be employed for absorption, adsorption, and distillation operations.

17. Suggest/develop a new support plate for an absorption column, adsorption column, and packed tower distillation unit.

18. List the various type of overhead separators employed in several unit operations of your choice.

19. Attempt to develop a new overhead separation device.

20. Develop a new liquid componential separation device.

21. Design a new membrane that could be employed in any of the four membrane separation processes discussed in the Overview section.

22. Develop a new method of drying solids.

23. Develop a new gas (or liquid) particulate separation device.

24. Describe the advantages and disadvantages of the various particulate separation devices that are employed in adsorption operations. Include those for both liquid and gaseous adsorption units.

25. Describe the various methods available to reduce the size of solids. Can a better method be developed?

26. Describe the various current methods employed for sedimentation operations.

27. Develop a generalized equation to describe the performance of a distillation column when one or more plates within the unit fail to perform any separation.

28. Outline a procedure to determine the optimum feel plate location(s) in a distillation tower. Compare your procedure to those employed in the past.

29. Discuss the problems associated with the various plate efficiencies employed in distillation calculations.

30. Convert the O'Connell correlation for overall distillation tray efficiency into equation form.
    Comment: Refer to L. Theodore and F. Ricci's *Mass Transfer Operations* text [1].

31. List some advantages and disadvantages of employing distillation versus another mass transfer operation to separate a two-component liquid stream.

32. Discuss the calculational problems that arise in absorption with chemical reaction.

33. Describe the difference between absorption and adsorption in both technical and layman terms.

34. Discuss the differences between drying, leaching, and extraction.

35. A dated plate distillation column is no longer delivering the degree of separation required for a process. Rather than replace the unit, you have been asked to recommend what other possible steps can be taken to the existing unit to get it back "on line".
    Comment: Refer to Problem 3 in the previous section.

36. An adsorber is designed to operate with a specified maximum discharge concentration. Once the unit is installed and running, the unit operated with a higher discharge concentration. Rather than purchase a new unit, what options are available to bring the unit into compliance with the specified design concentration?
    Comment: Refer to Problem 3 in the previous section.

37. You are an engineer at a water purification plant. The water is purified using a reverse osmosis (RO). One of the units at your plant is no longer operating at the required "separation" efficiency. What steps should be taken to avoid purchasing a new system?
    Comment: Refer to Problem 3 in the previous section.

38. Woist Engineers designed a cooling tower to operate at or below a maximum water discharge temperature. Once the tower was installed and running, the unit operated with a slightly higher temperature. Rather than purchase a new

tower, what options are available to bring the unit into compliance with the specified design temperature?

Comment: Refer to Problem 3 in the previous section.

39. A liquid-liquid extraction unit normally operates at a particular discharge concentration. Over time, the operating efficiency of the extractor has decreased. Rather than purchase a new unit, what options are available to bring the unit into compliance with the specified design concentration?

    Comment: Refer to Problem 3 in the previous Section.

40. Fitzmaurice Engineers designed a rotary dryer to operate at or below a solids water discharge concentration. Once the dryer was installed and running, the unit operated with a higher concentration. Rather than purchase a new dryer, what options are available to bring the unit into compliance with the specified design solids concentration?

    Comment: Refer to Problem 3 in the previous section.

41. You have been requested, as a part of a drying operation project, to convert the standard humidity chart into equation form.

    Comment: Good luck.

42. Develop a design procedure for each of the various classes of dryers.

43. Develop a crystallization design procedure.

References

1. L. Theodore and F. Ricci, *Mass Transfer Operations for the Practicing Engineer,* John Wiley & Sons, Hoboken, NJ, 2010.
2. R. Treybal, *Mass Transfer Operations,* McGraw-Hill, New York City, NY, 1955.
3. L. Theodore and J. Barden, *Mass Transfer,* A Theodore Tutorial, Theodore Tutorials, East Williston, NY, originally published by the USEPA/APTI, RTP, NC, 1990.
4. L. Theodore, *Engineering Calculations: Sizing Packed-Tower Absorbers Without Data,* CEP, New York City, NY, pp. 18-19, May 2005.
5. Personal notes: L. Theodore, East Williston, NY, 1991.
6. Personal notes: L. Theodore, East Williston, NY, 1996.
7. L. Theodore, *Engineering Calculations: Adsorber Sizing Made Easy,* CEP, New York City, NY, March 2005.
8. L. Theodore, *Air Pollution Control Equipment Calculations,* John Wiley & Sons, Hoboken, NJ, 2008.
9. W. McCabe and E. Thiele, *Ind Eng. Chem.,* New York City, NY, 17, 605, 1925.

10. E. Gilliland, *Ind. Eng. Chem.*, New York City, NY, 32, 1220, 1940.
11. R. Treybal, *Liquid Extraction*, McGraw-Hill, New York City, NY, 1951.
12. S. Slater, *Membrane Technology*, NSF Workshop Notes, Manhattan College, Bronx, NY, 1991 (adapted with permission).
13. P.C. Wankat, *Rate-Controlled Separations*, Chapter 12, Chapman & Hall, Boston, MA, 1990.
14. M.C. Porter, *Handbook of Industrial Membrane Technology*, Chapter 2, Noyes Publications, Park Ridge, NJ, 1990.
15. Adapted from L. Theodore, *Chemical Engineering: The Essential Reference*, McGraw-Hill, New York City, NY, 2014.
16. N. Chen, *New Equation Gives Tower Diameter*, Chem. Eng., New York City, NY, May 2, 1962.

8

Chemical Reactors

This chapter is concerned with chemical reactors. As with all the chapters in Part II, there are several sections: overview, several technical topics, illustrative open-ended problems open-ended problems. The purpose of the first section is to introduce the reader to the subject of chemical reactors. As one might suppose, a comprehensive treatment is not provided although several technical section are also included. The next section contains three open-ended problems; the authors' solution (there may be other solutions) are also provided. The last section contains 36 problems; *no* solutions are provided here.

8.1 Overview

This overview section is concerned—as can be noted from the chapter title—with chemical reactors. As one might suppose, it was not possible to address all topics directly or indirectly related to chemical reactors. However, additional details may be obtained from either the references provided at the end of this Overview section and/or at the end of the chapter.

Note: Those readers already familiar with the details associated with this subject may choose to bypass this Overview.

Almost every chemical process is designed to produce a desired product economically from a variety of starting materials through a succession of treatment steps. The raw materials may first undergo a number of physical treatment steps to put them in the form in which they can be reacted chemically. They then pass through a reactor. The discharge, or products of the reaction then usually undergoes additional physical treatment processes; i.e., separations, purifications, etc., for the final desired product to be obtained. The physical treatment processes were discussed in the earlier chapters. This chapter, however, is concerned with the chemical treatment processes that involve chemical reactors.

A major objective of this chapter is to prepare the reader to solve real-world engineering and design problems that involved chemical reactors. There are several classes of reactors. The three that are most often encountered in practice are batch (B), continuous stirred tank (CSTR), and tubular flow (TF). As such they receive the bulk of the treatment to follow. Another reactor reviewed is the semi-batch unit. Other topics reviewed include catalytic reactors and thermal effects.

The remainder of the chapter addresses 6 topics that are either directly or indirectly concerned with chemical reactors. They include:

1. Chemical Kinetics
2. Batch Reactors (B)
3. Continuous Stirred Tank Reactors (CSTR)
4. Tubular Flow Reactors (TF)
5. Catalytic Reactors
6. Thermal Effects

The reader should note that the bulk of the material in this chapter has been from L. Theodore, "Chemical Reaction Kinetics," A Theodore Tutorial, Theodore Tutorials, East Williston, NY, originally published by the USEPA/APTI, RTP, NC, 1995 [1]. Also note that in an attempt to be consistent with the chemical reactor literatures, the volumetric flow rate is represented by Q (not q, as in most of the chapters in this text).

8.2   Chemical Kinetics

Chemical kinetics involves the study of reaction rates and the variables that effect these rates. It is a topic that is critical for the analysis of reacting systems and chemical reactors. The rate of a chemical reaction can be described in any of several different ways. The most commonly used definition involves the time rate of change in the amount of one of the components participating in the reaction; the rate is also based on some arbitrary factor related to the reacting system size or geometry, such as volume, mass, and interfacial area.

Based on experimental evidence, the rate of reaction is a function of:

1. the concentration of components existing in the reaction mixture (this includes reacting and inert species);
2. temperature;
3. pressure; and (if applicable)
4. catalyst variables.

This may be put in equation form as:

$$r_A = r_A(C, P, T, catalyst\ variables) \tag{8.1}$$

or simply

$$r_A = \pm k_A f(C) \tag{8.2}$$

where k_A incorporates all the variables other than concentration. The $\pm$ notation is included to account for the reaction or formation of A. One may think of k_A as a constant of proportionality. It is defined as the *specific reaction rate* or more commonly the *reaction velocity constant*. It is a "constant" which is *independent* of concentration but *dependent* on the other variables. This approach has, in a sense, isolated one of the variables. The reaction velocity constant, k, like the rate of reaction, *must* refer to one of the species in the reacting system. However, k almost always is based on the same species as the rate of reaction.

Consider, for example, the reaction

$$aA + bB \rightarrow cC + dD \tag{8.3}$$

The notation $\rightarrow$ represents an irreversible reaction; i.e., if stoichiometric amounts of A and B are initially present, the reaction will proceed to the right until all the A and B have reacted (disappeared) and C and D have been formed. If the reaction is *elementary*, the rate of the above reaction is given by

$$r_A = -k_A C_A^a C_B^b \tag{8.4}$$

where the negative sign is introduced to account for the disappearance of A, and the product concentrations do not affect the rate. For elementary reactions, the reaction mechanism for r_A is simply obtained by multiplying the molar concentrations of the reactants raised to powers of their respective stoichiometric coefficients (power law kinetics). For *non-elementary* reactions, the mechanism can take any form.

The order of the above reaction with respect to a particular species is given by the exponent of that concentration term appearing in the rate of expression. The above reaction is, therefore of order a with respect to A and of order b with respect to B. The *overall order* n, usually referred to as "the order", is the sum of the individual orders; i.e.,

$$n = a + b \tag{8.5}$$

All real and naturally occurring reactions are *reversible*. A reversible reaction is one in which products react to form reactants back. Unlike irreversible reactions, which proceed to the right until completion, reversible reactions achieve an equilibrium state, in which rates of forward and reverse reactions are nearly the same for an infinite period of time. Reactants and products are still present in the system. At this (equilibrium) state, the reaction rate is zero. For example, consider the following reversible reaction:

$$aA + bB \rightleftharpoons cC + dD \tag{8.6}$$

where the notation $\rightleftharpoons$ is a reminder that the reaction is reversible $\rightarrow$ represents the forward reaction contribution to the total net rate, while $\leftarrow$ represents the contribution of the reverse reaction. The notation = is employed if the reaction system is at equilibrium. The rate of the reaction (see Equation 8.6) is then given by

$$r_A = \underset{\substack{\text{forward} \\ \text{reaction}}}{-k_A C_A^a C_B^b} + \underset{\substack{\text{reverse} \\ \text{reaction}}}{k_A{}' C_C^c C_D^d} \tag{8.7}$$

and

$$K_A = \frac{k_A}{k_A'} \qquad (8.8)$$

With regard to chemical reactions, there are two important questions that are of concern to the chemical engineer: (1) how far will the reaction go, and (2) how fast will the reaction go? Chemical thermodynamics (reaction equilibrium principles provides the answer to the first question; however, it provides no information regarding the latter question, which is concerned with kinetics. Reaction rates fall within the domain of chemical kinetics.

The two most common conversion variables employed by practitioners are denoted by α and X. The term α is employed to represent the change in the number of moles of a particular species due to chemical reaction. However, the most commonly used conversion variable is X; it is used to represent the change in the number of moles of a particular species (say A) relative to the number of moles of A *initially present* or *initially introduced* (to a flow reactor). Thus,

$$X_A = \frac{moles\ of\ A\ reacted}{initial\ moles\ of\ A} = \frac{N_A}{N_{A_0}} \qquad (8.9)$$

Other conversion (related) variables include: N_A, the number of moles of species A at some later time (or position); C_A, the concentration of A at some later time (or position); and x_A, the moles of A reacted/ total moles initially present. Also note that all of the above conversion variables can be based on mass, but this is rarely employed in practice. The conversion variable of choice is almost always X. In addition, the *yield* of a reaction is a measure of how much of the desired product is produced relative to how much would have been produced if only the desired reactions occurred, and if that reaction went to completion. Alternatively, *selectivity* is a measure of how a desired reaction performs relative to one of the side reactions.

Equations to describe the rate of reaction at the macroscopic level have been developed in terms of meaningful and measurable quantities. As noted earlier the reaction rate is affected not only by the concentration of species in a reacting system but also by the temperature. An increase in temperature will almost always result in an increase in the rate of reaction; in fact, the literature states that, as a general rule, a 10°C increase in reaction temperature will double the reaction velocity constant. However, this is generally no longer regarded as a truism, particularly at elevated temperatures.

The Arrhenius equation relates the reaction velocity constant with temperature, T. It is given by

$$A_{\exp}\left(\frac{-E_a}{RT}\right) \tag{8.10}$$

where A = frequency factor constant and is usually assumed to be independent of temperature

R = universal gas constant

E_a = activation energy and is usually assumed independent of temperature.

8.3   Batch Reactors

Batch reactors are commonly used in experimental studies. Their industrial applications are somewhat limited. They are used for gas phase, e.g. combustion reactions since small quantities (mass) of product are produced with even a very large-sized reactor. It is used for liquid phase reactions when small quantities of reactants are to be processed. It finds its major application in the pharmaceutical industry. As a rule, batch reactors are less expensive to purchase but more expensive to operate than either continuous stirred tank or tubular flow reactors.

The extent of a chemical reaction and/or the amount of product produced can be affected by the relative quantities of reactants introduced to the reactor. For two reactants, each is normally introduced through separate feed lines normally located at or near the top of the reactor. Both are usually fed simultaneously over a short period of time. Mixing is accomplished with the aid of a turning/spinning impeller. (See also Figure 8.1). The reaction is assumed to begin after both reactants are in the reactor. No spatial variations in concentration, temperature, etc., are generally assumed.

In most liquid-phase batch systems, the reactor volume is considered to be constant. However, in many systems, these parameters may vary, depending on the phase and reaction stoichiometry. Such variance must be included in the analysis of gas phase reaction systems [1–4].

As noted above, a batch reactor is normally used for small-scale operation, for testing new processes that have not been fully developed, for the manufacture of expensive products, and for processes that are difficult to convert to continuous operation. The batch reactor has the advantage of high conversions that can be obtained by leaving the reactants in the reactor

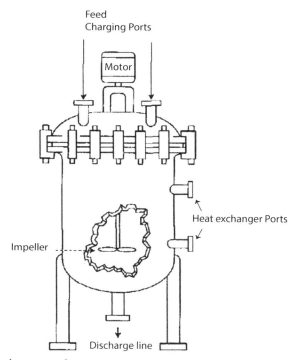

Figure 8.1 Batch reactor schematic.

for long periods of time, but it also has the disadvantage of high labor costs per unit production, and large-scale production, which is usually difficult.

A batch reactor is a solid vessel or container (see Figure 8.1). It may be open or closed. As noted, reactants are usually added to the reactor simultaneously. The contents are then mixed (if necessary) to ensure no spatial variations in the concentration of the species present. The reaction then proceeds. There is no transfer of mass into or out of the reactor during this period. The concentration of reactants and products changes with time; thus, this is a *transient* or *unsteady state* operation. The reaction is terminated when the desired chemical change has been achieved. The contents are then discharged and sent elsewhere, usually for further processing.

The describing equation for chemical reaction mass transfer is obtained by applying the conservation law for either mass or moles on a time-rate basis to the contents of a batch reactor. It is best to work with moles rather than mass since the rate of reaction is most conveniently described in terms of molar concentrations. The describing equation for species A in a batch reactor takes the form

$$\frac{dN_A}{dt} = -r_A V \tag{8.11}$$

where N_A = moles A at time t
r_A = rate of reaction of A; change in moles A/timevolume
V = reactor volume *contents*.

The above equation may also be written in terms of the conversion variable X since

$$N_A = N_{A_0} - N_{A_0} X \tag{8.12}$$

where N_{A_0} = initial moles of A.

Thus (setting X = X$_A$),

$$N_{A_0}\left(\frac{dX}{dt}\right) = -r_A V \tag{8.13}$$

The integral form of this equation is

$$t = N_{A_0} \int_0^X \left(\frac{-1}{r_A V}\right) dX \tag{8.14}$$

If V is constant (as with most liquid phase reactions)

$$t = \frac{N_{A_0}}{V} \int_0^X \left(\frac{-1}{r_A}\right) dX$$
$$= C_{A_0} \int_0^X \left(\frac{-1}{r_A}\right) dX \tag{8.15}$$

8.4   Continuous Stirred Tank Reactors (CSTRs)

Another reactor where mixing is important is the tank flow or continuously stirred tank reactor (CSTR). This type of reactor, like the batch reactor, also consists of a tank or kettle equipped with an agitator. It may be operated under steady or transient conditions. Reactant(s) are fed continuously, and the product(s) are withdrawn continuously (see Figure 8.2). The

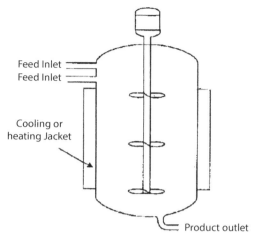

Figure 8.2 Continuously stirred tank reactor schematic. (CSTR)

reactant(s) and product(s) may be liquid, gas, or solid, or a combination of these. If the contents are perfectly mixed, the reactor design problem is greatly simplified for steady conditions because the mixing results in uniform concentrations, temperature, etc., throughout the reactor. This means that the rate of reaction is constant and the describing equations are not differential, and therefore, do not require integration. In addition, since the reactor contents are perfectly mixed, the concentration and/or conversion in a CSTR in exactly equal to the concentration and/or conversion *leaving* the reactor. The describing equation for the CSTR can then be shown to be:

$$V = \frac{F_{A_0} X_A}{(-r_A)} \tag{8.16}$$

where V = volume of reacting mixture
F_{A_0} = inlet molar feed rate of A
X_A = conversion of A
$-r_A$ = rate of reaction of A.

If the volumetric flow rate Q entering and leaving the CSTR are constant (this is equivalent to a constant density system), the above equation becomes

$$\frac{V}{Q} = \frac{C_{A_1} - C_{A_0}}{(r_A)}$$

$$= \frac{C_{A_0} - C_{A_1}}{(-r_A)} \tag{8.17}$$

where Q = total volumetric flow rate through the CSTR
C_{A_0} = inlet molar concentration of A
C_{A_1} = exit molar concentration of A.

In general, CSTRs are used for liquid phase reactions. High reactant concentrations can be maintained with low flow rates so that conversion approaching 100% can often be achieved. However, the overall economics of the system is reduced because of the low throughput.

CSTRs (as well as the tubular flow reactors described next) are often connected in series in such a manner that the exit stream of one reactor is the feed stream for another reactor. Under these conditions, it is convenient to define the conversion at any point downstream in the battery of CSTR reactors in terms of *inlet* conditions, rather than with respect to any one of the reactors in the series. The conversion X is then the total moles of A that have reacted up to that point per mole of A fed to the *first* reactor. However, this definition should only be employed if there are no side stream withdrawals and the only feed stream enters the first reactor in the series. The conversion from reactors 1, 2, 3,… in the series are usually defined as X_1, X_2, X_3,…, respectively, and effectively represent the overall conversion for that reactor relative to the feed stream to the *first* reactor.

As indicated earlier, for liquid phase reactions the design equation can be written as

$$\theta = \frac{V}{Q} = \frac{C_{A_1} - C_{A_0}}{r_A} \tag{8.18}$$

The term on the left-hand side (LHS) of the above equation has the units of time and represents the average holdup (residence) time in the reactor. It is usually denoted by the symbol θ. The reciprocal of θ is defined as the space velocity (SV) and finds wide application with tubular flow reactors. However, there is a distribution [5] around this average, it is often important enough that this distribution effect be included in the analysis of certain type systems.

8.5   Tubular Flow Reactors

The last "traditional" reactor to be examined is the tubular flow reactor. The most common type is the single-pass cylindrical tube. Another type is one that consists of a number of tubes in parallel. The reactor(s) may be vertical or horizontal. The feed is charged continuously at the inlet of the tube, and the products are continuously removed at the outlet. If heat exchange with surroundings is required, the reactor setup includes a jacketed tube (see Figure 8.3). If the reactor is empty, a homogenous reaction – one phase is present – usually occurs. If the reactor contains catalyst particles, the reaction is said to be heterogeneous; this type will be considered in the next section.

Tubular flow reactors are usually operated under steady conditions so that physical and chemical properties do not vary with time. Unlike the batch and tank flow reactors, there is no mechanical mixing. Thus, the state of the reacting fluid will vary spatially from point to point in the system, and this variation may be in both the radial and axial directions in tubular reactors. The describing equations are then differential, with position as the independent variable.

The reacting system for the describing equations presented below is assumed to move through the reactor in plug flow (no velocity variation through the cross-section of the reactor). It is further assumed that there is no mixing in the axial direction so that the concentration, temperature, etc., do not vary through the cross-section of the tube. Thus, the reacting fluid flows through the reactor in an undisturbed *plug* of mass. For these conditions, the describing equation for a tubular flow reactor is

$$\theta = \frac{V}{Q} = \frac{C_{A_1} - C_{A_0}}{r_A} \tag{8.19}$$

Since $F_{A_0} = C_{A_0} Q_0$

$$\frac{V}{Q_0} = C_{A_0} \int \left(\frac{-1}{r_A} \right) dX_A \tag{8.20}$$

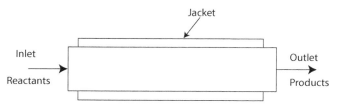

Figure 8.3 Tubular flow reactor; single pass.

The RHS (right hand side) of the above represents the residence time in the reactor based on inlet conditions. If Q does not vary through the reactor, then

$$\theta = \frac{V}{Q_0} \qquad (8.21)$$

where θ is once again the residence time in the reactor.

In actual practice, tubular flow reactors deviate from a plug flow model because of velocity variations in the radial direction. For this condition, the residence time for annular elements of fluid within the reactor will vary from some minimal value at a point where the velocity approaches zero. Thus, the concentration and temperature profiles, as well as the velocity profile, are not constant across the reactor and the describing equations based on the plug flow assumption are then not applicable. This situation can be further complicated if the reaction occurs in the gas phase. Volume effect changes that impact on the concentration term(s) in the rate equation need to be taken into account [1,2,4].

From a qualitative point of view, as the length of the reactor approached infinity, the concentration of a (single) reactant approaches zero for irreversible reactions (except zero order) and the equilibrium concentrations for reversible reactions. Since infinite time is required to achieve equilibrium conversion, this value is approached as the reactor length approaches infinity. For reactors of finite length, where the reaction is reversible, some fraction of the equilibrium conversion is achieved.

Note that the time for a hypothetical plug of material to flow through a tubular flow reactor is the same as the contact or reaction time in a batch reactor. Under these conditions, the same form of the describing equations for batch reactors will also apply to tubular flow reactors.

Another design variable is pressure drop. This effect is usually small for most liquid and gas phase reactions conducted in short or small reactors. This effect is usually estimated using Fanning's Equation (see also Chapter 5) [6].

$$\Delta P = \frac{4 f L V^2}{2 g_c D} \qquad (8.22)$$

where ΔP = pressure drop
f = Fanning friction factor
L = reactor length

V = average flow velocity
g_c = gravity conversion constant
D = reactor diameter

8.6   Catalytic Reactors

Metals in the platinum family are recognized for their ability to promote reactions at low temperatures. Other catalysts include various oxides of copper, chromium, vanadium, nickel, and cobalt. These catalysts are subject to poisoning, particularly from halogens, sulfur compounds, zinc, arsenic, lead, mercury, and particulates. High temperatures can reduce catalyst activity. It is therefore important that catalyst surfaces are clean and active to insure optimum performance. For example, catalysts can be regenerated with superheated steam.

Catalyst may be porous pellets, usually cylindrical or spherical in shape, ranging from 1/16 to ½ inch in diameter. Small sizes are recommended, but the pressure drop through the reactor increases. Other shapes include honeycombs, ribbons, wire mesh, etc. Since catalysis is a surface phenomena, an important physical property of these particles is that the total internal pore surface be many magnitudes greater than the outside surface. The reader is referred to the literature for more information on catalyst preparation, properties, comparisons, costs, and impurities [1,3]

Some of the advantages of catalytic reactors are:

1. low fuel requirements;
2. lower operating temperatures;
3. little or no insulation requirements;
4. reduced fire hazards; and
5. reduced flashback problems.

The disadvantages include:

1. high initial cost;
2. catalyst poisoning;
3. large particles must first be removed;
4. some liquid droplets must first be removed; and
5. catalyst regeneration problems.

Catalytic or heterogeneous reactors are an alternative to homogenous reactors. If a solid catalyst is added to the reactor, the reaction is said to be heterogeneous. For simple reactions, the effect of the presence of a catalyst is item to: increase the rate of reaction, permit the reaction to occur at a lower temperature, permit the reaction (usually) to occur at a more favorable pressure, reduce the reactor volume, and increase the yield of a reactant(s). The basic problem in the design of a heterogeneous reactor is to determine the quantity of catalyst and/or reactor size required for a given conversion and flow rate. In order to obtain this, information on the rate equation(s) and their parameter(s) must be made available. A rigorous approach to the evaluation of reaction velocity constants, etc., for many industrial applications has yet to be accomplished for catalytic reactions. At this time, industry still relies on simple procedures set forth in the literature. (1, 3)

In many catalytic reactions, the rate equation is extremely complex and cannot be obtained either analytically or experimentally. A number of rate equations may result and some simplification is warranted. As mentioned earlier, it is safe in many cases to assume that the rate expression may be satisfactorily expressed by the rate of equation of a single step.

It is common practice to write the describing equations for mass and energy transfer for homogeneous and heterogeneous flow reactors in the same way. However, the (units of the) rate of reaction may be expressed as either

$$\left(\frac{moles\ reacted}{time} \right) (volume\ of\ reactor) \tag{8.23}$$

or

$$\left(\frac{moles\ reacted}{time} \right) (mass\ of\ catalyst) \tag{8.24}$$

The latter is normally the preferred method employed in industry since it is the mass of catalyst present in the reactor that significantly impacts the reactor design. As noted, the rate expression is often more complex for a catalytic reaction than for a non-catalytic (homogeneous) one, and this can make the design equation of the reactor difficult to solve analytically. Numerical solution of the reactor design equation is usually required when designing tubular flow reactors for catalytic reactions.

As indicated above, the principal difference between reactor design calculations involving homogeneous reactions and those involving catalytic (fluid-solid) heterogeneous reactions is that the reaction rate for the latter is based on the mass of solid catalyst, W, rather than on the reactor volume

V. The reaction of a substance A for a fluid-solid heterogeneous system is then defined as

$$-r_A' = \left(\frac{mols\ A\ reacted}{mass\ catalyst} \right) (time)$$

(8.25)

A brief discussion of the two major classes of catalytic reactors follows.

8.6.1   Fluidized Bed Reactors

One type of catalytic reactor in common use is the fluidized-bed. The fluidized-bed reactor is analogous to the CSTR in that its contents, though heterogeneous, are well mixed, resulting in a uniform concentration and temperature distribution through the bed. The fluidized-bed reactor can therefore be modeled, as a first approximation, as a CSTR. For the ideal CSTR, the reactor design equation based on volume is

$$V = \frac{F_{A_0} X_A}{(-r_A)}$$

(8.26)

The companion equation for catalytic or fluid-solid reactor, with the rate based on the mass of solid, is

$$W = \frac{F_{A_0} X_A}{(-r_A')}$$

(8.27)

The reactor volume is simply the catalyst weight, W, divided by the fluidized bed density, ρ_{fb}, of the catalyst in the reactor.

$$V = \frac{W}{\left(\rho_{fb} \right)}$$

(8.28)

The fluid bed catalyst density is normally expressed as some fraction of the catalyst bulk density ρ_B.

8.6.2   Fixed Bed Reactors

A fixed-bed (packed-bed) reactor is essentially a tubular flow reactor that is packed with solid catalyst particles. This type of heterogeneous reaction system is most frequently used to catalyze gaseous reactions. The design equation for tubular flow reactor was previously shown to be

$$V = F_{A_0} \int_0^X \left(\frac{-1}{r_A} \right) dX; \; X = X_A \qquad (8.29)$$

The companion equation based on the mass of catalyst for a fixed-bed reactor is

$$W = F_{A_0} \int_0^X \left(\frac{-1}{r_A'} \right) dX \qquad (8.30)$$

The volume of the reactor, V, is then

$$V = \frac{W}{(\rho_B)} \qquad (8.31)$$

where ρ_B = bulk density of the catalyst.

The aforementioned Ergun equation [7] is normally employed to estimate the pressure drop for fixed bed units.

8.7   Thermal Effects

It was shown in the Chemical Kinetics section that the rate of reaction, r_A, is a function of temperature and concentration. The application of r_A to the reactor equations is simplest for isothermal conditions since r_A is generally solely a function of concentration. If non-isothermal conditions exist, another equations must be written to describe temperature variations with time or position within the reactor. Details of this effect for batch, continuous stirred tank and tubular flow reactors follows.

8.7.1   Batch Reactors

The equation describing temperature variations in batch reactors is obtained by applying the conservation law for energy on a time-rate basis to the reactor contents. Since batch reactors are stationary (fixed in space), kinetic and potential effects can be neglected. The equation describing the temperature variation in reactors due to energy transfer, subject to the assumptions in its development, is [2]

$$mC_p \left(\frac{dT}{dt} \right) = -UA_e \left(T - T_a \right) + V \left(-\Delta H_A \right) \left| -r_A \right|$$

(8.32)

where m = mass of the reactor contents
C_p = heat capacity
V = reactor volume
$-\Delta H_A$ = enthalpy of reaction of species A
$\left| -r_A \right|$ = absolute value of the rate of reaction of A
T = reactor temperature.

In addition (8-10) (see Chapter 6)

$$\dot{Q} = UA \left(T - T_a \right)$$

(8.33)

where Q = heat transfer rate across the walls of the reactor
U = overall heat transfer coefficient
A = area available for heat transfer
T_a = temperature surrounding the reactor walls

For non-isothermal reactors, one of the reactor design equations—the energy transfer equation (see above), and an expression for the rate in terms of concentration and temperature—must be solved simultaneously to give the conversion as a function of time. Note that the equations may be interdependent; i.e., each can contain terms that depend on the other equation(s). These equations, except for simple systems, are usually too complex for analytical treatment.

8.7.2   CSTR

If the conservation law for energy is applied to a CSTR, information on temperature changes and variations within the reactor can be obtained. For constant heat capacity, C_p, and enthalpy of reaction, ΔH_A, the describing equation becomes

$$FC_p \left(T_o - T_1 \right) + \left(-r_A \right) V \left(-\Delta H_A \right) = 0$$

(8.34)

where F = flow rate through the reactor, units consistent with C_p
   T_o = inlet temperature
   T_1 = outlet temperature

For adiabatic conditions, and noting that

$$(-r_A)V = F_{A_0}X_A; \quad F_{A_0} = feed\ rate\ of\ A \tag{8.35}$$

the above equation becomes (with $T_1 = T$)

$$FC_p(T_o - T) = F_{A_0}X_A(-\Delta H_A) = 0 \tag{8.36}$$

The equation may be rearranged to give

$$T = T_o + \left[\frac{F_{A_0}(-\Delta H_A)}{FC_p}\right]X_A \tag{8.37}$$

Thus, the need for simultaneous solution of the mass and energy transfer equations is removed. Fogler [4] has also accounted for enthalpy of reaction variation with temperature in a slightly different form.

8.7.3   Tubular Flow Reactions

The temperature in a tubular flow reactor can vary with position (volume) due to enthalpy of reaction effects or transfer of energy in the form of heat across the walls of the reactor. The reactor design equation must then include temperature variations before being solved. In order to obtain information on the temperature at every point in the reactor, the conservation law for energy is applied to a system. For adiabatic operation, heat transfer with the surroundings is zero. The energy equation reduces to [2]

$$FC_p dT = |-r_A|(-\Delta H_A)dV \tag{8.38}$$

Since

$$|-r_A|dV = F_{A_0}dX_A \tag{8.39}$$

for a tubular flow reaction, the above equation becomes

$$FC_p dT = F_{A_0}\left(-\Delta H_A\right)dX_A \qquad (8.40)$$

which may be integrated to give the previously developed equation for CSTRs:

$$T = T_o + \left[\frac{F_{A_0}\left(-\Delta H_A\right)}{FC_p}\right]X_A \qquad (8.41)$$

The term in brackets is a constant if the enthalpy of reaction and the average heat capacity are assumed independent of temperature. The temperature is then a linear function of conversion [1]. Fogler [4], also provides an equation describing temperature variation within a tubular flow reactor. The equation also takes the same form as that provided for CSTRs.

A detailed and expanded treatment of chemical reactors is available in the following three references

1. L. Theodore, *Chemical Reactor Analysis and Applications for the Practicing Engineer*, John Wiley & Sons, Hoboken, NJ, 2012 [21].
2. D. Green and R. Perry editors, *Perry's Chemical Engineers' Handbook*, 8th edition, McGraw-Hill, New York City, NY, 2008 [9].
3. L. Theodore, *Chemical Engineering: The Essential Reference*, McGraw-Hill, New York City, NY, 2014 [10].

8.8   Illustrative Open-Ended Problems

This and the last section provides open-ended problems. However, solutions *are* provided for the three problems in this section in order for the reader to hopefully obtain a better understanding of these problems, which differ from the traditional problems/illustrative examples. The first problem is relatively straightforward while the third (and last problem) is somewhat more difficult and/or complex. Note that solutions are not provided for the 36 open-ended problems in the next section.

Problem 1: Discuss rate versus equilibrium considerations as they apply to chemical reactors.

Solution: With regard to chemical reactions, there are two important questions that are of concern to the chemical engineer:

1. How *far* will the reaction go?
2. How *fast* will the reaction go?

Chemical thermodynamics provides the answer to the first question; however, it provides nothing about the second [9]. What follows the development presented in chapter 4.

To illustrate the difference and importance of both of the above questions in an engineering analysis of a chemical reaction, consider the following process [1–4]. Substance A, which costs 1 cent/ton, can be converted to B, which costs $1 million/lb, by the reaction A→B. Chemical thermodynamics will provide information on the maximum amount of B that can be formed. If 99.99% of A can be converted into B, the reaction would then appear to be economically feasible, from a *thermodynamics* point of view. However, a *kinetic* analysis might indicate that the reaction is so slow that, for all practical purposes, its rate is vanishingly small. For example, it might take 10^9 years to obtain a 10^{-9}% conversion of A. The reaction is then economically unfeasible. Thus, it can be seen that both equilibrium and kinetic effects must often be considered in an overall engineering analysis of a chemical reaction.

Equilibrium and rate are therefore both important factors to be considered in the design and prediction of the performance of equipment employed for chemical reactions. The rate at which a reaction proceeds will depend on the displacement from equilibrium, with the rate at which equilibrium is established essentially dependent on a host of factors discussed in the previous section. As noted, this rate process ceases upon the attainment of equilibrium.

Problem 2: [1,2] The following *gas* phase reaction is carried out *isothermally* in a constant volume batch reactor:

$$2A \rightarrow B + C + D$$

TABLE 8.1 Time – Pressure Data for Problem 2

Time (min)	0	1.20	1.95	2.90	4.14	5.70	8.10	∞
Total pressure (atm)	1	1.10	1.15	1.20	1.25	1.30	1.35	1.51

Pure A is initially in the reactor at STP (32°F, 1 atm). The following pressures in Table 8.1 were recorded at subsequent times during the reaction.

1. Calculate δ and ε.
2. Develop an expression for the conversion in terms of the total pressure.
3. Is the reaction reversible? Provide your explanation.
4. Verify that the order of the reaction is approximately 1.4, and then calculate the specific reaction rate constant using any convenient method of your choice.
5. Comment on the results.

Solution:

1. Rewrite the gas phase reaction

$$2A \rightarrow B + C + D$$

as

$$A \rightarrow (1/2)B + (1/2)C + (1/2)D$$

Preliminary calculations:

$$y_{A_0} = 1.0$$

$$\delta = (3/2) - 1 = \tfrac{1}{2}$$

$$\varepsilon = (1.0)(1/2) = \tfrac{1}{2} = 0.5$$

$$V = V_0(1 + \varepsilon X)(T_0/T)(P/P_0); \quad T_0/T = 1.0$$

$$= V_0(1 + \varepsilon X)(P/P_0)$$

2. Since the volume is constant, $V = V_0$, so that

$$X = \frac{P - P_0}{\varepsilon P_0}; \quad P_0 = 1.0$$

$$= 2P - 2$$

(8.42)

For a batch reactor,

$$r_A = \frac{1}{V}\frac{dN_A}{dt} = \frac{dC_A}{dt}; \text{ constant volume} \tag{8.43}$$

Assume the mechanism to be of the form

$$r_A = kC_A^n \tag{8.44}$$

Since

$$C_A = C_{A_0}(1-X)$$
$$-C_{A_0}\frac{dX}{dt} = kC_A^n \tag{8.45}$$

and

$$C_A = C_{A_0}\left[1-\left(\frac{P-P_0}{\varepsilon P_0}\right)\right] = \frac{C_{A_0}}{\varepsilon P_0}\left[(1+\varepsilon)P_0 - P\right] \tag{8.46}$$

Therefore

$$-r_A = k\left(\frac{C_{A_0}}{\varepsilon P_0}\right)^n\left[(1+\varepsilon)P_0 - P\right]^n = kC_A^n \tag{8.47}$$

And

$$X = \frac{P-P_0}{\varepsilon P_0} \tag{8.48}$$

3. The reaction equation indicates that 2 moles are converted to 3 moles, an increase of 50%. The pressure data indicates that the pressure increases from 1.0 atm to 1.5 atm (a 50% increase when the reaction goes to completion). Therefore the reaction is *irreversible*.

4. The derivative of the above Equation is

$$\frac{dX}{dt} = \frac{1}{\varepsilon P_0}\frac{dP}{dt} \tag{8.49}$$

Combining this equation with the rate equation yields

$$\frac{dP}{dt} = k\left(\frac{C_{A_0}}{\varepsilon P_0}\right)^{n-1}\left[(1+\varepsilon)P_0 - P\right]^n \tag{8.51}$$

One may use any approach to generate dP/dt information. One approach requires a log – log plot of rate in terms of dP/dt vs. $[(1 + \varepsilon)P_0 - P]$; i.e., $[1.5 - P]$. The slope is then n. The plot yields $n = 1.4$.

5. Two concerns arise in integrating the results:
   a. The assumed mechanism
   b. Ideal gas law behavior.

Problem 3: A vertical fixed bed catalytic reactor pictured in Figure 8.4 is designed to operate at a specified conversion. Once the unit is installed and running, the unit operates with a lower conversion. What options would you recommend to bring the unit into compliance with the specified design conversion? [2]

Solution: Some possible solutions are provided below:

1. Use a smaller catalyst size. A different size may produce a higher conversion.

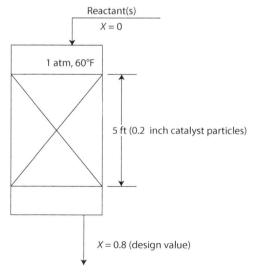

Figure 8.4 Vertical fixed-bed reactor.

2. Use a different type of catalyst. A different type may produce a higher conversion.
3. Use a different physical shaped catalyst.
4. Make sure there is no channeling inside the reactor. The catalyst should be randomly distributed and there should be no open spots.
5. Consider increasing or decreasing the flow rate entering the reactor.
6. Check to see if the pressure drop is excessive.
7. Lower the initial concentration of the reactant (if possible).
8. Increase the height of the reactor bed.
9. Increase the temperature.
10. Modify the process that may be producing the reactant(s); since the details of this process are not included in the problem statement, no specific recommendations can be made.
11. Finally, before considering changes to the system, one should undertake a thorough inspection of the reactor and surrounding components.

The reader should note that some of the recommendations above could lead to higher pressure drops and potential problems with the flow. An example of this problem would be the implementation of suggestion (1). The reader is left the exercise of determining what other steps could lead to flow/pressure drop problems. *Hint*: There are at least six suggestions that fall into this category, including a consideration to fire the (chemical?) engineer who designed the reactor [12].

8.9   Open-Ended Problems

This last section of the chapter contains open-ended problems as they relate to chemical reactors. No detailed and/or specific solution is provided; that task is left to the reader, noting that each problem has either a unique solution or a number of solutions or (in some cases) no solution at all. These are characteristics of open-ended problems described earlier.

There are comments associated with some, but not all, of the problems. The comments are included to assist the reader while attempting to solve the problems. However, it is recommended that the solution to each problem should initially be attempted *without* the assistance of the comments.

There are 36 open-ended problems in this section. As stated above, if difficulty is encountered in solving any particular problem, the reader should next refer to the comment, if any is provided with the problem. The reader should also note that the more difficult problems are generally located at or near the end of the section.

1. Describe the early history associated with chemical reactors.
2. Discuss the recent advances in chemical reactor technology.
3. Select a referred, published article on chemical reactors from the literature and provide a review.
4. Develop an original problem that would be suitable as an illustrative example in a book on chemical reactors.
5. Prepare a list of the various books which have been written on chemical reactors. Select the three best (hopefully the a choice will include the text written by one of the authors) and justify your answer. Also select the three weakest books and, once again, justify your answer.
6. Describe what, if any, advantages are derived from writing chemical reactions on a mole basis.
7. Generate a time-concentration (or conversion) solution to a reaction that is described by a complex rate equation of your choice.
8. Describe the relationships and differences between conversion, yield, and selectivity.
9. Describe in general terms how to increase conversion in a
   - batch reactor,
   - CSTR, and
   - tubular flow reactor.
10. Reactions are often described by rate equations that are of a complex nature. Discuss the impact that these complex equations can have on designing and predicting the performance of:
    - batch reactors,
    - CSTRs, and
    - tubular flow reactors.
11. List and discuss the advantages and disadvantages of batch, CSTR, and tubular flow reactors.
12. List and describe all the classes of reactors that have been employed in the past.
13. Develop and/or design a *new* reactor.

14. A batch reactor is no longer delivering the degree of conversion required for a process. Rather than replace the unit, you have been asked to recommend what other possible steps can be taken to the existing reactor to get it back "on line".
Comment: Refer to Problem 3 in the previous section.

15. A CSTR is designed to operate at a specified conversion. Once the unit is installed and running, the unit operates at a lower conversion. What options would you recommend to bring the unit into compliance with the specified design conversion?
Comment: Refer to Problem 3 in the previous Section.

16. Over time, the heat required to keep a catcracker in refinery at its operating temperature has increased. You are asked to recommend possible ways to correct the problem.
Comment: Refer to Problem 3 in the previous Section.

17. Discuss the options that are available to increase/decrease the temperature to a satisfactory value in a
   • batch reactor
   • CSTR
   • tubular flow reactor
   • fixed bed reactor
   • fluid bed reactor.

18. Design and/or develop a new catalyst.

19. Design and/or develop a new support plate for catalysts in a reactor.

20. Develop a general equation to describe temperature variation and its effect on conversion for a
   • batch reactor
   • CSTR
   • tubular flow reactor.

21. Describe the advantages and disadvantages of both the integration and differentiation method of analysis of experimental rate data.
Comment: Refer to the literature [2].

22. There are numerous approaches that have been employed to describe the mechanism associated with catalytic reactors. Compare the various approaches.
Comment: Refer to the literature [4]

23. Develop a simple model to describe the mechanism associated with a catalytic reactor.
Comment: Refer to the literature [4].

24. Develop a method to design and/or predict the performance of fluid bed reactors.

25. Develop a method to design and/or predict the performance of fixed bed reactors.

26. Discuss the problems that can arise if ideal flow patterns are assumed in designing and/or predicting the performance of the various classes of flow reactors.

27. Discuss the calculational problems that can develop for reactions with phase change(s).

28. Discuss the various experimental approaches that can be employed to determine the mechanism of a chemical reactor.

29. Would incorrect chemical reaction kinetic data result in the under or over design of a reactor? Explain.

30. Describe the problems associated with scaleup of the various classes of chemical reactors.

31. You have been assigned the task of developing an operation, maintenance, and inspection (OM&I) program for the reactors at your site. Also discuss the differences that would exist for the different class of reactors.

32. Describe the process of a catalytic reaction in technical terms.

33. Describe the process of a catalytic reaction in layman terms.

34. Consider the following two chemical reactions

$$C_2H_4 + 2H_2O \rightarrow C_2H_6 + O_2 + H_2$$

$$C_2H_4 + 2H_2O \rightarrow 2CO + 4H_2$$

Although both equations balance stoichiometrically, explain in layman terms why the equations are different. (See also Chapter 3).

35. An auditorium room has a volume of 100,000 ft^3. The CO_2 concentration in the air before a meeting is 0.04 per cent. The CO_2 concentration is 0.14 per cent one hour after the beginning of the session. The room is ventilated by 50,000 ft^3/min of fresh air (0.04 per cent CO_2). Describe how to calculate the CO_2 concentration 6 hr after the session has ended if the CO_2 is disappearing/reacting via

- a zero order reaction
- a first order reaction, and
- an nth order reaction.

36. One option available to a plant manager when a tube within a tubular flow reactor fails is to simply plug the inlet of the tube. Develop an equation to describe the impact on the performance of the reactor as a function of the number of tubes with in the reactor and the number of plugged tubes for a first-order reaction. Repeat the calculation for different reaction mechanisms.

References

1. L. Theodore, *Chemical Reaction Kinetics*, A Theodore Tutorial, Theodore Tutorials, East Williston, NY, orginally published in the USEPA/APTI, RTP, NC. 1995.
2. L. Theodore, *Chemical Reactor Analysis and Application for the Practicing Engineer*, John Wiley & Sons, Hoboken, NJ, 2012.
3. Personal notes: L. Theodore, East Williston, NY, 1968.
4. S. Fogler, *Elements of Chemical Reaction Engineering*, 4th edition, Prentice-Hall, Upper Saddle River, NJ, 2006.
5. S. shaefer and L. Theodore, *Probability and Statistics Applications in Environmental Science,* CRC Press/ Taylor & Francis Group, Boca Raton, FL, 2007.
6. P. Abulencia and L. Theodore, *Fluid Flow for the Practicing Chemical Engineer*, John Wiley & Sons, Hoboken, NJ, 2009.
7. S. Ergun, *Fluid Flow Though Packed Columns*, CEP, 48:49, New York City, NY, 1952.
8. L. Theodore, *Heat Transfer for the Practicing Engineer*, John Wiley & Sons, Hoboken, NJ, 2011.
9. D. Green and R. Perry (editors), *Perry's Chemical Engineers' Handbook,* 8th edition, McGraw-Hill, New York City, NY, 2008.
10. L. Theodore, *Chemical Engineering: The Essential Reference*, McGraw-Hill, New York City, NY, 2014.
11. L. Theodore, F. Ricci, and T. VanVliet, *Thermodynamics for the Practicing Engineer*, John Wiley & Sons, Hoboken, NJ, 2009.
12. Personal notes: L. Theodore, East Williston, NY, 2006.

9

Process Control and Instrumentation

This chapter is concerned with process control and instrumentation (PCI). As with all the chapters in Part II, there are several sections: overview, several technical topics, illustrative open-ended problems, and open-ended problems. The purpose of the first section is to introduce the reader to the subject of PCI. As one might suppose, a comprehensive treatment is not provided although several technical topics are included. The next section contains three open-ended problems; the authors' solution (there may be other solutions) is also provided. The final section contains 40 problems; *no* solutions are provided here.

9.1 Overview

This overview section is concerned—as can be noted from its title—with process control and instrumentation (PCI). As one might suppose, it was not possible to address all topics directly or indirectly related to PCI. However, additional details may be obtained from either the references provided at the end of this Overview section and/or at the end of the chapter.

Note: Those readers already familiar with the details associated with this subject may choose to bypass this Overview.

Chemical processes can be controlled to yield not only more products but also more uniform and higher quality products, usually resulting in a profit increase. In addition, there are processes which respond so rapidly to change(s) in the system that they cannot be properly controlled by a plant operator; these systems are eligible candidates for some form of automatic control. However, the decision to apply automatic control(s) should be based on an applicable and appropriate cost-effective economic analysis that includes process objectives.

In most modern chemical and petrochemical plants, computers are used for many data acquisition and control operations. Computers can collect data from instruments throughout a plant. The data can be used as input to plant or equipment models in order to provide information to control elements. The data can also be recorded for future analysis. Operators typically watch video display screens which contain information about the plant rather than watch panel boards of actual instruments.

Computer control has led to more optimum control of plants, to better integration of various process units, to make data storage easier, and to provide more up-to-date data for the operators. Complete computer control was slow to be adopted in the 1960's, 1970's and 1980's due to reliability problems, both real and perceived, of the computer systems. Any individual who works extensively with computers knows that interconnected networks often "crash." Highly reliable computer systems with back-up power supplies and data links have been developed for the process industries. In many cases, two fully-redundant systems are installed so that one can be used anytime the other requires service.

In plants using modern computer control, many computers are linked in a hierarchical fashion; single mainframe systems are rarely used. At the lowest level, small *control* computers, with roughly the capability of PCs, are used to record data from and provide control information to individual pieces ot process equipment or small groups of related equipment. These computers are linked to *process unit* computers which contain models of the overall process. Process unit computers periodically send updated instructions to the lower level machines and receive updated process information. The process unit computers throughout a plant are linked to a *supervisory* computer. It periodically provides updated information on such things as feed and product values, inventories, utility system conditions, etc. The *process unit* computers, in turn, provide the supervisory computer with operating information about each unit in the plant.

Full redundancy, either with a second computer system or with conventional control system backup, is needed at the control computer level. The plants can operate for hours at a time, however, if the process unit computers are down, and, perhaps, for days at a time if the supervisory computer is down. Nevertheless, full redundancy is often used since it is relatively inexpensive. Software and training costs are usually greater than hardware costs.

Much of the material to flow has been drawn from the work of Vasudevan [1]; numerous excellent illustrative examples are available in this reference. A number of topics (e.g., Routh criteria for stability, root locus analysis, Bode plots, etc.) are not treated in this chapter. In addition, a decision was made by the authors not to provide any detailed mathematics and complex describing equations. Topics addressed in the remainder of this chapter include the following:

1. Process Control Fundamentals
2. Feedback Control
3. Feedforward Control
4. Cascade Control
5. Alarms and Trips.

As noted above, the reader should note that nearly all of the material in this chapter has been drawn from P. Vasudevan, "Process Control," A Theodore Tutorial, Theodore Tutorials, East Williston, NY, originally published by the USEPA/APTI, RTP, NC, 1996 [1]. Also note that Vasudevan's notation has been retained in the presentation.

9.2   Process Control Fundamentals

Automatic control can perhaps be best described via a continuous stirred tank reactor [2] (CSTR) example, as pictured in Figure 9.1. The contents in the reactor are heated to a design temperature by the steam flowing through a heat exchanger, e.g., heating coils [3]. The temperature of the product flow (the variable controlled) and the CSTR mixture are affected by the flow rate and inlet temperature of the reactant(s), the temperature, pressure and flow ratio of the steam, the degree of mixing, and (any) heat losses to the surrounding environment.

Certain process control terms may now be introduced and defined. Figure 9.1 represents an *open-loop* system since the output temperature is not employed to adjust/change any of the reactors variables, i.e., the system cannot compensate/change any of the reactor valuables. A *closed-loop* system is one where the measured value of the temperature (the system variable to be controlled) is used to compensate/change one or more reactor variables, e.g., the system temperature.

In *feedback control* (see Figure 9.2), the temperature is compared to a particular value – often referred to as a *set point* or *design value*. The degree of displacement of the temperature from the set point provides a correction to one of the reactor variables in a manner to reduce the displacement (often referred to as the *error*).

Feedforward control (see Figure 9.3) allows a compensation for any reactor disturbance prior to a change to the controlled variable, i.e., the product temperature. This type of control has an obvious advantage if the controlled variable cannot be measured.

Feedforward and feedback control are often combined (see Figure 9.4) in certain systems.

The employment of *block diagrams* (see Figure 9.5) is the standard method of pictorially representing the controlled system with its variables. The block diagram is obtained from the physical system by dividing it into

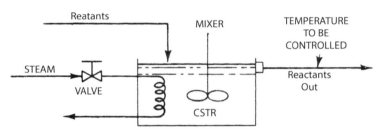

Figure 9.1 Continuous stirred tank reactor.

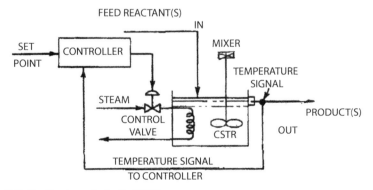

Figure 9.2 Feedback control of a CSTR process.

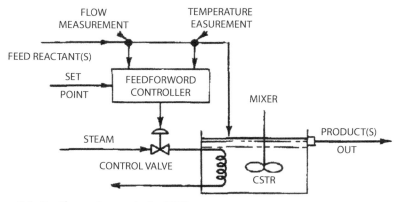

Figure 9.3 Feedforward control of a CSTR process.

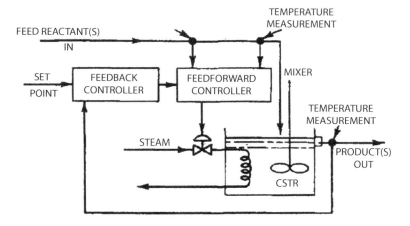

Figure 9.4 Feedforward and feedback control of a CSTR exchanger.

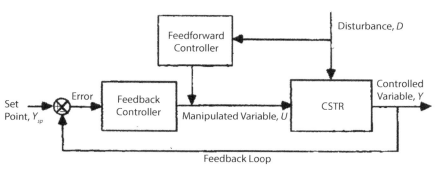

Figure 9.5 Block line diagram for feedforward and feedback control.

functional, non-interacting sections whose inputs and outputs are readily identifiable.

Four parameters that are employed in process control are

1. *Transfer function*
2. *Steady-state gain*
3. *Time constant*
4. *Dead-time or lag.*

These four terms are briefly discussed below:

1. The transfer function relates two variables in a process; one of these is the forcing function or input variable, and the other is the response of the output variable. The input and output variables are usually expressed in the Laplace [4] domain, and are in derivative form.
2. The steady-state gain gives an indication of how much the output variable changes (in the time-domain) for a given change in the input variable (also in the time-domain). The gain is also dependent on the physical properties and operating parameter of the process. The term steady state is applied since a step-change in the input variable results in a change in the output variable which reaches a new steady-state that can be predicted by application of the final-value theorem [1]. Gain may or may not be dimensionless.
3. The process time-constant provides an indication of the speed of the response of the process, i.e., the speed of the output variable to a forcing function or change in the input variable. The slower the response, the larger the time-constant is, and vice versa. The time-constant; with units of time, is usually related to the different physical properties and operating parameters of the process.
4. Process dead-time refers to the delay in time before the process starts responding to a disturbance in an input variable. It is sometimes referred to as *transportation lag* or *time delay.* Dead-time or delays can also be encountered in measurement sensors such as thermocouples, pressure transducers, and in transmission of information from one point to another. In these cases, it is referred to as measurement *lag.*

9.3   Feedback Control

Feedback control is a very important aspect of process control. Its role is best described in terms of an example (see also Figure 9.2). Assume that one desires to maintain the temperature of a polymer reactor at 70°C. Temperature is thus the controlled variable, and the desired temperature level 70°C, is called the aforementioned set-point. In feedback control, the temperature is measured using a sensor (such as a thermocouple device). This information is then continuously relayed to a controller, and a device known as a *comparator*, compares the set-point with the measured signal (or variable). The difference between the set-point and the measured variable is the previously defined error. Based on the magnitude of the error, the controller element in the feedback loop takes corrective action by adjusting the value of a process parameter, known as the *manipulated* variable.

The controller logic (the manner in which it handles the error) is an important process control criterion. Generally, feedback controllers are either *proportional*, P (sends signals to the final control element proportional to the error), *proportional-integral*, PI (sends a signal to the final control element that is both proportional to the magnitude of the error at any instant and the sum of the error), and *proportional-integral-derivative*, PID (sends a signal that is also based on the slope of the error).

In the above example, the manipulated variable may be cooling water flow through the reactor jacket. This adjustment or manipulation of the flow rate is achieved by a *final-control-element*. In most chemical processes, the final-control-element is usually a pneumatic control valve. However, depending upon the process parameter being controlled, the final control element could very well be a motor whose speed is regulated. Thus, the signal from the controller is sent to a final control element which manipulates the manipulated variable in the process.

In addition to the controlled variable, there may be other variables that disturb or affect the process. In the reactor example above, a change in the inlet temperature of the feed or inlet flow rate are considered to be *load* disturbances. A *servo* problem is one in which the response of the system to a change in set-point is *recorded*, whereas a load or regulator problem is one in which the response of a system to a disturbance or load variable is *measured*.

Before selecting a controller, it is very important to determine its action. Consider another example—the heat exchanger shown in Figure 9.6. Steam is used to heat the process fluid. If the inlet temperature of the process fluid increases, an increase in the outlet temperature will result. Since the outlet temperature moves above the set-point (or desired temperature), the

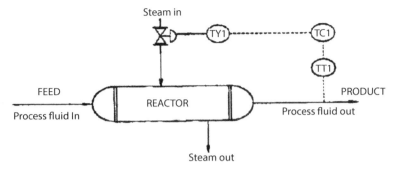

Figure 9.6 Controller actions.

controller must close the steam valve. This is achieved by the controller sending a lower output (pneumatic or current) signal to the control valve, i.e., an increase in the input signal from the controller to the valve. The action of the controller is considered to be *reverse*. If the input signal to the controller and the output signal from it act in the same direction, the controller is *direct acting*.

It is important to consider the process requirements for control to determine the action of the controller and the action of the final control element. The controller action is usually set by a switch on electronic and pneumatic controllers. On microprocessor–based controllers, the setting can be made by changing the sign of the scale factor in the software (which then changes the sign of the proportional gain of the controller).

The functions of a feedback controller in a process control loop is twofold: (1) to compare the process signal from the transmitter (the controlled or measured variable) with the set point, and (2) to send a signal to the final control element with the sole purpose of maintaining the controlled variable at its set point.

As noted above, the most common feedback controllers are proportionally controllers (P control), proportional-integral controllers (PI control), and proportional-integral-derivative (PID) controllers.

9.4   Feedforward Control

Feedforward control has several advantages. Unlike feedback control, a feedforward control measures the disturbance directly, and takes preemptive action before the disturbance can affect the process. Consider the heated tank example discussed earlier. A conventional feedback controller would entail measuring the temperature in the tank, the controlled variable, and maintaining it at the desired set-point by regulating the heat input, the manipulated variable. In feedforward control, the load disturbances would first be identified, i.e., the inlet flow rate, and the temperature of the inlet fluid. Any change in the inlet temperature, for example, would be monitored in a feedforward control system, and corrective action taken by adjusting the heat input (again the manipulated variable), before the process is affected. Thus, unlike feedback control, the error is not allowed to propagate through the system.

It is clear that a feed forward controller is not a PID controller, but a special computing or digital machine. Good feedforward control relies to a large extent on good knowledge of the process, which is often the biggest drawback. Finally, the stability of a feedforward-feedback system is determined by the roots of the characteristic equation of the feedback loop (feedforward control does not affect the stability of the system) [1].

Vasudevan provides additional developmental material and illustrative examples [1].

9.5   Cascade Control

The simple feedback control loop considered earlier is an example of a Single Input Single Output (SISO) system. In some instances, it is possible to have more than one measurement but one manipulated variable, or one measurement and more than one manipulated variable. *Cascade control* is an example where there is one manipulated variable but more than one measurement. Consider a slight modification of the continuous stirred tank reactor (CSTR) shown in Figure 9.7. The control objective is to maintain the reactor temperature at a set value by regulating the cooling water flow to the exchanger. The load disturbances to the reactor include changes in the feed inlet temperature

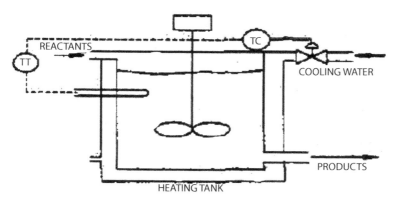

Figure 9.7 Simple feedback temperature control in a CSTR.

or in the cooling water temperature. In simple feedback control, any change in the cooling water temperature will affect the reactor temperature, and the disturbance will propagate through the system before it is corrected. In other words, the control loop will respond faster to changes in the feed inlet temperature compared to changes in the cooling water temperature.

Now consider the following example. In this case, any change in the cooling water temperature is corrected by a new controller added to the process control loop before its effect propagates through the reaction system. These are two different measurements – reaction temperature and cooling water temperature, but only one manipulated variable (cooling water flow rate). The loop that measures the reaction temperature is known as the *primary* control loop, and the loop that measures the cooling water temperature, the *secondary* control loop. The secondary control loop uses the output of the primary controller as its set-point, whereas, the set-point to the primary controller is supplied by the operator. Cascade control has wide applications in chemical processes. Usually, flow rate control loops are cascaded with other control loops.

9.6   Alarms and Trips

Alarms are used in process plants to inform operators that plant conditions are outside of the normal operating range. Trips are devices which sense operation outside of normal range and automatically shut off or turn on some device. Alarms are used when the time constants involved are large enough for an operator to make adjustments to the process and return it to normal conditions. Trips are used when process time constants are very short and immediate action may be needed to prevent a disaster. As might be expected, complex plants may have several different priority levels of alarms and trips.

An everyday example of an alarm would be a high-temperature warning light on the dash board of a car. If it goes on, it indicates that there is a problem and the driver should do something about it. If a trip were installed here, it would shut off the motor when a certain high temperature point was reached. Household gas water heaters have trips on the fuel supply linked to a temperature element which senses whether the flame has gone out.

A detailed and expanded treatment of process control and instrumentation is available in the following two references:

1. P.T. Vasudaven, *Process Dynamics and Control*, A Theodore Tutorial, Theodore Tutorials, East Williston, NY, originally published by the USEPA/APTI, RTP, NC, 1996 [1].

2. L. Theodore, *Chemical Engineering: The Essential Reference*, McGraw-Hill, New York City, NY, 2014 [5].

9.7   Illustrative Open-Ended Problems

This and the last section provide open-ended problems. However, solutions *are* provided for the three problems in this section in order for the reader to hopefully obtain a better understanding of these problems which differ from the traditional problems/illustrative examples. The first problem is relatively straightforward while the third (and last problem) is somewhat more difficult and/or complex. Note that solutions are not provided for the 39 open-ended problems in the next section.

Problem 1: Discuss the solution of linear differential equations using Laplace transforms, particularly as they apply to process control.

Solution: There are three basic steps involved in the solution of ordinary differential equations using Laplace transforms.

1. The first key step is to take the Laplace transform of the differential equation by applying the real differentiation theorem. This transforms the differential equation into an algebraic equation.
2. The algebraic equation is then rearranged in such a way that the output variable, $Y(s)$, is expressed as a function of the input variable, or forcing function, $X(s)$.
3. The resulting equation is inverted in order to obtain the output variable, $Y(t)$, in the real time-domain.

One of the problems associated with Laplace transforms is that they can only be applied to linear systems. The dynamic control responses of most industrial processes is non-linear. However, it is possible to linearize the equations describing a non-linear process.

Refer to Chapter 23, Illustrative Open-ended Problem 3 for additional details.

Problem 2: Process instruments and control devices make up the nerve system, or the "brains" of chemical or petrochemical plant. Control systems consist of sensing devices, models, and control elements. *Sensing devices* measure such things as temperature, flow rate, pressure, composition, level height, etc. *Models* consists of information about how a plant is expected to operate. They may be as simple as individual temperature or flow rate set points, or as complex as dynamic computer models of whole process units. *Control elements* are the actual devices that adjust the conditions of a plant, such as valves or speed controls on rotating machinery.

Standard symbols are used on process flow diagrams and piping instrumentation diagrams to represent various elements of control systems. They consist of small circles, or "bubbles," with key initials inside. List some of the most common initials.

Solution: Most of the initials are concerned with sensing devices and functions. Some of the more common ones are listed below:

Sensing devices:

F	Flow rate
FQ	Integrated flow rate
L	Level
P	Pressure
PD	Pressure difference
pH	pH ($-\log_{10}$ of hydrogen ion concentration)
S	Speed
T	Temperature
W	Weight
X	Chemical composition

Functions:

A	Alarm
AH	Alarm, high reading
AL	Alarm, low reading
C	Controller
I	Indicator
R	Recorder
S	Switch
V	Valve

Problem 3: Describe the role several forcing functions play in process control.

Solution: Commonly encountered forcing functions (or input variables) in process control are step-input (positive or negative), pulse function, impulse function and ramp function. For example, consider a step-input of magnitude 1. The step-function may be represented by $u(t)$, where

$$u(t) = 1 \text{ for } t \geq 0$$

and

$$u(t) = 0 \text{ for } t > 0$$

Substituting the value of this function into the definition of the Laplace transform and solving yields

$$L(u(t)) = 1/s$$

Now consider a rectangular pulse whose magnitude is H and duration is t units of time. In the time interval $0 \le t < T$, the function $f(t) = H$. When $t < 0$ or when $t > T$, the function $f(t) = 0$. Substituting the value of the pulse function into the definition of the Laplace transform, and integrating between the limits 0 to T yields:

$$L(f(t)) = F(s) = H/s\left(1 - e^{(-T_s)}\right)$$

It is very important for the chemical engineer to understand how a system responds to various forcing-functions. The forcing functions considered above can be used in a number of examples. The effect of a delay in time is also important since processes generally experience a time-lag or 'dead-time.'

9.8   Open-Ended Problems

This last section of the chapter contains open-ended problems as they relate to process control and instrumentation. No detailed and/or specific solution is provided; that task is left to the reader, noting that each problem has either a unique solution or a number of solutions or (in some cases) no solution at all. These are characteristics of open-ended problems described earlier.

There are comments associated with some, but not all, of the problems. The comments are included to assist the reader while attempting to solve the problems. However, it is recommended that the solution to each problem should initially be attempted *without* the assistance of the comments.

There are 40 open-ended problems in this Section. As stated above, if difficulty is encountered in solving any particular problem, the reader should next refer to the comment, if any is provided with the problem.

The reader should also note that the more difficult problems are generally located at or near the end of the section.

1. Describe the early history associated with process control.
2. Discuss the recent advances in process control technology.
3. Select a refereed, published process control article from the literature and provide a review.
4. Provide some normal everyday domestic applications involving the general topic of process control.
5. Develop an original problem that would be suitable as an illustrative example in a book on process control.
6. Prepare a list of the various books which have been written on process control. Select the three best and justify your answer. Also select the three weakest books and, once again, justify your answer.
7. Discuss the advantages and disadvantages of employing process control in the chemical industry.
8. Discuss the advantages and disadvantages of employing instrumentation in the chemical industry.
9. Discuss the role measurement plays in process control.
10. Discuss the advantages and disadvantages of computer process control of plants.
11. Describe and discuss the difference between instrument accuracy and precision.
12. Discuss the general subject of control stability from both a technical perspective and a layman's perspective.
13. Describe the differences between an alarm and a trip. Also provide an everyday example of both.
14. Describe how a pneumatic control element works.
15. Provide a list of the common initials/abbreviations that are employed for a variety of sensing devices and functions.
16. Discuss the role instrumentation flow sheets play in process control.
17. Describe the difference between open and closed loop systems.
18. Provide a technical description of feedforward control.
19. Provide a technical description of feedback control.
20. Provide a layman's definition of feedforward and feedback control.
21. Describe the role block diagrams play in process control.

22. Discuss the difference between process control and instrumentation.
23. Provide definitions for proportional, reset, response lag, and tuning.
24. Discuss the advantages and disadvantages of feedforward control.
25. Discuss the advantages and disadvantages of feedback control.
26. Discuss the role that location plays in process control.
27. List all the process variables that you believe can be controlled.
28. List all the process variable that you believe cannot be controlled... and explain why.
29. Rank (and justify) the order of importance of controlling various process variables.
30. Is it possible to develop a multi-purpose controller? Explain your answer.
31. Discuss the role of (process) control as it applies to heat exchangers.
32. Discuss the role of (process) control as it applies to fluid flow systems.
33. Discuss the role of (process) control as it applies to mass transfer equipment, e.g., distillation columns.
34. Discuss the role of (process) control as it applies to chemical reactors.
35. Discuss the role of (process) control as it applies to drying operations.
36. Which operation in the chemical process industries is most significantly impacted by process control. Justify your answer.
37. Develop an experiment for a Unit Operations Laboratory that would (best) illustrate the role of process control.
38. A controller is *not* operating according to norm. Indicate what steps you would take before purchasing a new unit.
39. As noted in the Overview, trips and alarms are used extensively in chemical and petrochemical processing plants. A "trip" is an automatic control function, typically the opening or closing of a particular valve or the turning on or off a device such as a pump. Trips are used when a very rapid response is needed when a process upset occurs. An alarm is usually an audible and/or visual signal to the operator in

the control room that something is not proper. Alarms are used when a process is drifting away from normal conditions, but there is still adequate time for a human operator to make adjustments and/or shut the process down. Discuss these two factors in layman terms.

40. Logic diagrams, similar to computer program flow charts, are often used to lay out trip and alarm systems. They graphically illustrate what must be built into the control system hardware and software. Lay out a logic diagram for the temperature controller on an exothermic continuous stirred tank reactor. The system should have the following characteristics:

   a   If the reactor temperature is below the set-point, T_s, reduce the coolant flow.
   b   If the reactor temperature is above the set-point, T_s, increase the coolant flow.
   c   If the reactor temperature exceeds the first limit, T_1, sound an alarm in the control room. (ALARM)
   d   If the reactor temperature exceeds the second limit, T_2, shut off the feed to the reactor. (TRIP)
   e   If the reactor temperature exceeds the third limit, T_3, sound a high priority alarm in the control room. (ALARM)

References

1. P.T. Vasudaven, *Process Dynamics and Control*, A Theodore Tutorial, Theodore Tutorials, East Williston, NY, originally published by the USEPA/APTI, RTP, NC, 1996.

2. L. Theodore, *Chemical Reactor Analysis and Applications for the Practicing Engineer*, John Wiley & Sons, Hoboken, NJ, 2012.

3. L. Theodore, *Heat Transfer for the Practicing Engineer*, John Wiley & Sons, Hoboken, NJ, 2011.

4. M. Speigel, *Laplace Transforms*, Schaum's Outline Series, Shaum Publishing Co., New York City, NY, 1965.

5. L. Theodore, *Chemical Engineering: The Essential Reference*, McGraw-Hill, New York City, NY, 2014.

10

Economics and Finance

This chapter is concerned with process economics and finance. As with all the chapters in Part II, there are several sections: overview, several technical topics, illustrative open-ended problems, and open-ended problems. The purpose of the first section is to introduce the reader to the subject of economics and finance. As one might suppose, a comprehensive treatment is not provided although several technical topics are also included. The next section contains three open-ended problems; the authors' solutions (there may be other solutions) are also provided. The final section contains 43 problems; *no* solutions are provided here.

10.1 Overview

This overview section is concerned with—as can be noted from its title—economics and finance. As one might suppose, it was not possible to address all topics directly or indirectly related to economics and finance. However, additional details may be obtained from either the references provided at the end of this Overview section and/or at the end of the chapter.

Note: Those readers already familiar with the details associated with this subject may choose to bypass this Overview.

An understanding of the economics and finances involved in chemical engineering applications is important in making decisions at both the engineering and management levels. Every chemical engineer should be able to execute an economic evaluation, or the equivalent, of a proposed project. If the project is not profitable, it should obviously not be pursued, and the earlier such a project can be identified, the fewer resources will be wasted.

The fundamental purpose of any process plant is to convert one or more feeds into products of greater value. In order to predict whether a new process will be economically feasible, it is necessary to estimate the costs of feeds, the value of products, the capital investment required to build the plant, and the cost of operating the plant.

Capital costs are usually estimated once a fairly detailed material balance, flow sheet and equipment list are worked out. The flow sheet used at this stage is referred to as a Process Flow Diagram, or PFD (see next Chapter on Plant Design). It is necessary to know key dimensions of major pieces of equipment, but it is not necessary to have a complete, detailed design of every item in the plant. Environmental control systems located within the main processing plant, such as an electrostatic precipitator, must also be included. The costs of auxiliary, or "off-site", systems are often added in using rules-of-thumb rather than engineering calculations. These systems include cooling towers, steam boiler plants, in-plant roads, shipping and storage facilities, and perhaps some of the aforementioned environmental control systems, such as air and wastewater treatment facilities.

Operating costs can also be estimated once a preliminary flowsheet and material balance have been developed. The largest costs are usually those for feed materials and for utilities (electricity, steam, fuel, etc.). Waste treatment costs are occasionally a major factor in overall plant operations. The costs for labor, plant maintenance, insurance and property taxes are also included; but these are often estimated using common rules-of-thumb rather than detailed engineering calculations. Likewise, common rule-of-thumb factors are usually used for such hard-to-calculate overhead costs as sales promotion, workmen's compensation, accident management, public relations, employee morale, etc.

The overall economics of a plant venture are determined using *present-worth* or *discounted-cash-flow* methods that take into account the time value of money, income taxes, and other factors. Often a simpler method, a *payout time*, can be used for preliminary estimations. In general, investment in a new plant or modification to an existing one must have a simple pay-off (increased annual revenue or decreased annual operating costs, before taxes) of three years or less to be economically attractive.

Before the cost of a unit or process or facility can be evaluated, the various factors contributing to the cost must be recognized. As noted above, there are two major contributing factors: capital costs and operating costs; these are discussed in the next two sections. Once the total cost of the facility has been estimated, the engineer must determine whether or not the project will be profitable. This may involve converting all cost contributions to an annualized basis, a method favored by one of the authors that is discussed in the following subsections. If more than one project proposal is under study, this method provides a basis for comparing alternate proposals and for choosing the best proposal. Project optimization is the subject of a later section, where a brief description of a perturbation analysis is presented.

Detailed cost estimates are beyond the scope of this section and chapter. Such procedures are capable of producing accuracies in the neighborhood of ±5%; however, such estimates generally require many months of engineering work. This section is primarily designed to give the reader a basis for a *preliminary cost analysis* only, with an expected accuracy of approximately ±20%. See also Chapter 11 – Plant Design and Chapter 13 – Project Management for additional detail.

The reader should note that the material for this chapter was adapted primarily from the following four sources:

1. T. Shen, Y. Choi, and L. Theodore, *EPA Manual Hazardous Waste Incineration*, USEPA/APTI, RTP, NC, 1985[1].
2. L. Theodore and K. Neuser, *Engineering Economics and Finance*, A Theodore Tutorial, Theodore Tutorials, East Williston, NY, originally published by the USEPA/APTI, RTP, 1997 [2].
3. L. Theodore and E. Moy, *Hazardous Waste Incineration*, USEPA/APTI, RTP, NC, 1992[3].
4. J. Santoleri, J. Reynolds and L. Theodore, *Introduction to Hazardous Waste Incineration*, 2nd edition, John Wiley & Sons, Hoboken, NY, 2009[4].

The remaining five Sections in this chapter include:

1. Capital Costs
2. Operating Costs

3. Project Evaluation
4. Perturbation Studies in Optimization
5. Principles of Accounting.

10.2   Capital Costs

Equipment cost is a function of many system variables, one of the most significant of which is capacity. There are other important variables that vary with the cost of equipment or process. Preliminary estimates are often made from simple cost-capacity relationships that are valid when the other variables are assumed constant or confined to narrow ranges of values; these relationships can be represented by approximate linear (on log-log coordinates) cost equipment, equation of the form [5]

$$C = a\,q^{\beta} \tag{10.1}$$

where C = cost
   q = some measure of equipment capacity
   α, β = empirical "constants" that depend mainly on equipment type

It should be emphasized that this procedure is suitable for rough estimation only; actual estimates from vendors are more preferable. Only major pieces of equipment are included in this analysis; small peripheral equipment such as pumps and compressors may not be included. Similar methods for estimating costs are available in the literature [5]. If greater accuracy is needed, however, actual quotes from vendors should be used.

Again, the equipment cost estimation model just described is useful for a very preliminary estimation. If more accurate values are needed and old price data is available, the use of an indexing method may be better, although a bit more time consuming. The method consists of adjusting earlier cost data to present values using factors that correct for inflation. A number of such indices are available; some of the most commonly used are the Chemical Engineering Fabricated Equipment Cost Index (FECI), the Chemical Engineering Plant Cost Index, and the Marshall and Swift (M&S) Equipment Cost Index, all three of which are available in the magazine *Chemical Engineering*.

The usual technique for determining the *capital costs* (i.e., total capital costs, which include equipment design, purchase, and installation) for a facility is based on the *factored method* of establishing direct and indirect installation costs as a function of the known equipment costs. This is basically a *modified Lang method*, whereby cost factors are applied to known equipment costs [6,7].

The first step is to obtain directly from vendors (or, if less accuracy is acceptable, from one of the estimation techniques previously discussed) the purchase prices of the primary and auxiliary equipment. The total base prices designated by X, which should include instrumentation, control, taxes, freight costs, etc., serves as the basis for estimating the direct and indirect installation costs (ICF). The installation costs are obtained by multiplying X by the cost factors, which can be adjusted to more closely model the proposed system by using adjustment factors that may be available in order to take into account for the complexity and sensitivity of the system [6–8].

The second step is to estimate the *direct installation cost* by summing all the cost factors involved in the direct installation costs, which can include piping, insulation, foundation, and supports, etc. The sum of these factors is designated as the DCF (*direct installation cost factor*). The direct installation costs are then the product of the DCF and X.

Once the direct and indirect installation costs have been calculated, the total *capital cost* (TCC) may be evaluated as

$$TCC = X + (DCF)(X) + (ICF)(X) \qquad (10.2)$$

This cost is then converted to *annualized* capital costs with the use of the *capital recovery factor* (CRF), which is described in a later section. The *annualized capital cost* (ACC) is the product of the CRF and TCC and represents the total installed equipment cost distributed over the lifetime of the facility.

10.3   Operating Costs

Operating costs can vary from site to site, plant to plant, and equipment to equipment, since these costs, in part, reflect local conditions, e.g., staffing practices, labor, and utility costs. Operating costs, like capital costs, may also be separated into two categories: direct and indirect costs. *Direct* costs are those that cover material and labor and are directly involved in operating the facility. These can include labor, materials, maintenance labor and maintenance supplies, replacement parts, waste (e.g., residues) disposal fees, utilities, and laboratory costs. *Indirect* costs are those operating costs associated with, but not directly involved in, operating the unit or facility in question; costs such as overhead (e.g., building-land leasing and office supplies), administrative fees, local property taxes, and insurance fees fall into this category [9,10].

The major direct operating costs are usually those associated with labor and materials. *Material* costs usually involve the cost of chemicals needed for the operation of the system(s). *Labor* costs differ greatly, depending on whether or not the costs are located on-site or off-site and the degree of controls and/or instrumentation. Typically, there are three working shifts per day plus the standby shift used for rotation on a weekly basis, with one supervisor per shift. On the other hand, it may be manned by a single operator for only one third or one half of each shift; usually only an operator, supervisor, and site manager are necessary to run a facility. The cost of *utilities* generally consists of that for electricity, water, fuel, compressed air, and steam. The annual costs are estimates and can be described as a percentage of the capital equipment costs. Typical life expectancies can be found in the literature. Laboratory costs (if applicable) depend on the number of samples tested and the extent of these tests; these costs may be estimated as 10-20% of the operating labor costs.

The indirect operating costs consist of overhead, local property tax, insurance, administration, less any credits. The overhead can comprise payroll, fringe benefits, social security, unemployment insurance, and other compensation that is indirectly paid to the plant personnel. This cost can be estimated as 70-90% of the operating labor, supervision, and maintenance costs [11,12]. Local property costs may be estimated as 2% of the TCC.

The total operating costs is the sum of the direct operating costs and the indirect operating costs less any credits that may be recovered (e.g., the value of recovered by-products such as steam). Unlike capital costs, operating costs are usually calculated on an annual basis.

10.4   Project Evaluation

Although this section primarily deals with a plant or process, it may be applied to equipment and/or other economic issues. In comparing alternate processes or different options of a particular process from an economic point of view, it is recommended that the total capital cost be converted to an annual basis by distributing it over the projected lifetime of the facility. The sum of both the annualized capital costs (ACC) and the annual operating costs (AOC) is known as the total annualized cost (TAC) of the facility. The economic merit of the proposed facility, process or scheme can be examined once the total annual cost is available. Alternate facilities, processes, or options may also be compared. Note, a small flaw in this procedure is the assumption that the operating costs remain constant throughout the lifetime of the facility. However, since the analysis is geared to comparing different alternatives, the changes with time is often uniform among the various alternatives, resulting in little loss of accuracy.

The conversion of the total capital cost to an annualized basis involves an economic parameter known as the capital recovery factory (CRF) an approach routinely employed by one of the authors in the past. These factors can be found in any standard economics text [13–15] or can be calculated directly from

$$CRF = \frac{i(1+i)^n}{(n+i)^n - 1} \tag{10.3}$$

where i = annual interest rate (expressed as a fraction)

n = projected lifetime of the system (years)

The CRF calculated from Equation (10.3) is a positive, fractional number. The ACC is computed by multiplying the TCC by the CRF. The annualized capital cost reflects the cost associated in recovering the initial capital outlay over the depreciable life of the system.

Investment and operating costs can be accounted for in other ways, such as the popular aforementioned *present worth* analysis. However, the capital recover method is preferred because of its simplicity and versatility. This is especially true when comparing systems having different depreciable lives. There are usually other considerations in such decisions besides the economics, but if all the other factors are equal, the proposal with the lowest total annualized cost should be the most attractive.

10.5   Perturbation Studies in Optimization

Once a particular process scheme (or project) has been selected, it is common practice to optimize the process from a capital cost and O&M (operation and maintenance) standpoint. There are many optimization procedures available, most of them too detailed for meaningful application to some studies. These sophisticated optimization techniques, some of which are routinely used in the design of conventional chemical and petrochemical plants invariably involve computer calculations. Occasionally employed by one of the authors, the use of these techniques in many chemical engineering applications is usually not warranted.

One simple optimization procedure that is recommended is the perturbation study (see also Chapter 2). This involves a systematic change (or perturbation) of variables, one by one, in an attempt to locate the optimum design from a cost and operation viewpoint. To be practical, this often means that the chemical engineer must limit the number of variables by assigning constant values to those process variables that are known beforehand to play an insignificant role. Reasonable guesses and simple or short-cut mathematical methods can further simplify the procedure. Much information can be gathered from this type of study since it usually identifies those variables that significantly impact on the overall performance of the process and also helps identify the major contributors to the total annualized cost.

10.6   Principles of Accounting

Accounting is the science of recording business transactions in a systematic manner. Financial statements are both the basis for and the result of management decisions; practicing chemical engineers are rarely involved. Such statements can tell a manager or a chemical engineer a great deal about a company, provided that one can interpret the information correctly.

Accounting has also been defined by accountants as the language of business. The different departments of management use it to communicate within a broad context of financial and cost terms. The chemical engineer who does not take the trouble to learn the language of accountancy denies him/herself the most important means available for communicating with top management. He/she may be thought by them to lack business acumen. Some chemical engineers have only themselves to blame for their lowly status within the company hierarchy since they seem determined to hide themselves from business realities behind the screen of their specialized technical expertise. However, more and more chemical engineers are becoming involved in decisions that are business related. In addition to understanding the principles of accountancy and obtaining a working knowledge of its practical techniques, chemical engineers should be aware of possible inaccuracies of accounting information in the same way that he/she allows for errors in any technical information data.

A detailed and expanded treatment of economics and finance is available in the following two references:

1. L. Theodore and K. Neuser, *Engineering Economics and Finance*, A Theodore Tutorial, Theodore Tutorials, East Williston, NY, originally published by the USEPA/APTI, RTP, NC, 1997 [2].
2. L. Theodore, *Chemical Engineering: The Essential Reference*, McGraw-Hill, New York City, NY, 2014 [16].

10.7   Illustrative Open-Ended Problems

This and the last section provide open-ended problems. However, solutions *are* provided for the three problems in this section in order for the reader to hopefully obtain a better understanding of these problems which differ from the traditional problems/illustrative examples. The first problem is relatively straightforward while the third (and last problem) is somewhat more difficult and/or complex. Note that solutions are not provided for the 43 open-ended problems in the next section.

Problem 1: To what does the term "due diligence process" refer? How would this process reveal any environmental liabilities a corporation may have?

Solution: As defined by law, the term "due diligence" refers to "such measure of prudence, activity, or assiduity, as is properly to be expected from, and ordinarily exercised by, a reasonable and prudent man under the particular circumstances; not measured by any absolute standard, but depending on the relative facts of the special case."

If, for example, a corporation is seeking a loan from a bank to purchase a piece of property to expand its operations, the bank would be required by FDIC regulations to include checks for any liabilities associated with the property or with the corporation seeking the loan. This check is required in order to be sure that neither the borrower nor the property has any potential liabilities which could make the bank responsible for any liabilities.

Problem 2: An investor may invest $60,000 in either Option A or Option B. The return on the investment for each option is given in Table 10.1 below. The investor wished to earn the highest rate of return possible. What is the present value of each option? Select various end-of-year discounting rates.

Table 10.1 Return on Investment for Investment Options A and B

Year	Annual Income Option A	Annual Income Option B
1	$10,000	$10,000
2	$15,000	$10,000
3	$10,000	$15,000
4	$10,000	$15,000
5	$15,000	$10,000
Total	$60,000	$60,000

Solution: One approach is to apply the *present value* formula.

$$PV = (AI)\frac{1.0}{\left(1.0+i\right)^n}$$ (10.4)

where PV = present value of annual income for period n in $, AI = annual income in $, i = annual interest factor or discount rate on a fractional bases, and n = number of annualized periods.

Select a discount rate of 0.10 (10%). For option A, year one, the present value of the income is calculated as follows:

$$PV = \$10,000\frac{1.0}{\left(1.0+0.10\right)^1} = \$9091$$

For Option A, year two, the present value of the income is:

$$PV = \$15,000\frac{1.0}{\left(1.0+0.10\right)^2} = \$12,397$$

The results for both options for the 5 years are given in the Table 10.2. It can be seen that Option A yields a higher present value than Option B ($45,145 versus $45,079). Option A is marginally the better investment even though both options earn the same total undiscounted income.

Table 10.2 Return on Investment for Investment Options A and B Expressed in Present Value Terms. Present Value Calculated Results

Year	Annual Income Option A	Present Value Option A	Annual Income Option B	Present Value Option B
1	$10,000	$9,091	$10,000	$9,091
2	$15,000	$12,397	$10,000	$8,264
3	$10,000	$7,513	$15,000	$11,270
4	$10,000	$6,830	$15,000	$10,245
5	$15,000	$9,314	$10,000	$6,209
Total	$60,000	$45,145	$60,000	$45,079

The above calculational procedure can now be extended to include other discount rates. The analysis of these calculations is left as an exercise for the reader.

Problem 3: Plans are underway to construct and operate a commercial waste treatment facility in Dumpsville in the state of Egabrag. The company is still undecided as to whether to install a liquid injection or rotary kiln incinerator at the waste site. The liquid injection unit is less expensive to purchase and operate than a comparable rotary kiln system. However, projected waste treatment income from the rotary kiln unit is higher since it will handle a larger quantity and different varieties of waste.

Based on economic and financial data provided below in Table 10.3, select the incinerator that will yield the higher annual profit

Table 10.3 Costs/Credits for Liquid Injection and Rotary Kiln Incinerators

Costs/Credits	Liquid Injection	Rotary Kiln
Capital ($)	2,625,000	2,975,000
Installation ($)	1,575,000	1,700,000
Operation ($/yr)	400,000	550,000
Maintenance ($/yr)	650,000	775,000
Income ($/yr)	2,000,000	2,500,000

Calculations should be based on interest rates in the 2 – 18% range and process lifetime of 8 – 15 years for the both incinerators [2-4].

Solution: The solution for $i = 0.12$ (12%) and $n = 12$ follows

1. Calculate the capital recovery factor, CRF.

$$CRF = i/[1 - (1+i)^{-n}]$$
$$= 0.12/[1 - (1+0.12)^{-12}]$$
$$= 0.1614$$

2. Determine the annual capital and installation costs for the liquid injection (LI) unit.

LI costs = (Capital + Installation)(CRF)
$$= (2625000 + 1575000)(0.1614)$$
$$= \$677,880 / yr$$

3. Determine the annual capital and installation costs for the rotary kiln (RK) unit.

RK costs = (Capital + Installation)(CRF)
$$= (2975000 + 1700000)(0.1614)$$
$$= \$754,545 / yr$$

4. Complete the following (see Table 10.4) which provides a comparison of the costs and credits for both incinerators.

Table 10.4 Cost Comparison for Problem 3

	Liquid Injection	Rotary Kiln
Total Installed ($/yr)	678 ,000	755,000
Operation ($/yr)	400,000	550,000
Maintenance ($/yr)	650,000	775,000
Total annual cost ($/yr)	1,728,000	2,080,000
Income credit ($/yr)	2,000,000	2,500,000

5. Calculate the profit for each unit on an annualized basis. The profit is the difference between the total annual cost and the income credit.
LI (profit) = 2,000,000 – 1,728,000 = \$272,000 /yr
RK (profit) = 2,500,000 – 2,080,000 = \$420,000 /yr

A rotary kiln incinerator should be selected based on the above economic analysis.

10.8   Open-Ended Problems

This last section of the chapter contains open-ended problems as they relate to economics and finance. No detailed and/or specific solution is provided; that task is left to the reader, noting that each problem has either a unique solution or a number of solutions or (in some cases) no solution at all. These are characteristics of open-ended problems described earlier.

There are comments associated with some, but not all, of the problems. The comments are included to assist the reader while attempting to solve the problems. However, it is recommended that the solution to each problem should initially be attempted *without* the assistance of the comments.

There are 43 open-ended problems in this section. As stated above, if difficulty is encountered in solving any particular problem, the reader should next refer to the comment, if any is provided with the problem. The reader should also note that the more difficult problems are generally located at or near the end of the section.

1. Describe the early history associated with economics and finance.
2. Discuss the recent advances in economics and finance.
3. From an insurance perspective, why is it in a corporation's best interest to be ethically conscious?
4. The Securities and Exchange Commission (SEC) requires corporations to disclose their liabilities for accounting purposes. What happens to a corporation if it fails to properly report a liability?
5. Select a refereed, published economics and finance article from the literature and provide a review.
6. Provide some normal everyday domestic applications involving the general topics of economics and finance.
7. Develop an original problem that would be suitable as an illustrative example in a book on economics and finance.
8. Prepare a list of the various books which have been written on economics. Select the three best and justify your answer. Also select the three weakest books and, once again, justify your answer.

9. Prepare a list of the various books which have been written on finance. Select the three best and justify your answer. Also select the three weakest books and, once again, justify your answer.

10. Define and describe single interest in layman terms.

11. Define and describe compound interest in layman terms.

12. Define and describe present worth in layman terms.

13. Define and describe depreciation (straight line sinking fund) in layman terms.

14. Define and describe the capital recover factor (CRF) in layman terms.

15. Define and describe perpetual life in layman terms.

16. Define and describe break-even point in layman terms.

17. Define and describe rate of return in layman terms.

18. Define and describe bonds in layman terms.

19. Define and describe deferred investments in layman terms.

20. Define and describe amortization in layman terms.

21. Detail the role a journal plays in a company's economic and finance dealings.

22. Explain why the practicing chemical engineer should familiarize him/herself with the economics and finance details of their company.

23. Briefly discuss the various methods that industry employs in performing economic evaluations.

24. List and describe some intangible assets at the domestic level.

25. List and describe some intangible assets at the industrial level.

26. Compare and discuss the difference of comparing alternate proposals based on total cost or present worth.

27. What criteria should be employed in choosing between new and existing equipment?

28. Describe the relationship between depreciation and appraisal value.

29. Explain why break-even point calculations are often important in economic analysis.

30. Explain the difficulties that can arise in appraising a home for sale purposes.

31. Due to government cutbacks in funding, regulatory-driven programs for most federal agencies are feeling the pinch. With fewer and fewer enforcement personnel, many feel that

corporate behavior will take a turn for the worse. The traditional sources of pressure on corporations to perform in a responsible manner have come from federal and state regulatory agencies. However, with governmental cuts in funding, the role of these regulators is being diminished. It would be reasonable to expect, then, an increase in these activities by corporations. However, such has not been the case because of the increasingly important role of other entities whose impacts on the performance of corporations may be even more significant. These entities have come to be known as "surrogate regulators". Who and what are these entities? [2].

32. Why are banks interested in determining the potential liability of corporations that seek to borrow money from them?

33. The cost for an outside group to maintaining the air-conditioning equipment for your company's laboratory is under consideration. Discuss the advantages (and disadvantages) of accepting this offer.

34. Consider the following reaction:

$$aA \rightleftharpoons bB$$

Component A is 1¢/ton while B is worth 10^6/ng. Qualitatively describe how the conversion of A to B could be maximized if:

    1. a=1, b=10 and the reaction is exothermic
    2. a=1, b=10 and the reaction is endothermic
    3. a=10, b=1 and the reaction is exothermic
    4. a=10, b=1 and the reaction is endothermic

Comment: See also Chapters 3 and 8.

35. Which of the three major classes of reactors has the greatest OM&I problems from an economic perspective. Justify your answer. Comment: See also Chapter 8.

36. Set up a chart, figure and/or equations based solely on capital and operating costs that would allow a chemical engineer responsible for purchasing equipment to select the most economical choice.

37. There are a host of topics that reside under the economic and finance umbrella. List these topics in terms of importance. Justify your answer.

Table 10.6 Air Pollution Initial Equipment Data, Problem 39.

	ESP	VS	BH
Total capital cost	$17.5/acfm	$14.0/acfm	$16.0/acfm
Operating cost	$0.30/acfmyr	$0.35/acfmyr	$0.40/acfmyr
Maintenance costs	$120,000/yr	$150,000/yr	$130,000/yr

38. Prior to being processed in an absorber, a 20,000 acfm stream of particulate contaminated air is to be treated using one of three devices: an electrostatic precipitator (ESP), a venturi scrubber (VS), or a baghouse (BH). The following data were obtained for Theodore Consultants from a reliable vendor. Which air pollution control device should be selected for this operation?
    Perform the calculations based on equipment lifetime and interest in the 10 – 15 year range and 4 – 10% range, respectively [2].

39. As part of an energy conservation project, Abulencia Airlines has decided to reduce their planes speed from 530 mi/hr to 500 mi/hr. Estimate the annual reduction in gasoline used. Also estimate the annual savings.
    Comment: This will require a literature search to obtain gasoline usage vs. speed data as well as gasoline cost information.

40. It is desired to determine whether to use 1-inch-thick or 2-inches-thick insulation for a steam pipe. Theodore Consultants have determined that the following economic data need to be provided in order to perform a meaningful analysis.
    • cost of heat loss w/o insulation
    • cost of each insulation and the corresponding heat loss
    • interest rate
    • insulation lifetime
    • depreciation
    • salvage value
    Outline the details of the calculation.

41. A vendor has provided a total installed capital cost estimate for a particulate-ladened flue gas generated at a waste incineration facility of 25.36 $/acfm. A baghouse is to be installed for particulate control at the site employing Teflon felt bags, each with an area of 12 ft² ($75.00 per bag) at an air-to-cloth

ratio (acfm/ft^2) of 5.81 to 1. The total pressure drop across the system is 16.5 inches of water (in H$_2$0), including 3 in H$_2$O for the baghouse. The following economic factors exist at the time of purchase: the gas flow rate (at 350°F) is 70,000 acfm, the overall fan efficiency E$_f$ = 60% (350°F), and the operating time is 6,240 hours/year. Electrical power costs 0.15$/kW·h. Yearly maintenance cost is $50,000 per year plus replacing 25% of the bags each year. In addition, there is no salvage value for the bags at the end of useful life.

Assuming that there are no fuel requirements for the incineration of the waste, determine the installed capital, operating, and maintenance costs for the proposed facility on an annualized basis for a faculty lifetime in the 10 – 20 year range with corresponding interest rate in the 5 – 10% range [2]. Comment on the results.

42. Plans are underway to construct and operate some type of a mass transfer unit that will serve to purify a product stream from a chemical reactor. The company is still undecided as to whether to install an extraction unit or a distillation column. The extraction unit is less expensive to purchase and operate than a comparable distillation system, primarily because of energy costs. However, projected income from the distillation unit is higher since it will handle a larger quantity of process liquid and provide a purer product. Based on the economic and financial data provided below, select the mass transfer unit that will yield the higher annual profit. Select income ($/yr) for the extraction and

Table 10.7 Equipment Cost Data, Problem 42.

Costs/Credits	Extraction	Distillation
Mass Transfer Unit ($)	750,000	800,000
Peripherals ($)	1,875,000	2,175,000
Total Capital ($)	2,625,000	2,975,000
Installation ($)	1,575,000	1,700,000
Operation ($/yr)	400,00	550,000
Maintenance ($/yr)	650,000	775,000
Income ($/yr)	-	-

distillation unit in the 2 – 3 million dollar range and 2.4 – 4.0 million dollar range, respectively. The time value of money is 12% and both units have a process lifetime of 12 years [2].

43. Fourteen years ago, a 1200 kW steam electric plant was constructed at a cost of $2,200 per kW. Annual operating expenses had been $310,000 to produce the annual demand of 5,400,000 kW·h. It is estimated that the annual operating expenses and current demand will continue. The original estimate of a 20-year life with a 5% salvage value at that time is still expected to apply.

The company is contemplating the replacement of the old steam plant with a new diesel plant. The old plant can be sold now for $750,000, while the new diesel plant will cost $2,450 per kW to construct. The diesel plant will have a life of 25 years with a salvage value of 10% at the end of that time and will cost $230,000 annually to operate. Annual taxes and insurance are estimated to be 2.3% of the first cost of either plant. Using several interest rates in the 1 – 10% range, determine whether the company is financially justified in replacing the old steam plant now [2]. Comment on the results.

References

1. T. Shen, Y. Choi and L. Theodore, EPA Manual *Hazardous Waste Incineration*, USEPA/APTI, RTP, NC, 1985.
2. L. Theodore and K. Neuser, *Engineering Economics and finance*, A Theodore Tutorial, Theodore Tutorials, East Williston, NY, originally published by the USEPA/APTI, RTP, 1997.
3. L. Theodore and E. Moy, *Hazardous Waste Incineration*, USEPA/APTI, RTP, NC, 1992.
4. J. Santoleri, J. Reynolds and L. Theodore, *Introduction to Hazardous Waste Incineration*, 2nd edition, John Wiley & Sons, Hoboken, NY, 2009.
5. R. McCormick and R. DeRoiser, *Capital and O&M Cost Relationships for Hazardous Waste Incinerators*, Acurex Corp., Cincinnati, OH, EPA report, date unknown.
6. R. Neveril, *Capital and Operating Costs of Selected Air Pollution Control Sytstems*, Gard, Inc., Niles, IL, EPA Report 450/5-80-002, Dec., 1978.
7. W. Vatavuk and R. Neveril, *Factors for Estimating Capital and Operating Costs*, Chem. Eng., 157-162, New York City, NY, Nov. 3, 1980.

8. G. Vogel and E. Martin, *Hazardous Waste Incineration Part 1—Equipment Sizes and Integrated-Facility Costs*, Chem. Eng., 143-146, New York City, NY, Sept. 5, 1983.

9. G. Vogel and E. Martin, *Hazardous Waste Incineration Part 2—Estimating Costs of Equipment and Accessories*, Chem. Eng., 75-78, New York City, NY, Oct. 17, 1983.

10. G. Vogel and E. Martin, *Hazardous Waste Incineration Part 3—Estimating Capital Costs of Facility Components*, Chem. Eng., 87-90, New York City, NY, Nov. 28, 1983.

11. G. Ulrich, *A Guide to Chemical Engineering Process Design and Economics*, John Wiley & Sons, Hoboken, NJ, 1984.

12. G. Vogel and E. Martin, *Hazardous Waste Incineration Part 4—Estimating Operating Costs*, Chem. Eng., 97-100, New York City, NY, Jan. 9, 1984.

13. E. DeGarmo, J. Canada, and W. Sullivan, *Engineering Economy*, 6th ed., Macmillan, New York City, NY, 1979.

14. C. Hodgman, S. Selby, and R. Weast, ed., *CRC Standard Mathematical Tables*, 12th ed., Chemical Rubber Company, Cleveland, OH, 1961 (presently CRC Press, Boca Raton, FL).

15. T. Tielenberg, *Environmental and Natural Resource Economics*, Scott, Foreman, and Company, Glenview, IL, 1984.

16. L. Theodore, *Chemical Engineering: The Essential Reference*, McGraw-Hill, New York City, NY, 2014.

11

Plant Design

This chapter is concerned with process plant design. As with all the chapters in Part II, there are several sections: overview, several technical topics, illustrative open-ended problems, and open-ended problems. The purpose of the first section is to introduce the reader to the subject of plant design. As one might suppose, a comprehensive treatment is not provided although several technical topics are also included. The next section contains three open-ended problems; the authors' solution (there may be other solutions) is also provided. The final section contains 35 problems; *no* solutions are provided here.

11.1 Overview

This overview section is concerned—as can be noted from the chapter's title—with plant design. As one might suppose, it was not possible to address all topics directly or indirectly related to plant design. However, additional details may be obtained from either the references provided at the end of this Overview section and/or at the end of the chapter.

Note: Those readers already familiar with the details associated with this subject may choose to bypass this Overview.

Current process and plant design practices can fall into the category of state of the art and pure empiricism. Past experience with similar applications is commonly used as the sole basis for the design procedure. The vendor (seller) maintains proprietary files on past installations, and these files are periodically revised and expanded as new installations are evaluated. In designing in a new unit, the files are consulted for similar applications and old designs are heavily relied upon [1].

By contrast, the engineering profession in general, and the chemical engineering profession in particular, has developed fairly well-defined procedures for the design, construction, and operating of many chemical plants. These techniques are routinely used by today's chemical engineers. These same procedures may also be used in the design of other facilities [1].

The purpose of this chapter is to introduce the reader to some of the process and plant design fundamentals. Such an introduction to design principles, however sketchy, can provide the reader with a better understanding of the major engineering aspects of new or modified facilities, including some of the operational, economic, controls, and instrumentation for safety and regulatory requirements, and environmental factors associated with the process [1].

The author [1] has simplified the design process by keying in five topics that are summarized via the acronym SCORE; i.e.,

Safety
Costs
Operability
Reliability
Environment

It should also be noted that process plant location and layout considerations are not reviewed. However, in a general sense, physical plant considerations should include process flow, construction, maintenance, operator access, site conditions, space limitations, future expansion, special piping requirements, structural supports, storage space, utility requirements, instruments and controls, and (if applicable) energy conservation. A more detailed presentation is provided in a later section. Finally the reader should note that no attempt is made in the sections that follow to provide extensive coverage of this topic; only general procedures and concepts are presented and discussed.

The remaining chapter contents include:

1. Preliminary Studies
2. Process Schematics
3. Material and Energy Balances
4. Equipment Design
5. Instrumentation and Controls
6. Design Approach
7. The Design Report.

The bulk of the material in this chapter has been drawn from:

1. Personal notes of Theodore [1],
2. T. Shen, Y. Choi and L. Theodore, *EPA Manual Hazardous Waste Incineration*, USEPA/APTI, RTP, NC, 1985 [2], and
3. L. Theodore, *Hazardous Waste Incineration*, Instructor's Guide, USEPA/APTI, RTP, NC, 1986 [3].

11.2   Preliminary Studies

A process chemical engineer is often involved in one of two principle activities: designing/building a plant or deciding whether to do so. The skills required in both cases are often similar, but the money, time, and details involved are not as great in the latter situation. The authors estimate that only one out of 15 proposed new processes ever reaches the construction stage.

In general engineering design practice, there are usually five levels of sophistication for evaluating projects and estimating costs. Each is discussed in the following list, with particular emphasis on cost.

1. The first level of analysis requires little more than identification of products, raw materials, and utilities. This what is known as an *order of magnitude estimate* and is often made by extrapolating or interpolating from data on similar existing processes.
2. The next level of sophistication is called a *study estimate* and requires a preliminary process flow sheet (to be discussed in the next section) and a first attempt at identification of equipment, utilities, materials of construction, and other processing units.

3. A *scope* or *budget authorization* is the next level of evaluation; it requires a more defined process definition, detailed process flow sheets, and pre-final equipment design (discussed in a later section).
4. If the evaluation is positive at this stage, a *project control estimate* is then prepared. Final flow sheets, site analyses, equipment specifications, and architectural and engineering sketches are employed to prepare this estimate.
5. The fifth and final (economic) analysis is called a *firm* or *contractor's estimate*. It is based on detailed specifications and actual equipment bids. It is employed by the contractor to establish a project cost and has the highest level of accuracy.

11.3   Process Schematics

To the practicing engineer and particularly the chemical engineer, the process flow sheet is the key instrument for defining, refining, and documenting a chemical process. The process flow diagram is the authorized process blueprint, the framework for specifications used in equipment designation and design; it is the single, authoritative document employed to define, construct, and operate the chemical process [4].

Beyond equipment symbols and process stream flow lines, there are several essential elements contributing to a detailed process flow sheet. These include equipment identification numbers and names, temperature and pressure designations, utility designations, mass, molar and volumetric flow rates for each process steam, and a material balance table pertaining to process flow rates. The process flow diagram may also contain additional information such as energy requirements, major instrumentation, environmental equipment (and concerns), and physical properties of the process streams. When properly assembled and employed, a process schematic provides a coherent picture of the overall process; it can pinpoint some deficiencies in the process that may have been overlooked earlier in the study, e.g., instrumentation overkill, by-products (undesirable or otherwise), and recycle needs. Basically, the flow sheet symbolically and pictorially represents the interrelation between the various flow streams and equipment, and permits easy calculations of material and energy balances. These two topics are considered in the next section. Controls and instrumentation must also be considered in the overall requirements of the system; these concerns are covered later in this chapter as well as in a previous chapter.

As one might expect, a process flow diagram for a chemical or petroleum plant is usually significantly more complex than that for a simple facility. For the latter case, the flow sequence and determinations thus approach a "railroad" or sequential type of calculation that does not require iterative calculations.

The degree of sophistication and details of a flow sheet usually vary with both the preparer and time. The flow sheet may initially consist of a simple free-hand block diagram with limited information that includes only the equipment. Later versions may include line drawings with pertinent process data such as the overall and componential flow rates, utility and energy requirements, environmental equipment, and instrumentation. During the latter stages of the design project, the flow sheet will consist of a highly detailed P&I (piping and instrumentation) diagram; this aspect of the design procedure is beyond the scope of this text; the reader is referred to literature [2,5] for information on P&I diagrams.

In a sense, flow sheets are the international language of the engineer, particularly the chemical engineer. Chemical engineers conceptually view a (chemical) plant as consisting of a series of interrelated building blocks that are defined as *units* or *unit operations*. The plant essentially ties together the various pieces of equipment that make up the process. Flow schematics follow the successive steps of a process by indicating where the pieces of equipment are located and the materials streams entering and leaving each unit [6–8].

11.4   Material and Energy Balances

Overall and componential material balances have already been described in some detail earlier in Chapter 4. As noted at that time, material balances may be based on mass, moles, or volume, usually on a rate (time rate of change) basis. Care should be exercised here since moles and volumes are *not* conserved, i.e., the quantities may change during the course of a reaction. Thus, the initial material balance calculation should be based on *mass*. Mole balances and molar information are important in not only stoichiometric calculations but also in chemical reaction and phase equilibria calculations. Volume rates play an important role in equipment sizing calculations.

Some present-day design calculations in the chemical industry include transient effects that can account for *process upsets, startups, shutdowns*, etc. The describing equations for these time-varying (unsteady-state) systems are generally *differential*. The equations usually take the form of a first-order derivative with respect to time, where time is the independent variable. However, design calculations for most facilities assume steady-state conditions, with the ultimate design based on maximum flow conditions. This greatly simplifies the calculations, since the describing equations provide an accounting or inventory of all mass entering and leaving one or more pieces of equipment, or the entire process.

The number of material balance equations can significantly depend on a host of factors, including the number of components in the system, process chemistry, and pieces of equipment. These are critical calculations since (as noted earlier) the size of the equipment is often linearly related to the quantity of material being processes. This can then significantly impact – often linearly or even exponentially – capital and operating costs. In addition, componental rates can impact on (other) equipment needs, energy considerations, material of construction, etc.

Once the material balance(s) is/are completed, one may then precede directly to energy calculations, some of which play a significant role in the design of a facility. As indicated earlier, energy calculations are also usually based on steady-state conditions. An extensive treatment of this subject has already been presented earlier in this text, and need not be repeated here. However, a thorough understanding of thermodynamic principles – particularly the enthalpy calculations – is required for most of the energy (balance) calculations.

One of the principal jobs of a chemical engineer involved in the design of a facility is to account for the energy that flows into and out of each process unit and to determine the overall energy requirements for the process. This is accomplished by performing the aforementioned energy balances on each process unit and on the overall process. These balances play as important a role as material balances in facility design. They find particular application in determining fuel requirements, in heat exchanger design, in heat recovery systems, in specifying materials of construction, and in calculating fan and pump power requirements.

11.5 Equipment Design

As noted previously, chemical engineers describe the operation of any piece of equipment on the basis of mass, energy, and/or momentum transfer as a *unit operation*. A combination of two or more of these operations is defined as a *unit process*. A whole chemical process can be described as a coordinated set of unit operations and unit processes. This subject matter has received much attention over the years and, as a result, is adequately covered in the literature. From details on these unit operations and processes, it is therefore possible to design new plants more efficiently by coordinating a series of *unit actions*, each of which operates according to certain laws of physics regardless of the other operations being performed along with it. The unit operation of combustion is used in many different types of industries; many of the critical design parameters for the combustion processes, however, are common to all combustion systems and independent of the particular industry.

For example, in a hazardous waste incineration plant, the major pieces of equipment that must be considered are all of the following: [9]

1. Storage and handling facilities (feed and residuals)
2. Incinerator (rotary kiln or liquid injection)
3. Waste heat boiler (primary quench system or energy recovery if economically practical)
4. Quench system
5. Wet scrubber – venturi scrubber (particulate removal scrubber)
6. Absorber (packed tower for acid-gas absorption)
7. Spray dryer (quench and acid gas absorption)
8. Baghouse or electrostatic precipitators (ESP)
9. Peripheral equipment (cyclone)
10. Fan(s) and blower(s)
11. Stack
12. Pumps (feed, recycle, and scrubber)

Since design calculations are generally based on the maximum throughput capacity for the proposed process or for each piece of equipment, these calculations are never completely accurate. It is usually necessary to apply reasonable *safety factors* when setting the final design. Safety factors vary widely and are a strong function of the accuracy of the data involved, calculation procedures, and past experience. A process engineer's attempt to quantify these is a difficult task.

Unlike some of the problems encountered and solved by the chemical engineer, there is absolutely no correct solution to a design problem; however, there is usually a *better* solution. Many alternative designs, when properly implemented, will function satisfactorily, but one alternative will usually prove to be economically more efficient and/or attractive than the others. This leads to the general subject of optimization, a topic briefly addressed earlier in Chapters 2 and 10.

11.6   Instrumentation and Controls

The control of a system or process requires careful consideration of all operational and regulatory requirements. The system is usually designed to process materials. Safety [10] should be a primary concern of all individuals involved with the handling, treatment, or movement of the materials. The safe operation of any unit requires that the controls keep the system operating within a safe operating envelope. (See also Chapters 14 and 15). The envelope is based on many of the design, process, and regulatory constraints. These are placed on the unit to ensure proper operation. Additional controls may be installed to operate additional equipment needed for energy recovery, neutralization, or other peripheral operations.

The control system should also be designed to vary one or more of the process variables to maintain the appropriate conditions with the unit. These variations are programmed into the system based on the past experience of the unit manufactured. The operational parameters that may vary include the temperatures, system pressure, etc.

The purpose of a control system is to ensure that the system is operating in a reliable and safe manner, and within the guidelines of the design. The control system is responsible for all of the variables that occur during operation of the system. The reader is referred to Chapter 9 for additional details on instrumentation and controls.

11.7   Design Approach

Although chemical engineers approach design problems somewhat differently, six major steps are generally required. These six steps are briefly discussed below and may also be applied to the design of most facilities:

1. The first step is to conceptualize and define the process. A designer must know the assumptions that apply, the plant capacity, and the process time allowed. Some of the answers to a host of questions pertaining to the process operation will be known from past experience.

2. After the problem has been defined, a method of solution must be sought. Although a method is seldom obvious, a good starting point is the construction of a process flow sheet. This effort usually produces variable results. For example, it may suggest to the designer ways of reducing the complexity of the problem; it can allow for easier execution of material and energy balances, which in turn can point out the most important process variables. It is an efficient way to become familiar with the process; and, information that is initially lacking often becomes evident.

3. The third step is the actual design of the process equipment that involves the numerous calculations needed to arrive at specifications of operating conditions, equipment geometry, size, materials of construction, controls, instrumentation, monitors, safety equipment (automatic feed cut-off), etc. Equipment costs must be established part of this step. Cost-estimating precision is dependent on the desired accuracy of the estimate. If the decision based on an estimate is positive, a detailed project control or contractor's estimate will follow.

4. An overall economic analysis must also be performed in order to determine the process feasibility. The main purpose of this step is to answer the question of whether a process will ultimately be profitable or not. To answer this, raw material, labor, equipment, and other processing costs are estimated to provide an accurate economic forecast for the proposed operation.

5. In a case where alternate design possibilities exist, economics and engineering optimization are necessary. Since this is often the case, optimization calculations are usually applied several times during most design projects.

6. The final step of this design scheme is the compiling of a design report. A design report may represent the only relevant product of months or even years of effort. This is discussed in the next, and final, section.

These six activities are prominent steps in the traditional development of all modern chemical processes. Today, safety and regulatory (if applicable) concerns have also been integrated into the approach [10].

As noted earlier, the safe operation of equipment requires that some of the operational parameter be constrained within specific bounds. Each system has parameters that must remain within the appropriate bounds to assure that the system is stable. Most systems will have safety equipment to prevent the system from being operated at a condition outside of the safe limits. Insurance companies such as Industrial Risk Insurers (IRI), Factory Mutual (FM), and national groups such as National Fire Protection Agency (NFPA) have recommended specific requirements, the most important being to assume that all personnel are properly trained on operation limitations of the equipment.

Any environmental regulation requires that each of the operational limits be monitored to assure that the system has not been operated when the parameters have been exceeded. However, any of the permit parameters often cannot be monitored on a continuous basis. The design of a system must include both standard and nonstandard operational conditions.

11.8   The Design Report

As pointed out in Step 6 in the previous Section, a comprehensive plant design project report is often required. This material should be written up in a clear and concise fashion. In addition, the project leader might be requested to make informal and/or formal presentations to management on the study. The report and presentation should explain what has been accomplished and how it has been carried out. There are many different formats for design reports; the format will vary with the organization and the project. One possible outline employed by the authors for a project report is given in the following list:

1. Title page
2. Table of contents
3. Executive summary (abstract)
4. Prefatory comments (optional)
5. Introduction
6. Discussion
7. Laboratory studies (if applicable)
8. Pilot plant studies (if applicable)
9. Comprehensive process design, process flow diagrams, including mass and energy balances, and annualized costs
10. Calculation limitations
11. Design limitations
12. Optimization studies

13. Recommendations
14. Acknowledgements
15. Appendix.

Great care should be exercised with the preparation of the *executive summary*. It is recommended that the executive summary be no longer than one single-spaced typewritten page. It should contain a short introduction, important results, and pertinent recommendations and conclusions. It should *not* refer to the body of the report. In many instances, the executive summary is the only portion of the report that upper-level management will initially review; *this one section is unquestionably the most important part of the report.*

A detailed and expanded treatment of plant design is available in the following three references:

1. R. Perry and D. Green (editors), *Perry's Chemical Engineers' Handbook*, 7th edition, McGraw-Hill, New York City, NY, 1997.
2. D. Kauffman, *Process Design*, A Theodore Tutorial, Theodore Tutorials, East Williston, NY, originally published by the USEPA/APVI, RTP, NC 1992 [11].
3. L. Theodore, *Chemical Engineering: The Essential Reference*, McGraw-Hill, New York City, NY, 2014 [12].

11.9   Illustrative Open-Ended Problems

This and the last section provide open-ended problems. However, solutions *are* provided for the three problems in this Section in order for the reader to hopefully obtain a better understanding of these problems which differ from the traditional problems/illustrative examples. The first problem is relatively straightforward while the third (and last problem) is somewhat more difficult and/or complex. Note that solutions are not provided for the 35 open-ended problems in the next section.

Problem 1: Discuss the role utility systems play in the design and operation of a chemical plant.

Solution: Utility systems provide the infrastructure in which chemical and petrochemical process plants operate. These systems can include such things as steam boilers and distribution lines, fuel distribution systems, cooling towers and cooling water distribution systems, compressed air systems, electrical distribution systems, inert gas distribution systems, lighting, communications, fire water, etc. No plant could operate without at least some of these utility systems, and many plants require all of them.

Since plant utility systems are quite similar from one plant to the next, the chemical process engineers do not have to design them from scratch each time. It is customary to assume that the utility services will be available in whatever quantity is needed. They can be designed later. Or, in the case of installations at existing plants, they can be modified to supply the needs of a new process unit.

Problem 2: Discuss the difference between continuous and batch plant operation.

Solution: Most process plants can be categorized as either *continuous* or *batch* in nature. Continuous processes operate at steady state with respect to time; there is a continuous inflow of raw materials and a continuous outflow of products. Batch plants are characterized by process conditions that change with time; feeds are added at specific times, and products are removed at specific times.

Continuous plants have many distinct advantages. Once the plants are started up, all feed materials are processed in exactly the same way. Each step of the process takes place in a different piece of equipment, with the materials flowing continuously from one to the next. Uniform product quality can often be maintained easily. Storage requirements for feeds, products and intermediates are minimized. The quantity of material being actively processed at any one time is minimized. The task of the operators, once the plant is started up, is to maintain it in a single, steady operating mode. Small deviations from desired conditions can be easily detected and corrected. Continuous plants are most often used for production of large quantities of materials.

In batch plants, many distinct operations may take place in the same piece of equipment. Feeds are added at discrete times based on a "recipe" for the process. The operators must process each batch of material individually. Temperature and residence time control are more difficult than in continuous plants, but some difficult-to-control process features, such as extent of a polymerization reaction, can be followed and adjusted more closely. Batch plants are most often used when only small quantities of a product are needed or where many variations of a single basic product are needed, such as in the specialty chemical business. In addition, there are large-scale processes, such as manufacture of some plastic resins, for which continuous plants have proved impractical.

There are, of course, plants which are hybrid mixtures of continuous and batch processing. Many food processing plants fall in this category. "Cooking" operations are likely to be batch in nature; packaging operations are likely to be continuous.

Problem 3: A research chemical engineer has devised the processing scheme described below for the manufacture of a commodity chemical. The overall chemistry of the process can be represented by the equations:

$$A + 2B \rightarrow C + D$$

$$C + E \rightarrow F + G$$

A is an organic liquid. B is a gas. D is an unwanted gaseous byproduct which can be used as a fuel. E is an organic liquid. F is the desired product, an organic liquid. G is an unwanted organic byproduct which can be used as a fuel.

In the laboratory, the first reaction readily took place at 240°F and 150 psig, requiring about 30 minutes to go to completion. It was endothermic, with a heat of reaction of about 65 kcal/gmol of A reacted. The reactor used consisted of a one-liter stainless steel flask with the gases sparged in at the bottom. A reflux condenser was used to keep product C while venting D.

The second reaction takes place almost instantaneously upon addition of E to C at ambient conditions. It is exothermic, with a heat of reaction of about 25 kcal/gmol of C reacted. In the laboratory, this reaction was carried out by slowly adding E to the C, still contained in the stainless steel flask. Finally, F and G were separated by distillation in a lab column. F boils at about 240°F and G boils at 155°F at atmospheric pressure. Both remain liquid at ambient temperature.

Develop a preliminary flowsheet for a continuous plant to carry out the process.

Solution: One possible continuous plant for this process is sketched below in Figure 11.1.

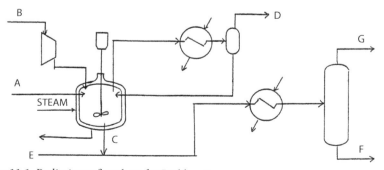

Figure 11.1 Preliminary flowsheet for Problem 2

The following 6 features of the system are described below.

1. Compressor for feed B: The compressor is needed to maintain the reactor at 150 psig. If the above supply of B is already at or above this pressure, the compressor would not be needed.
2. Reactor with steam heating jacket or coils: The reactor is designed to hold the liquid mixture for whatever reaction time is necessary to achieve sufficient conversion of A. It most likely will not be exactly the same as the 30 minutes used in the lab. Steam heating is provided by using steam coils in the reactor and/or a steam jacket surrounding the reactor.
3. Vent condenser: The vent condenser cools the gaseous by-product D and condenses out any residual A or C, retuning them to the reactor. D is sent for use as fuel elsewhere in the plant.
4. "Reactor" two: Where is it? There is no actual "reactor: for the second reaction. E is added directly to C and is mixed in the piping.
5. Intermediate cooler: This heat exchanger is designed to cool down the reaction products after the exothermic reaction between C and E. It will take more detailed analysis to determine whether or not it is actually required.
6. Distillation column: The column provides for separation of product F from by-product G.

In this process, undesired organic by-products are used as fuel elsewhere in the plant. A normal part of process design and analysis would be to see whether there were any higher-value uses for either of these by-products, or whether their formation could be eliminated by changing the overall process.

Note that this is but one possible scheme.

11.10   Open-Ended Problems

This last section of the chapter contains open-ended problems as they relate to plant design. No detailed and/or specific solution is provided; that task is left to the reader, noting that each problem has either a unique solution or a number of solutions or (in some cases) no solution at all. These are characteristics of open-ended problems described earlier.

There are comments associated with some, but not all, of the problems. The comments are included to assist the reader while attempting to solve the problems. However, it is recommended that the solution to each problem should initially be attempted *without* the assistance of the comments.

There are 35 open-ended problems in this section. As stated above, if difficulty is encountered in solving any particular problem, the reader should next refer to the comment, if any is provided with the problem. The reader should also note that the more difficult problems are generally located at or near the end of the section.

1. Describe plant design activities during the early years of the chemical engineering profession.
2. Discuss the recent advances in the design of plants.
3. Select a refereed, published article on plant design from the literature and provide a review.
4. Develop an original problem in plant design that would be suitable as an illustrative example in a book.
5. Prepare a list of the various books that have been written on plant design. Select the three best and justify your answer. Also select the three weakest books and justify your answer.
6. Describe the difference between plant design and project management. Also discuss how they are interrelated.
7. Explain why it is more difficult to design a plant/process than to predict performance.
8. Describe the difference between process synthesis for a known process as opposed to a new process.
9. Which equipment generally play the most important role in a plant design?
10. Describe the differences between informal and formal design reports.
11. "A generalization for equipment design is that *standard* equipment should be selected in the design of a plant". Comment on this statement.
12. Describe the differences between the following three classes of designs.
    - Preliminary designs
    - Detailed estimate designs
    - Firm process designs.
13. Describe in general terms the relationship between plant design and sales.

14. Describe the various procedures that are available in order to obtain the optimum design for a plant.
    Comment: Refer to Chapter 10, Economics and Finance.

15. Provide some general comments on flow diagrams.

16. Provide some general comments on process flow sheets.

17. There is much more variety in the selection and specification of equipment for handling solids than for liquids or gases, which are nearly always confined to pipes or ducts. Discuss the various methods of moving/conveying solids.

18. Describe how new processes impact the chemical engineer's approach to process and plant design.

19. Select a process of your choice and draw a flow diagram of the system.
    Comment: This will require the review of the literature.

20. Select a particular chemical process industry. Prepare a report using a tabulated format to describe the physical and chemical properties of the raw materials, intermediate products, by-products, and principal chemicals that will be encountered in the manufacturing process.

21. Refer to the previous problem. Prepare a material balance for individual units and the plant itself.

22. Refer to the previous problem. Prepare an energy balance for individual units and the plant itself.

23. Describe the advantages and disadvantages that pilot plant studies play in the design of a plant.

24. Suggest what steps can be taken to eliminate pilot plant studies in the design of a plant.

25. Discuss the general problems/associated with scale-up.

26. Discuss specific problems associated with scale-up as it applies to some of the major chemical process equipment, e.g., fluid bed systems, crystallizers, mixers, etc.

27. Describe the role the research department of a company plays in the design of a plant.

28. Describe the role the legal department of a company plays in the design of a plant.

29. Describe the differences between plant design drawings and construction and installation drawings.

30. Describe the logical evolution of a process (for a plant).
    Comment: Refer to F. Vilbrandt and C. Dryden's "Chemical Engineering Plant Design", McGraw-Hill, New York City, NY, 1934 [13].

31. Discuss the potential role the stock market plays in the design of a plant.

32. The analysis of chemical reactors is a basic feature of most chemical and petrochemical plant designs. It is the ability to operate large-scale chemical reactors that has made possible the efficient production of many chemicals employed in the world today. Chemical reactors in process plants can have many different shapes and sizes. Describe these units. Comment: Refer to the literature [14] for additional details and to Chapter 8.

33. Cooling towers operate by contacting warm water with ambient air. The air is heated and becomes saturated with water vapor at its exit temperature. The evaporation of a small portion of the warm water passing into the air cools the rest of the water as pictured in Figure 11.2 [11].

    A cooling tower at sea level is fed 10,000 gpm (5,000,000 lb/h) of water at 115°F and cools it to 90°F. On a particular day, the ambient air temperature is 70°F and the relative humidity is 50%. How much air flow is needed for the cooling tower and how much water is evaporated? [11] Finally, design the cooling tower. Specifically state the assumptions in your design.

34. Atmospheric pressure tanks are used to store most liquids in process plants. The vapor in an atmospheric pressure tank is connected to the atmosphere through a vent line so that there will not be any net pressure difference between the inside and outside of the tank roof. Vapor recovery systems are often put on the vent line to reduce either air pollution or material losses, or both [11].

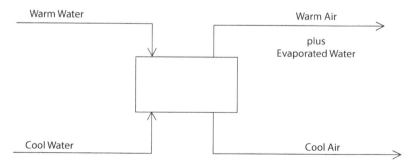

Figure 11-02 Cooling Tower Schematic

A 40,000-bbl atmospheric pressure tank is used to store kerosene. The tank is 30 ft tall and 98 ft in diameter. What is the maximum pressure that could be exerted by the kerosene on the tank wall? How much kerosene vapor would be vented to the atmosphere in day/night "tank breathing" if the tank has no vapor recovery system? [11] Discuss the validity and limitations of your calculations.

35. The basic goal of most chemical and petrochemical processing is to convert less valuable raw materials into more valuable products. A simple example is found in many refineries – the conversion of toluene, the most abundant aromatic compound in crude oil, to more valuable benzene and xylene products. The process is referred to by some as "toluene disproportionation." The overall chemical reaction involved is:

$$2C_6H_5CH_3 = C_6H_6 + C_6H_4(CH_3)_2$$

This reaction is usually carried out in the vapor phase at moderately high temperatures and moderate pressures. There are three isomers of xylene (para-xylene, meta-xylene, and ortho-xylene), and all three are formed in the reaction, which proceeds rapidly to chemical equilibrium [11]. The equilibrium mixture from the reactor, which consists of benzene, toluene and all three xylene isomers, is separated. The toluene is recycled to the front end of the process; therefore, the only end products are benzene and the xylenes.

Prepare a line flow diagram for this example. Also prepare a material balance in bbl/day for a toluene disproportionation unit which has a fresh feed rate of 1000 bbl/day. The reactor operates at 50 psia and 400°F. Under these conditions, one can assume an equilibrium chemical product mixture of 26.2 mol% benzene, 47.6 mol% toluene, 6.4 mol% para-xylene, 14.4 mol% meta-xylene and 5.4 mol% ortho-xylene.

References

1. Personal notes: L. Theodore, East Williston, NY, 1967.
2. T. Shen, Y. Choi, and L. Theodore, *EPA Manual Hazardous Waste Incineration*, USEPA/APTI, RTP, NC, 1985.

3. L. Theodore, *Hazardous Waste Incineration*, Instructor's Guide, USEPA/APTI, RTP, NC, 1986.

4. G. Ulrich, *A Guide to Chemical Engineering Process Design and Economics*, John Wiley & Sons, Hoboken, NJ, 1984.

5. Personal notes: L. Theodore, East Williston, NY, 1983.

6. R. Felder and R. Rousseau, *Elementary Principles of Chemical Processes*, 2nd ed., John Wiley & Sons, Hoboken, NJ, 1986.

7. R. Perry and D. Green (editors), *Perry's Chemical Engineers' Handbook*, 7th edition, McGraw-Hill, New York City, NY, 1997.

8. W. McCabe, J. Smith, and P. Harriott, *Unit Operations of Chemical Engineering*, 5th ed., McGraw-Hill, New York City, NY, 1993.

9. J. Santoleri, J. Reynolds and L. Theodore, *Introduction to Hazardous Waste Incineration*, 2nd edition, John Wiley & Sons, Hoboken, NJ, 2000.

10. L. Theodore and R. Dupont, *Environmental Health and Hazard Risk Assessment: Principles and Calculations*, CRC Press/Taylor & Francis Group, Boca Raton, FL, 2012.

11. D. Kauffman, *Process Design*, A Theodore Tutorial, Theodore Tutorials, East Williston, NY, originally published by the USEPA/APVI, RTP, NC 1992.

12. L. Theodore, *Chemical Engineering: The Essential Reference*, McGraw-Hill, New York City, NY, 2014.

13. F. Vilbrandt and C. Dryden, *Chemical Engineering Plant Design*, McGraw-Hill, New York City, NY, 1934.

14. L. Theodore, *Chemical Reactor Design and Applications for the Practicing Engineer*, John Wiley & Sons, Hoboken, NJ, 2012.

12

Transport Phenomena

This chapter is concerned with process transport phenomena. As with all the chapters in Part II, there are several sections: overview, several technical topics, illustrative open-ended problems, and open-ended problems. The purpose of the first section is to introduce the reader to the subject of transport phenomena. As one might suppose, a comprehensive treatment is not provided although numerous references are included. The second section contains three open-ended problems; the authors' solution (there may be other solutions) are also provided. The third (and final) section contains 31 problems; *no* solutions are provided here.

12.1 Overview

This overview section is concerned—as can be noted from its title—with transport phenomena. As one might suppose, it was not possible to address all topics directly or indirectly related to transport phenomena. However, additional details may be obtained from either the references provided at the end of this Overview section and/or at the end of the chapter.

Note: Those readers already familiar with the details associated with this subject may choose to bypass this Overview.

Transport phenomena deals with the transfer of certain quantities (momentum, energy, and mass) from one point in a system to another. Three basic transport mechanisms are involved in a process. They are:

1. Radiation
2. Convection
3. Molecular diffusion.

The first mechanism, radiative transfer, arises due to wave motion and is not considered, since it may be justifiably neglected in most engineering applications. Convective transfer occurs simply due to *bulk motion*. Molecular diffusion is defined as the transport mechanism arising due to *gradients*. For example, momentum is transferred in the presence of a *velocity* gradient; energy in the form of heat is transferred due to a *temperature* gradient and mass is transferred in the presence of a *concentration* gradient. These molecular diffusion effects are described by *phenomenological* laws. These laws have been defined as mathematical models which happen to be obeyed within experimental precision by most media. Each of the laws described below reduces to the product of an appropriate transport coefficient and a gradient.

1. Newton's second law serves to define the viscosity—the transport coefficient for momentum transfer.
2. Fourier's law defines the thermal conductivity—the transport coefficient for heat transfer.
3. Fick's law serves to define the diffusivity—the transport coefficient for mass transfer.

The aforementioned transport coefficient is almost always determined by experiment, although it can be predicted theoretically from knowledge at the molecular level. The methods of evaluating and correlating these coefficients is not presented. Instead, the reader is referred to any standard text on physical properties for this information.

The remaining sections of this chapter are concerned with

1. Development of Equations
2. The Transport Equations
3. Boundary and Initial Conditions
4. Solution of Equations

5. Analogies.

It should be noted that the bulk of the material to follow has been drawn from the work of Theodore [1].

12.2   Development of Equations

Much of the materials in this section was presented in Chapter 3, Momentum, energy and mass are all conserved. As such, each quantity obeys the conservation law within a system.

$$
\left\{ \begin{array}{c} quantity \\ into \\ system \end{array} \right\} - \left\{ \begin{array}{c} quantity \\ out\ of \\ system \end{array} \right\} + \left\{ \begin{array}{c} quantity \\ generated \\ in\ system \end{array} \right\} = \left\{ \begin{array}{c} quantity \\ accumulated \\ in\ system \end{array} \right\} \quad (12.1)
$$

This equation may also be written on a *time* rate basis:

$$
\left\{ \begin{array}{c} rate\ of \\ quantity \\ into \\ system \end{array} \right\} - \left\{ \begin{array}{c} rate\ of \\ quantity \\ out\ of \\ system \end{array} \right\} + \left\{ \begin{array}{c} rate\ of \\ quantity \\ generated \\ in\ system \end{array} \right\} = \left\{ \begin{array}{c} rate\ of \\ quantity \\ accumulated \\ in\ system \end{array} \right\} \quad (12.2)
$$

The conservation law may be applied at the macroscopic, microscopic or molecular level. One can best illustrate the differences in these methods with an example. Consider a system in which a fluid is flowing through a cylindrical tube (see Figure 12.1). Define the system as the fluid contained within the tube between points 1 and 2 at any time. If one is interested in determining changes occurring at the inlet and outlet of the system, the conservation law is applied on a "macroscopic" level to the entire system. The resultant equation describes the overall changes occurring *to* the system without regard for internal variations *within* the system. This approach is usually applied in the Unit Operations (or its equivalent) courses. The microscopic approach is employed when detailed information concerning the behavior *within* the system is required, and this is often requested of and provided by the engineer. The conservation is then applied to a *differential* element within the system which is large compared to an individual molecule, but small compared to the entire system. The resultant equation

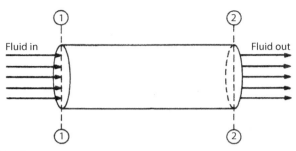

Figure 12.1 Pipe flow

is then expanded, via integration, to describe the behavior of the entire system. This is defined as the *transport phenomena* or *microscopic approach*. The molecular approach involves the application of the conservation law to individual molecules. This leads to a study of statistical and quantum mechanics—both of which are beyond the scope of this text. In any case, the description of individual matter at the molecular level is of little value to the engineer. However, the statistical averaging of molecular quantities in either a differential or finite element within a system leads to a more meaningful description of the behavior of a system.

Traditionally, the applied mathematician has developed the differential equations describing the detailed behavior of systems by applying the appropriate conservation law to a differential element or shell within the system. Equations were derived with each new application. The engineer later removed the need for these tedious and error-prone derivations by developing a general set of equations that could be used to describe systems. These are referred to as the *transport equations*. Since they are so general, they may be used rather indiscriminately to describe the infinite variety of specific problems confronting engineers. Needless to say, these transport equations have proven to be an invaluable asset in describing the behavior of many systems, operations and processes.

A complete description of the transport process requires certain additional information. The pressure, temperature, and composition dependence of viscosity, thermal conductivity, and diffusivity, must be made available from thermodynamics data. Chemical reaction systems require kinetic data.

12.3   The Transport Equations

The aforementioned transport equations are available in the literature [1,2], the details of which are significantly beyond the purpose and scope of this text. Theodore [1] developed the equations in vector form and then expanded them into the following coordinate system:

1. rectangular (Cartesian);
2. cylindrical; and
3. spherical.

Theodore's development included material concerned with the continuity, momentum transfer, energy transfer in solids, energy transfer, mass transfer in solids, and the mass transfer equations. The classic work of Bird, et al. [2] provides additional and a more expansive treatment of this subject.

12.4   Boundary and Initial Conditions

In order to solve the differential transport equation(s) so that one may obtain a complete description of the pressure, temperature, composition, etc., of a system, it is necessary to specify boundary and/or initial conditions (BC/IC) for the system. This information arises from a description of the problem or the physical situation. The number of boundary conditions (BC) that must be specified is the sum of the highest order derivative for each independent position variable appearing in the differential equation. A value specified at the boundary of the system is one type of boundary condition. The number of initial conditions (IC) that must be specified is the highest order time derivative appearing in the differential equations. This condition is used only if time is a variable. The value of the solution at time equal to zero constitutes an IC.

For example, the equation

$$\frac{d^2 v_y}{dz^2} = 0 \tag{12.3}$$

requires two BC. The equation

$$\frac{dT}{dt} = 0; \qquad t = \text{time} \tag{12.4}$$

requires one IC. And finally, the equation

$$\frac{\partial c_A}{\partial t} = D \frac{\partial^2 c_A}{\partial y^2} \tag{12.5}$$

requires one IC and two BC.

12.5   Solution of Equations

This section is introduced by outlining the general procedure that engineers should follow in solving problems in transport phenomena. The procedure is as follows.

1. Draw a line diagram representing the physical system.
2. List all pertinent variables and dimensions on the diagram.
3. Select the most convenient coordinate system.
4. Obtain the mathematical equations (in the chosen coordinates) describing the behavior of the system. This information can be "extracted" from the transport equations.
5. Specify the BC/IC.
6. Solve the equations.
7. Check to see if the solution satisfies *both* the differential equation and the BC/IC

As stated earlier, combining the conservation and phenomenological laws leads to a set of partial differential equations that can usually be solved subject to the system's BC/IC. In principle, this approach leads to a complete solution. In practice, two major difficulties may arise:

1. There is insufficient knowledge of the transport coefficients appearing in the equations.
2. The complexity of the differential equation and the accompanying BC/IC prohibits solution.

The reader is referred to other texts dealing exclusively with item (1). Information (data) on the coefficients is general available. One can then focus attention on item (2). These solutions may be obtained by:

1. Intuition
2. Graphical methods
3. Analytical methods
4. Analog methods
5. Numerical methods.

12.6   Analogies

There are certain common principles and laws that apply to the transport processes. Because of this, many similarities and analogies exist between the transport mechanisms discussed in the preceding sections; these are outlined and discussed in the present section.

Four common subject areas are discussed below. These include:

1. Conservation law
2. Phenomenological law
3. Units of molecular diffusion coefficient
4. Ratio of molecular diffusion coefficients

Details of each follows

1. *Conservation law.* Each of the equations describing the transfer of momentum, energy, and mass are developed by application of the conservation law on a rate basis to momentum, energy, and mass, respectively. The general form of the equation is

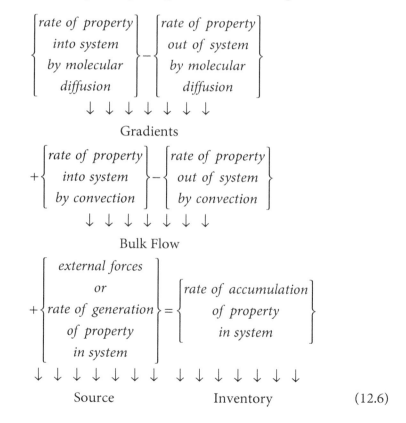

$$
\begin{Bmatrix} rate\ of\ property \\ into\ system \\ by\ molecular \\ diffusion \end{Bmatrix} - \begin{Bmatrix} rate\ of\ property \\ out\ of\ system \\ by\ molecular \\ diffusion \end{Bmatrix}
$$

$$\downarrow \downarrow \downarrow \downarrow \downarrow \downarrow \downarrow$$

Gradients

$$
+ \begin{Bmatrix} rate\ of\ property \\ into\ system \\ by\ convection \end{Bmatrix} - \begin{Bmatrix} rate\ of\ property \\ out\ of\ system \\ by\ convection \end{Bmatrix}
$$

$$\downarrow \downarrow \downarrow \downarrow \downarrow \downarrow \downarrow$$

Bulk Flow

$$
+ \begin{Bmatrix} external\ forces \\ or \\ rate\ of\ generation \\ of\ property \\ in\ system \end{Bmatrix} = \begin{Bmatrix} rate\ of\ accumulation \\ of\ property \\ in\ system \end{Bmatrix}
$$

$$\downarrow \downarrow \downarrow \downarrow \downarrow \downarrow \downarrow \quad \downarrow \downarrow \downarrow \downarrow \downarrow \downarrow \downarrow$$

Source   Inventory   (12.6)

2. *Phenomenological law.* The laws governing the molecular diffusion of momentum, energy, and mass were developed

by Newton, Fourier, and Fick, respectively. These phenomenological laws express the corresponding fluxes for momentum, energy, and mass in terms of measurable quantities, i.e.,

$$\begin{Bmatrix} molecular \\ diffusion \\ flux \end{Bmatrix} = -\begin{Bmatrix} transport \\ coefficient \end{Bmatrix} \times \{gradient\} \qquad (12.7)$$

or more specifically,

$$\tau_{zy} = -\frac{\mu}{g_c}\frac{dv_y}{dz}; \quad \text{Newton's Law} \qquad (12.8)$$

$$q_z = -k\frac{dT}{dz}; \quad \text{Fourier's Law} \qquad (12.9)$$

$$J_{A_z} = -D_{AB}\frac{dc_A}{dz}; \quad \text{Fick's Law} \qquad (12.10)$$

with the standard notation employed in the transport
field [1,2].

3. *Units of molecular diffusion coefficient.* The molecular diffusion for momentum, energy, and mass are defined

$$v = -\frac{\mu}{\rho}; \quad \text{kinematic viscosity, momentum} \qquad (12.11)$$

$$a^2 = \frac{k}{\rho C_p}; \quad \text{thermal diffusivity, energy} \qquad (12.12)$$

$$D_{AB} = D_{AB}; \quad \text{diffusion coefficient, mass} \qquad (12.13)$$

A dimensionless analysis of these three coefficients produces
an interesting result. The units of v, a^2, and D_{AB} are the same
and given in ft²/s (English units).

4. *Ratio of molecular diffusion coefficients.* The ratio of the molecular diffusion coefficients can play an important role in the analysis of a system undergoing the simultaneous transfer of any combination of momentum, energy, and/or mass. It is a measure of the relative magnitude of these effects. The three coefficients corresponding to momentum, energy, and mass previously have been defined as v, a^2, and D_{AB}, respectively. Three dimensionless ratios can be generated from these coefficients.

a. Ratio of momentum to energy; i.e.,

$$\frac{v}{a^2} = \frac{\mu / \rho}{k / \rho C_p} \tag{12.14}$$

This may be written as

$$\frac{\mu C_p}{k} \tag{12.15}$$

and is defined as the Prandtl number. It finds application in fluid flow and heat transfer processes.

b. Ratio of momentum to mass; i.e.,

$$\frac{v}{D_{AB}} \tag{12.16}$$

or

$$\frac{\mu}{\rho D_{AB}} \tag{12.17}$$

This term is defined as the Schmidt number and finds application in systems undergoing momentum and mass transfer.

c. Ratio of energy to mass; i.e.,

$$\frac{a^2}{D_{AB}} = \frac{k / \rho C_p}{D_{AB}} \tag{12.18}$$

or

$$\frac{k}{\rho C_p D_{AB}} \qquad (12.19)$$

This term is defined as the Lewis number. It finds application in heat and mass transfer operations.

12.7   Illustrative Open-Ended Problems

This and the last section provide open-ended problems. However, solutions *are* provided for the three problems in this section in order for the reader to hopefully obtain a better understanding of these problems which differ from the traditional problems/illustrative examples. The first problem is relatively straightforward while the third (and last problem) is somewhat more difficult and/or complex. Note that solutions are not provided for the 31 open-ended problems in the next section.

Problem 1: Describe the differences between the macroscopic, microscopic, and molecular approaches (as they apply to the conservation laws) from a technical perspective.

Solution: Refer to the development provided earlier in the Development of Equations section.

Problem 2: Outline the general procedure that chemical engineers should follow in solving problems in transport phenomena.

Solution: Solutions can be obtained almost immediately by inspection or intuition for a few of these examples. However, the majority of the differential equations encountered are solved by well-known standard analytical methods. These include:

1. Separation of variables
2. Fourier series
3. Bessel functions
4. Laplace transforms
5. Error functions.

Complet programs now permits solution to some of the more formidable problems. The more complex equations, encountered in practice, can be solved by numerical methods with a computer.

The reader is referred to the Solution of Equations section for additional details see Chapter 2.

Problem 3: A moving fluid enters the reaction zone of a tubular reactor given at concentration c_{A_0} and undergoes chemical reaction. Obtain the steady-state equations describing the concentration in the reaction zone if the flow is either laminar or plug. Assume various reaction mechanisms in generating the solutions. Do not neglect diffusion effects [1].

Solution: The problem is solved using cylindrical coordinates. Based on the problem statement

$$c_A = c_A(r,z), \quad \text{laminar flow}$$

$$c_A = c_A(z), \quad \text{plug flow}$$

and

$$v_r = 0$$
$$v_\phi = 0$$
$$v_z = 2v_z \left[1 - \left(\frac{r}{a} \right)^2 \right], \quad \text{laminar flow}$$
$$v_z = v_z, \quad \text{plug flow}$$

The partial differential equation describing this system for a *first order* reaction is given by

$$v_z \frac{\partial c_A}{\partial z} = D_{AB} \left[\frac{1}{r} \frac{\partial}{\partial r} \left(r \frac{\partial c_A}{\partial r} \right) + \frac{\partial^2 c_A}{\partial z^2} \right] - k_A c_A$$

For laminar flow, this equation becomes

$$\left(2v_z \right) \left[1 - \left(\frac{r}{a} \right)^2 \right] \frac{\partial c_A}{\partial z} = D_{AB} \left[\frac{1}{r} \frac{\partial}{\partial r} \left(r \frac{\partial c_A}{\partial r} \right) + \frac{\partial^2 c_A}{\partial z^2} \right] - k_A c_A$$

For plug flow,

$$v_z \frac{\partial c_A}{\partial z} = D_{AB} \frac{\partial^2 c_A}{\partial z^2} - k_A c_A$$

One can obtain the concentration profile in the reactor for laminar and plug flow. Neglecting axial and radial diffusion leads to

$$2v_z \left[1 - \left(\frac{r}{a} \right)^2 \right] \frac{\partial c_A}{\partial z} = -k_A c_A$$

for laminar flow. This many now be rewritten as

$$2v_z \left[1 - \left(\frac{r}{a} \right)^2 \right] \frac{dc_A}{dz} = -k_A c_A$$

The BC for this problem is

$$c_A = c_{A_0} \qquad \text{at} \qquad z = 0$$

The solution to the above equation is

$$c_A = c_{A_0} e^{-k_A z / 2 v_z \left[1 - (r/a)^2 \right]}$$

This approach may be applied to reactions of other/different orders.

12.8   Open-Ended Problems

This last Section of the chapter contains open-ended problems as they relate to transport phenomena. No detailed and/or specific solution is provided; that task is left to the reader, noting that each problem has either a unique solution or a number of solutions or (in some cases) no solution at all. These are characteristics of open-ended problems described earlier.

There are comments associated with some, but not all, of the problems. The comments are included to assist the reader while attempting to solve the problems. However, it is recommended that the solution to each problem should initially be attempted *without* the assistance of the comments.

There are 31 open-ended problems in this section. As stated above, if difficulty is encountered in solving any particular problem, the reader should next refer to the comment, if any is provided with the problem. The reader should also note that the more difficult problems are generally located at or near the end of the section.

1. Discuss the recent advances in transport phenomena education.
   Comment: Refer to the literature for details [2].

2. Describe the early history associated with transport phenomena education.
   Comment: Refer to the literature for details [2].

3. Describe the differences between the macroscopic, microscopic, and molecular approaches from a layman's perspective.

4. Select a refereed, published transport phenomena article from the literature and provide a review.

5. Develop an original problem that would be suitable as an illustrative example in a book on transport phenomena.

6. Prepare a list of the various books that have been written on transport phenomena. Select the three best and justify your answer. Also select the three weakest books and, once again, justify your answer.

7. Provide in layman terms, the Boltzmann equation describing the kinetic theory of gases.

8. Describe and discuss the limitations associated with Boltzmann's kinetic theory of gases.

9. Attempt to improve on Boltzmann's kinetic theory of gases.

10. Describe Newton's Law of viscosity in layman terms.

11. Describe Fourier's Law in layman terms.

12. Describe Fick's Law in layman terms.

13. Discuss the differences between macroscopic and microscopic coefficients.

14. Describe the various classes of polymeric liquids. Also discuss the differences.

15. Describe the various velocity distributions that can arise for flowing fluids in conduits.

16. Discuss the differences between free and forced convection at the microscopic level.

17. Describe the complications that arise in describing multi-component systems at the microscopic level.

18. Describe the problems associated with applying the microscopic approach to turbulent flow systems.

19. Discuss the molecular theory of predicting the viscosity of both liquids and gases.

20. Discuss the molecular theory of predicting the thermal conductivity of solids, liquids and gases.

21. Discuss the molecular theory of predicting the diffusivities of liquids, colloidal suspensions and gases.

22. Energy is being absorbed in a long solid cylinder of radius a. The temperature at the outer surface of the cylinder is maintained at a constant value T_0. Calculate the temperature profile in the solid at steady-state conditions. Assume the energy generation term A is a
   - linear,
   - quadratic, and
   - cubic

   function of the temperature. Comment on the results.

23. A long hollow *cylinder* has its inner and outer surfaces maintained at constant temperatures. Calculate the temperature profile in the solid section of the cylinder and determine the flux at both surfaces for different temperatures. Comment on the results. Assume steady-state conditions.

24. A component is reacting uniformly in a batch reactor of arbitrary shape. Obtain the concentration as a function of position and time for various reaction mechanisms if the initial concentration is everywhere constant. Assume no mass transfer across the surface of the solid.

25. An incompressible fluid enters the reaction zone of an insulated tubular (cylindrical) reactor at temperature T_0. The chemical reaction occurring in the zone causes a rate of energy per unit volume to be liberated. Obtain the steady-state equation describing the temperature in the reactor zone if the flow is laminar and the rate of energy generation is a:
   - linear
   - parabolic
   - cubic

function of temperature. Neglect axial diffusion. Also comment on the results.

Comment: Refer to the literature [3].

26. Two different viscosity fluids are contained between the region bounded by two infinite parallel horizontal plates separated by a finite distance. The volumes occupied by each fluid are equal. The upper plate is moving with a velocity that varies with time. Set up the describing equation(s) and calculate the velocity profile of both fluids for different velocity variations.

27. Consider an insulated cylindrical copper rod. If the rod is initially at a constant temperature and the ends of the rod are maintained at a temperature that varies with time, provide an equation that describes the temperature (profile) in the rod as a function of both position and time.

28. Refer to the previous problem. Calculate the temperature profile for different temperature variations.

29. Some have argued (including the senior – in terms of age – author of this book) that transport phenomena has outlived its usefulness for the chemical engineer. Comment on this statement.

30. Prepare a detailed review of the second edition of the Bird, et al. book.

31. Prepare a detailed review that highlights the differences between the first [4] and second [2] editions of the Bird, et al books.

References

1. L. Theodore, *Transport Phenomena for Engineers*, Theodore Tutorials, East Williston, NY, originally published by International Textbook CO., Scranton, PA, 1971.

2. R. Byrd, W. Stewart, and Lightfoot, *Transport Phenomena*, 2nd edition, John Wiley & Sons, Hoboken, NJ, 2002.

3. L. Theodore, *Heat Transfer for the Practicing Engineer*, John Wiley & Sons, Hoboken, NJ, 2011.

4. R. Byrd, W. Stewart, and Lightfoot, *Transport Phenomena*, John Wiley & Sons, Hoboken, NJ, 1960.

13

Project Management

This chapter is concerned with process project management. As with all the chapters in Part II, there are several sections: overview, several technical topics, illustrative open-ended problems, and open-ended problems. The purpose of the first section is to introduce the reader to the subject of project management. As one might suppose, a comprehensive treatment is not provided although numerous references are included. The second section contains three open-ended problems; the authors' solution (there may be other solutions) are also provided. The third (and final) section contains 37 problems; *no* solutions are provided here.

13.1 Overview

This overview section is concerned—as can be noted from its title—write project management. As one might suppose, it was not possible to address all topics directly or indirectly related to project management. However, additional details may be obtained from either the references provided at the end of this Overview section and/or at the end of the chapter.

Note: Those readers already familiar with the details associated with this subject may choose to bypass this Overview.

One of the authors [1] has estimated that better than 75% of chemical engineers are involved with project management activities… and this percentage will continue to increase in the future. The need for the traditional chemical engineer who designed heat exchangers, specified pumps, predicted the performance of multi-component distillation columns, etc., has all but disappeared. This is a fact that the profession has difficulty in accepting [2,3]. But, the reality is that chemical engineers in the future will require some understanding of project management.

Project management has come to mean different things to different people in the technical community. The term *project* is a time-constrained endeavor undertaken to create a unique product, service, result or activity; it is temporary in duration, with a defined beginning and end. The term *management* is an activity concerned with bringing a group of people (generally more than two) together to accomplish a desired or set goal. Management can include a host of activities, some of which are discussed in subsequent sections. *Project management* may be viewed as the process of "managing" multiple related projects. Alternately, project management has been defined as a discipline involved with the planning, organizing, securing, and managing resources to achieve specified goals. Thus, the challenge of project management is to achieve all (where possible) of the project goals and objectives. Perhaps key to project management is selecting the right/best projects, the right personnel and then applying appropriate project management techniques and approaches.

Project participants are referred to as the project team. In effect, it is the management team that leads the project and provides all the necessary services to the project. The project manager should be a professional in the field of project management; he/she has the responsibility of the planning, execution, and closing of the project. This individual is primarily responsible for the project plan, defined as the formal, approved document used to guide both *project execution* and *project control*. It documents planning assumptions and decisions, facilitates communication among team members, and documents approved scope, cost, and schedule *baselines*.

Another important element in project management is *risk management*. Its objective is to reduce different risks related to various activities to an acceptable level. It may be related to numerous other types of risks arising because of the environment, technology, humans, organizations, social constraints, politics, etc. Another important element in project management is the risk management associated with the probability of specific eventualities [1]. The risk may be related to financial or environmental considerations in order to reduce different risks related to a preselected activity to a level accepted by society.

It should be noted that there have been several attempts to develop international management standards. The International Organization for Standardization (ISO) develops ISO standards; ISO 9000 are a family of standards for quality management systems. This group was founded in 1947 and is responsible for standardizing everything from paper size to film speeds. (The American National Standards Institute, ANSI, is the US representative to ISO.) The standards are a voluntary series of guidelines designed to address management. They focus on management on the belief that correct management will lead to better performance.

The chapter consists of 8 additional sections. The presentation to follow will address the following (some of which are noted above) project management topic areas.

1. Managing Project Activities
2. Initiating
3. Planning/Scheduling
4. Gantt Charts
5. Executing/Implementing
6. Monitoring/Controlling
7. Completion/Closing
8. Reports.

13.2   Managing Project Activities

There are a number of approaches to managing project activities, including agile, interactive, incremental, and phased approaches. Each requires careful consideration with respect to overall project objectives, timeline, and cost, as well as the roles and responsibilities of all participants and stakeholders. The traditional phase activity involves a "railroad" approach that consists of 6 developmental components of the project:

1. Initiation
2. Planning and design
3. Execution and construction

4. Monitoring/controlling
5. Completion/closing
6. Reports.

Not all projects will involve every stage, as some projects can be terminated before they reach completion. Some projects may not follow a simple structured (or railroad according to one of the authors) planning and/or monitoring process. Some projects are interactive and will go through steps 2, 3, and 4 multiple times. It should be noted that many industries use variations of this approach. For example, these stages may be supplemented with decisions (go/no go) at which point the project's continuation is defined and decided.

Another relatively simple project management approach could take the following form [1].

1. Develop an overall project schedule, work plan, and task outline(s)
2. Develop project budget estimates
3. Solicit and evaluate equipment vendor proposals
4. Compare capital and operating costs
5. Prepare cost/benefit analyses and understand the sensitivity of these analyses relative to the factors involved
6. Monitor and report project status with respect to both schedule and budget
7. Prepare a (final) report.

13.3   Initiating

The initiating process generally determines the nature and scope of the project. If this stage is not performed well, it is unlikely that the project will be successful in meeting its objectives. Following the selection of the project manager and team members, this stage should include a plan that addresses the following areas:

1. Project objectives
2. Analyzing the business needs/requirements in terms of measurable goals
3. Reviewing present operations
4. Financial analysis of the costs and benefits
5. Stakeholder analysis, including users, and support personnel for the project.

Projects have beginnings and ends. The project schedule or cycle is often a reflection of what should be accomplished and what options are available. Since projects arise out of need, the whole project management process begins when someone or something has a need to be fulfilled. When it comes to project selection, choices have to be made by the group. Some projects are selected, while others are rejected. Decisions are made on the basis of available resources, the number of needs which must be addressed, the cost of fulfilling those needs, and the relative importance of satisfying one set of needs and ignoring others.

The key player during this initiating stage is the aforementioned project manager. Since (in a general sense) project management is the planning, organizing, directing, and controlling of company resources for a relatively short-term objective that has been established to complete specific goals and objectives, the project manager is responsible for coordinating and integrating activities related to the project in question. The two most important roles of the project manager are to select the team and complete the project on time, within budget, and according to any specification(s). His/her basic roles are:

1. Technical supervision
2. Initial project planning
3. Project team organizing
4. Project team direction
5. Controlling technical quality, schedule and costs
6. Early communication with client(s) and vendor(s).

The strong project manager is the nucleus of the project and can ensure the success of the project. However, individual experts must be coordinated accordingly. Team managers are often asked to deal with specified constraints of time and dollars, sometimes under great stress. The project manager needs to give technical guidance, management expertise, plus enthusiasm and support to the team. Interestingly, the project manager often does not have the authority to pick his or her own team, but it is the authors' contention that he/she should be given this directive.

13.4   Planning/Scheduling [1–4]

This obviously is the most important stage of the project, and understandably receives the bulk of the treatment. Planning is conducted throughout the duration of the project. Project milestones are identified, and tasks are laid out. Many tools exist to assist the project manager in devising the formal project plan: work-breakdown structures, Gantt charts (see next section), network diagrams, resource allocation charts, cumulative cost distributions, etc. As the project is carried out, the plan may undergo considerable modification as it adapts to unanticipated circumstances. Project management plans are *time-variable*, allowing the project staff to manage change in an orderly and systematic fashion.

The primary objective of/at this stage is to plan time, cost, and resources adequately in order to estimate the work required and to effectively manage any potential risk(s) during project execution. As with the initiation step, a failure to adequately plan greatly reduces the project's chances of successfully accomplishing its goals.

Project planning generally includes the following 10 steps:[4]

1. Approval from management to initiate the project
2. Identifying deliverables and creating the work breakdown structure
3. Identifying the activities needed to complete deliverables
4. Networking the activities in a logical order
5. Estimating the resource requirements for the project, including time and cost
6. Developing the project schedule
7. Developing the project budget
8. Investigating potential risks
9. Identifying all roles and responsibilities
10. Conducting an initial get-together meeting.

Any work plan should also include:

1. What is to be done?
2. Who will do it?
3. When is it to be done?
4. How much will it cost?

Time, money, human and material resources all come into play at this time. Time (management) is handled normally through the use of schedules. Money is handled by budgets which indicate how project funds are to be allocated, while human and material resources are concerned with how best to allocate (potentially limited) resources on projects.

13.5   Gantt Charts

An integral part of project planning involves scheduling the project tasks in such a way that the project is carried out both logically and efficiently. The schedule serves as a master plan from which the client, as well as management, can have an up-to-date picture of the progress of the project. Schedules (as noted earlier) normally include lists of tasks and activities, dates when those tasks are to be performed, duration of those tasks, and other information related to the timing of project activities.

The information discussed above can be displayed in several ways. The most popular form is the bar chart, originally developed by Gantt (which carries his name). The Gantt chart is a type of bar chart that illustrates a project schedule. It includes the beginning and ending dates of the terminal elements and summary elements of a project. Terminal elements and summary elements comprise the work breakdown structure of the project. Figure 13.1 is a crude type of Gantt chart. By reading from the horizontal axis, the project staff will know the planned start and finish dates for different tasks. The chart is also useful for project control when actual achievements vary, thereby enabling the determination of schedule variance. Figure 13.1 shows, for example, that the project is off schedule from the very beginning, when Task 1 begins later than planned. Note that the actual duration of Task 1 is equal to the planned duration, so the schedule slippage is entirely accounted for, despite the fact that it started late. With Task 2, it is clear that not only did the task begin late, but it took longer to accomplish than planned. Schedule slippage here could be caused by

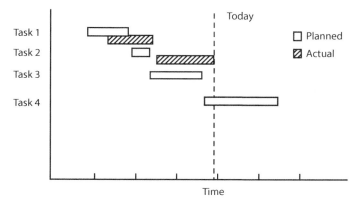

Figure 13.1 Gantt chart

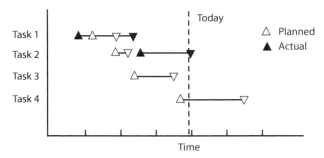

Figure 13.2 Gantt chart, alternate form

both the late start and sluggish performance that stretched out the task's planned duration.

Figure 13.2 presents a slightly different approach to the employment of a Gantt chart. The basic facts here are *identical* to those given in Figure 13.1. However, they are presented in a different manner. With this approach, planned start dates are pictured as hollow upright triangles, and actual finish dates are solid downward-pointing triangles. A comparison of the two charts indicates that they both give the same information.

These Gantt charts are widely used for the planning and control of schedules on projects. Their past and present popularity lies in their simplicity [5].

13.6   Executing/Implementing [1,4]

When a formal plan has been devised, the project is ready to be carried out. In a sense, implementation is at the heart of a project, since it entails doing the things that need to be done, as spelled out in the aforementioned project plan. The executing consists of the processes used to complete the work defined in the project plan in order to accomplish the project's objectives. The execution process involves coordinating people and resources, as well as integrating and performing the activities of the project management plan. The deliverables are produced as outputs from the process(es) performed as defined in the project management plan and other frameworks that might be applicable to the type of project at hand.

This stage also involves direct (or indirect) communication between all members of the project team and with upper-level management. The procurement process must be clearly spelled out and adhere to. Quality assurance/quality control (QA/QC) needs to be implemented and this involves both this and the next stage to be discussed.

The implementation phase of a project essentially consists of performing the actual work to complete the project. It is usually the project manager who performs a variety of functions to ensure the project is effectively implemented, considering the constraints of schedule and budget.

Successful project completion is dependent on a number of factors, including, but not limited to the following eight:

1. Updating the project plan, as necessary
2. Staying within the scope of work specified by the project plan
3. Getting authorization for changes
4. Providing deliverables in stages
5. Conducting project review meetings
6. Frequently checking the technical work being performed in order to provide guidelines and direction as needed
7. Keeping up-to-date project information
8. Reviewing the performance of the project staff in order to assist if help is necessary.

Other techniques to assist in effective project implementation is to conduct regular meetings. Periodic review meetings are an integral component of project implementation and maintaining control over project execution. The general purposes for these project meetings are quality review, schedule review, cost review, and keeping the project team informed. Review meetings with appropriate corporate and functional managers as well as the customer (if applicable) to review the current status of the project are essential.

13.7   Monitoring/Controlling [1,4]

Monitoring and controlling consists of those processes performed to observe project implementation so that any problems can be identified at an early stage and in a timely manner so that corrective action can be taken. The main objective is that project performance is observed and measured regularly to identify any variances from the project management plan. These control systems are required not only for cost, communication, time changes and procurement issues, but also for risk, quality, and human resources. Thus, this stage may be viewed as an independent element in the overall plan. In multi-phase projects, the monitoring and control process also provides feedback between project phases in order to implement corrective or preventive actions to bring the plan into compliance with the objective of the project management plan.

Project control is based on establishing an effective project plan with cost, time, and technical baselines. During the aforementioned implementation phase, control is accomplished by measuring, evaluating, and reporting actual performance against these baselines. It is important for the chemical engineer to understand and realize that project management is a team effort, and must rely on the project team to successfully complete their project activities; this will help maintain control of the project.

There are many functions that the project manager may perform to help keep project control. These include the following five assignments:

1. Frequently review project team activities
2. Provide the support the team needs to complete assigned activities
3. Inform the project team of the status of the project, and what needs to be done next
4. If necessary, prioritize tasks for team members
5. Communicate problems or changes to the customer, management and/or the project team as soon as possible.

While the basis of project control is based on good project planning and monitoring, project execution and control is facilitated by using effective control methods such as:

1. Updating schedule and cost estimates
2. Conducting project team meetings
3. Using work authorizations.

13.8   Completion/Closing

Projects ultimately come to an end. Sometimes this end is abrupt and premature, as when it is decided to kill a project before its scheduled termination. However, it is always hoped that the project will meet a more positive ending. In any case, when projects end, the project manager's responsibilities almost always continue. There are various closeout duties to be performed. For example, if equipment was used, this equipment must be accounted for and possibly reassigned to new uses. On contracted projects, a determination must be made as to whether the project deliverables satisfy the contract. Final reports may have to be written, a topic discussed in the last section of this chapter.

The closing procedure includes the formal acceptance of the project and the ending thereof. Administrative activities include the archiving of the files and documenting any lessons learned. In fact, this final phase consists of three closures.

1. Finalize all activities across all of the process groups in order to formally close the project or project phase.
2. Complete and settle each contract (including the resolution of any unresolved items) and terminate each contract applicable to the project (or project phase).
3. Release all members of the team from their assignments.

13.9   Reports

The most important project management report is the *final* report. These reports can take any of several forms. As noted in Chapter 11, the most important part of a final report is the ABSTRACT or what some refer to as EXECUTIVE SUMMARY. *This is the most important part of the report.* It should briefly summarize (without referring the body of the report) the project's description and objectives, project activities and the results/findings/inclusions. However, there are other earlier reports, and consideration should be given to appending them to the final report. Those other reports generally fall into three categories: *progress reports*, *contact reports*, and *project status reports*.

Project progress reports are one of the best ways to track the progress of a project; these reports are used to monitor the activities and compare them to schedule and budget. Periodic progress reports should contain the information on the status of project activities, schedule, budget, goals achieved, goals not achieved, goals due, important meetings, correspondence, release or delivery of deliverables, other reports, and (if applicable) equipment specifications and designs.

As noted earlier, contact reports are used to document project related communication with the client. They are particularly useful in documenting any changes in project schedule, feedback on deliverables, and changes in project scope. Contact reports are normally sent to project team members and functional managers. They provide an excellent opportunity to keep everyone abreast of project progress and project status. Contact reports are also used to confirm telephone conversations with clients. When sent to the client, they place in writing an understanding of the client's comments and planned actions. When used in this manner, contact reports minimize misunderstandings. However, it is not necessary or desirable to document everything in contact reports; to do so can undermine their value. Routine project communication can also be recorded on a contact/telephone log.

A detailed and expanded treatment of project management is available in the following two references.

1. L. Theodore, *Chemical Engineering: The Essential Reference*, McGraw-Hill, New York City, NY, 2014.
2. H. Kerzner, *Project Management*, 2nd edition, Van Nostraud Reinhold, New York City, NY, 1984.

13.10   Illustrative Open-Ended Problems

This and the last section provide open-ended problems. However, solutions *are* provided for the three problems in this section in order for the reader to hopefully obtain a better understanding of these problems which differ from the traditional problems/illustrative examples. The first problem is relatively straightforward while the third (and last problem) is somewhat more difficult and/or complex. Note that solutions are not provided for the 37 open-ended problems in the next section.

Problem 1: Provide a brief description of the potential role(s)/activities of a chemical engineer in a management project.

Solution:

1. Selects the personnel for the technical portion of the proposal work
2. Develops a general scope of work assignment
3. Assigns individuals to specific tasks
4. Develops man-hour estimates
5. Prepares the technical portion of any initial proposal
6. Prepares the technical portion of intern reports
7. Prepares the technical portion of the final report
8. Ensures the completeness and accuracy of any technical responses
9. Reviews cost estimates
10. Assists the project manager in coordinating all technical activities of the project
11. Suggests candidates for any additional technical work
12. Reviews the cost estimate of the project
13. Reviews inquiries from a legal or contractual perspective.

Problem 2: Conflicts. Conflicts. Conflicts. They are almost always an integral part of the project management process. List some of the potential conflicts.

Solution: Kerzner [6] provides a list of some of these conflicts.

1. Technical abilities
2. Managerial abilities
3. Manpower availability
4. Equipment and facilities availability
5. Capital expenditures
6. Peripheral costs
7. Technical opinions
8. Technical trade-offs
9. Project priorities
10. Administrative procedures
11. Schedule
12. Responsibilities
13. Personality clashes
14. Financial compensation
15. Recognition of efforts.

Problem 3: Estimating the time requirements for a project activity occasionally is a difficult task. The general procedure to provide a best-estimate for each activity is relatively simple. This best-estimate is almost always based on past experience and (what has come to be defined as) good engineering judgement. One of the authors [7] has advocated, and previously employed, a Delphi Panel method. (This same author has also modestly referred to this method as the Theodore Panel Approach (TPA).) Describe this approach to estimating risks and suggest a method that *quantitatively* provides information on the variance of the estimate.

Solution: At the simplest level, a group of project members or experts are brought together to discuss a risk valuation in order to reach a consensus as to its most appropriate value. The procedure is iterative, with feedback between iterations, and involves 6 steps once the experts have been chosen. These six steps are as follows: [7]

1. Select, in isolation, independent estimates of the risk *and* reasons/justification for the selected value.
2. Provide these initial results and reasons to the other experts.

3. Allow each expert to reverse his or her initial estimate and to provide the reasoning for any change to the initial value.
4. Repeat Steps 1 through 3 until a "consensus" value is approached.
5. Use the average of the final estimate as the best estimate of the risk.
6. Use the standard deviation of estimates as a measure of the uncertainty.

In effect, the experts get locked in separate rooms, providing independent judgments, until some approach to convergence is achieved. Naturally, the experts (panelists) must be willing to share their knowledge, experience, and information with each other if this effort is to be successful.

Kerzner [6] suggest applying (perhaps in conjunction with the method described above) the following procedure to generate time information

1. *The most optimistic completion time.* That should occur approximately 1 percent of the time.
2. *The most pessimistic completion time.* That should also occur approximately 1 percent of the time.
3. *The most likely completion time.*

The expected time between any project event period can then be found from the expression:

$$\bar{t} = \frac{a + 4t + b}{6} \tag{13.1}$$

where $\bar{t}$ = expected time
  a = most optimistic time
  b = most pessimistic time
  t = most likely time

In order to calculate the probability of completing the project on time, the standard deviation of each activity in the project must be known. This can be found from the expression:

$$s = \frac{b - a}{6} \tag{13.2}$$

where s is the standard deviation associated of the expected time for a particular activity, $\bar{t}$. Another useful expression is the variance, s^2, which is the square of the standard deviation [8]. For a multi-time phase one must calculate the square root of the sum of the squares of each of the time activity standard deviations. For example, assume there are four phases—A, B, C, D—to a project with corresponding time activities (low, medium, high) in months:

A: 1 – 2 – 3

B: 3 – 4 – 5

C: 6 – 10 – 14

D: 2 – 4 – 6

The best estimate for the project is obtained by first calculating s_i for each activity employing Equation (13.2) and applying Equation (13.3)

Thus

$$s_A = \frac{3-1}{6} = 0.33$$

$$s_B = \frac{5-3}{6} = 0.33$$

$$s_C = \frac{14-6}{6} = 1.33$$

$$s_D = \frac{6-2}{6} = 0.66$$

$$s = \sqrt{\sum (A_i)^2}$$

and

$$s = \sqrt{(0.33)^2 + (0.33)^2 + (1.33)^2 + (0.66)^2}$$
$$= \sqrt{0.111 + 0.111 + 1.769 + 0.444} = \sqrt{2.435}$$
$$= 1.560$$

If the project time is normally distributed, there exists a 68 percent chance of completion within on standard deviation, 95 percent within two standard deviations, and 99 percent within three standard deviations. This is presented in Table 13.1, noting that the initial average estimate is 20, i.e., (2+4+10+4). If there appears the possibility that the project might be extended beyond the expected 20 months, then additional costs will arise with accompanying cost overruns.

Table 13.1 Statistical Estimates of Total Time in Months

	PERCENTAGE	EXPRESSION	RANGE
s	68	20 ± 1.56	18.44 – 21.56
$2s$	95	20 ± 3.12	16.88 – 23.12
$3s$	99	20 ± 4.68	15.32 – 24.68

13.11   Open-Ended Problems

This last section of the chapter contains open-ended problems as they relate to project management. No detailed and/or specific solution is provided; that task is left to the reader, noting that each problem has either a unique solution or a number of solutions or (in some cases) no solution at all. These are characteristics of open-ended problems described earlier.

There are comments associated with some, but not all, of the problems. The comments are included to assist the reader while attempting to solve the problems. However, it is recommended that the solution to each problem should initially be attempted *without* the assistance of the comments.

There are 37 open-ended problems in this section. As stated above, if difficulty is encountered in solving any particular problem, the reader should next refer to the comment, if any is provided with the problem. The reader should also note that the more difficult problems are generally located at or near the end of the section.

1. Describe the early history associated with project management.
2. Discuss the recent advances in project management.
3. Discuss how project management can impact the career of a chemical engineer.
4. Select a refereed, published article on project management from the literature and provide a review.
5. Provide some normal everyday domestic applications involving the general topic of project management.
6. Develop an original problem in project management that would be suitable as an illustrative example in a book.
7. Prepare a list of the various books which have been written on project management. Select the three best and justify your answer. Also select the three weakest books and, once again, justify your answer.

8. One of the major concerns with project management is cost control, particularly as it applies to cost overruns. Cost control is important to all projects, regardless of size. Small projects generally have tighter monetary controls, mainly because of the smaller risk associated with the failure of the project. Large companies may have the luxury to distribute project losses over several projects. Cost control is primarily concerned with the monitoring of costs and it often involves the recording of massive quantities of data. It is recommended that cost control be performed by all personnel who incur costs—not only the project officer.
Cost control activities generally include:
   - Cost estimating
   - Cost accounting
   - Project cash flow
   - Company cash flow
   - Direct labor and any overhead costing
   - Miscellaneous costs.
Detail and describe some of these activities

9. Financial compensation and/or rewards are important to the morale and motivation of people in any project management endeavor. However, there are several issues that often make it necessary to treat compensation issues related to project personnel separately from the rest of the organization. Explain why this dilemma exists and outline how it can be best addressed.

10. Refer to Problem 2 in the previous section. Provide procedures/suggestions that can reduce or eliminate the listed conflicts.

11. Refer to Problem 2 in the previous section. As one might suppose, there are numerous other categories of conflict. List some of the other potential conflicts and provide procedures/suggestions that can reduce their impact. In effect, how can conflicts be best managed?

12. Four of the major causes for the failure of a proposed project management include:
   - Selection of a project that is not appropriate for the organization and/or team.
   - Selection of the wrong individual as project manager.
   - Selection of the wrong individual(s) for management staff.
   - Upper management is not supportive.

Discuss how these potential failures can be reduced.

13. Refer to the previous problem. There are obviously other causes for a project to fail. Detail at least four other causes and suggest approaches that can be implemented to reduce the impact of these causes.

14. Kerzner [6] and Avots [9] have indicated that more can often be learned from failure than from success. They list the following 10 lessons that can be learned from project failure.

- When starting off in project management, plan to go all the way.
- Don't skimp on the project manager's qualifications.
- Do not spare time and effort in laying out the project groundwork and defining work.
- Ensure that the work packages in the project are of proper size.
- Establish and use network planning techniques, having the network as the focal point of project implementation.
- Be sure that the information flow related to the project management system is realistic.
- Be prepared to adjust assignments continually to accommodate frequent changes on dynamic program
- Whenever possible, tie together responsibility, performance, and rewards.
- Long before a project ends, provide some means for accommodating the employees' personal goals.
- If mistakes in project implementation have been made, make a fresh try.

Provide additional explanatory detail(s) on these lessons.

15. Project management analysis via the matrix approach has received significant attention in this field. Discuss the advantages of employing a matrix-type approach.
Comment: Refer to the analysis provided by Kerzner [6].

16. Project management analysis via the matrix approach has received significant attention in this field. Discuss the disadvantages of employing a matrix-type approach.
Comment: Refer to the analysis provided by Kerzner [6].

17. One of the main problems that arises during a project is that a team member (including the project manager) may have more "loyalty" elsewhere within the organization. Discuss how this can best be addressed.

18. Most project managers often request that the project evaluation process be different from how they are usually evaluated. Is this a reasonable response?

19. ISO 21500:2012, *Guidance on Project Management*, was published in 2012. Although ISO (International Organization for Standardization) 10006, *Quality Management Systems—Guide—lines for Quality Management in Projects*, has been around since 2003, it focuses only on the application of quality management practices and principles to project management. ISO 21500 is the first ISO standard to address the complete project management process. Discuss the key features of ISO 21500.

20. The Project Management Institute (PMI) released the 5th Edition of the *A Guide to the Project Management Body of Knowledge* (PMBOK, an ANSI Standard) in early 2013. Although there are some differences between this and earlier editions of PMBOK, this edition is not a rework of the practices detailed in the fourth edition. It has been reported [10] that the fifth edition builds on the fourth edition to add some new material while consolidating and reorganizing select existing material. Provide you review of the 5th edition.

21. Elam [10] has indicated that there is excellent agreement between the 5th edition of PMBOK and ISO 21500:2012, the result of PMI's active involvement in shaping the new ISO standard on project management, and that project managers will find little difference between the two standards. Are his comments reasonable?

22. The company president has requested that you provide a two hour seminar to all technical employees on the merits of the traditional project management approach. Provide a PowerPoint (P2) presentation that you feel would not only satisfy the president but also be beneficial to the organization's employees.

23. There are various codes of ethics for chemical engineers, engineers in general, and scientists. (See also Chapter 22). Prepare a code of ethics for project managers.

24. Describe the difference between project management and product management. Also comment on how these differences will/may change in the future.

25. Suggest how the Gantt "chart" approach can be improved.

26. It has been reported that lifestyles in the U.S. will change drastically in the future. Can this change impact project management? Justify your answer.

27. Describe the role government and government-related activities has, is, and will influence project management.

28. One of the authors [11] has recommended that the following approach be employed in purchasing equipment.

- Refrain from purchasing any equipment without reviewing *certified independent test data* on its performance under a similar application. Request the manufacturer to provide performance information and design specifications.
- In the event that sufficient performance data are unavailable, request that the equipment supplier provide a small pilot model for evaluation under existing conditions.
- Consider requesting participation of an "authority" in the decision-making process.
- Prepare a good set of specifications. Include a *strong performance guarantee* from the manufacturer to ensure that the control equipment will meet all applicable local, state, and federal codes at specific process conditions.
- Closely review the process and economic fundamentals.
- Make a careful material balance study before authorizing or purchasing equipment.
- Refrain from purchasing any equipment until *firm* installation cost estimates have been added to the equipment cost. *Escalating installation costs are the rule rather than the exception.*
- Give operation and maintenance costs high priority on the list of equipment selection factors.
- Refrain from purchasing any equipment until a solid commitment from the supplier(s) is obtained. Make every effort to ensure that the new system will utilize equipment fuel, controllers, filters, motors, etc, that are compatible with those already available at the plant.
- The specification should include written assurance of *prompt* technical assistance from the equipment supplier. This, together with a complete operating manual

(with parts list and full schematics), is essential and is too often forgotten in the rush to get the equipment operating.

- Schedules, particularly on projects being completed under a court order or consent judgment, can be critical. In such cases, delivery guarantees should be obtained from the manufacturers and penalties identified.
- If applicable, equipment should be of fail-safe design with built-in indicators to show when performance is deteriorating [7].
- Withhold 10 to 15% of the purchase price until satisfaction is clearly demonstrated.

The above procedure was developed nearly a half century ago. You have been assigned the task of both updating and improving the above procurement procedure.

29. As a purchasing agent for D'Aquino Industries, you have been requested to develop a procedure to justify decisions regarding the purchases of equipment. The procedure should be of a *quantifiable* nature. See also the previous problem.

30. As a plant manager, would you consider maintenance or inspection more important?

31. Discuss the role economics plays on OM&I activities.

32. Tiffany, a consultant for Theodore Partners, was recently hired by a cosmetic firm to develop an OM&I procedure for their recently designed and constructed nanotechnology facility. The plant generates nanoparticles for several of their cosmetic products. (See also Chapter 20). Outline a procedure.

33. Discuss the role operation and maintenance plays in the purchase of equipment and equipment parts. (See also Chapter 10).

34. Which of the various mass transfer devices has the greatest OM&I problems? Justify you answer (See also Chapter 7).

35. ATE (Abulencia-Theodore Enterprises) consultants have been assigned to a management project concerned with the development and marketing of a new nanotechnology [12,13] product. After setting the schedule for the project, the consultants are informed that the due date for the two year project has been reduced to one year. Describe in qualitative terms the impact this change could have on the overall success/performance of the project.

36. One of the authors [11] has indicated the following. Many design/procurement/construction/startup problems can be compounded by any one or combination of the following:
    - Unfamiliarity of process engineers with engineering
    - New and changing codes
    - New suppliers with frequently unproven equipment
    - Lack of industry standards in some key areas
    - Interpretations of control agency field personnel
    - Compliance schedules that are too tight
    - Vague specifications
    - Weak guarantees for the new equipment
    - Unreliable delivery schedules
    - Process unreliability problems.

    You have been assigned the task of outlining how to remove or eliminate some of these project-related problems.

37. ATE consultants has developed three possible approaches to satisfying the objectives of a project. The four scenarios from a cost-schedule perspective are presented in Figure 13.3. Quantitatively discuss the pros and cons associated with approaches (A), (B), (C), and (D).

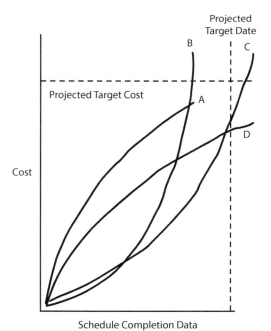

Figure 13.3 Cost-schedule analysis, problem 37.

References

1. Personal notes, L. Theodore, East Williston, NY, 2006.
2. L. Theodore, *The Challenges of Change*, CEP, New York City, NY, January 2007.
3. L. Theodore, Letter to the Editor (author response*), Changing the ChE Curriculum: How Much Is Too Much?* , CEP, New York City, NY, March 2007.
4. Adapted from: L. Theodore, *Chemical Engineering: The Essential Reference*, McGraw-Hill, New York City, NY, 2014.
5. J. Davidson, *Managing Projects in Organizations*, Jossey-Bass Publishers, location unknown, 1998.
6. H. Kerzner, *Project Management*, 2nd edition, Van Nostraud Reinhold, New York City, NY, 1984.
7. L. Theodore and R. Dupont, *Environmental Health and Hazard Risk Assessment: Principles and Calculations*, CRC Press/Taylor & Francis Group, Boca Raton, FL, 2012.
8. S. Shaefer and L. Theodore, *Probability and Statistics Applications in Environmental Science,* CRC Press/ Taylor & Francis Group, Boca Raton, FL, 2007.
9. I. Avots, *Why Does Project Management Fail?*, California Management Review, Vol. 12, pp 78-82, location unknown, 1969.
10. D. Elam, *Standardized Project Management*, EM, Pittsburgh, PA, March 2013.
11. L. Theodore, *Air Pollution Control Equipment Calculations*, John Wiley & Sons, Hoboken, NJ, 2008.
12. L. Theodore and R. Kunz, *Nanotechnology: Environmental Implications and Solutions*, John Wiley & Sons, Hoboken, NJ, 2005.
13. L. Theodore, *Nanotechnology: Basic Calculations for Engineers and Scientists*, John Wiley & Sons, Hoboken, NJ, 2007.

14

Environmental Management

This chapter is concerned with process environmental management. As with all the chapters in Part II, there are several sections: overview, several technical topics, illustrative open-ended problems, and open-ended problems. The purpose of the first section is to introduce the reader to the subject of environmental management. As one might suppose, a comprehensive treatment is not provided although several technical topics are also included. The next section contains three open-ended problems; the authors' solution (there may be other solutions) are also provided. The final section contains 32 problems; *no* solutions are provided here.

14.1   Overview

This overview section is concerned with —as can be noted from its title— environmental management. As one might suppose, it was not possible to address all topics directly or indirectly related to environmental management. However, additional details may be obtained from either the references provided at the end of this Overview and/or at the end of the chapter.

Note: Those readers already familiar with the details associated with this subject may choose to bypass this Overview.

For better than four decades there had been an increased awareness of a wide range of environmental issues covering all sources: air, land, and water. More and more people are becoming aware of these environmental concerns, and it is important that chemical engineers, many of whom do not possess an understanding of environmental problems or have the proper information available when involved with environmental issues, develop capabilities in this area. All professionals, not only chemical engineers, should have a basic understanding of the technical and scientific terms related to these issues as well as the regulations involved. Hopefully this chapter will serve the needs of the chemical engineer—increasing his or her awareness of (and help solve) the environmental problems facing society.

The current human population on Earth is approximately 7 billion, and it almost certainly will increase in the future. The influence and effects of human activities on the environment have become increasingly evident at the local, state, national, and international levels, while issues of environmental degradation, health, safety, green chemistry and engineering, sustainability, etc., have become more pervasive and pressing. The net result is that there has been a more significant increase in both awareness and interest about the environment since the turn of this century.

Section titles and contents of this chapter are:

1. Environmental Regulations
2. Classification, Sources, and Effects of Pollutants
3. Multimedia Concerns
4. ISO 14000
5. The Pollution Prevention Concept
6. Green Chemistry and Green Engineering
7. Sustainability

Note that parts of the material in this Chapter have been adapted from:

1. M.K. Theodore and L. Theodore, *Introduction to Environmental Management* CRC Press/Taylor & Francis Group, Boca Raton, FL, 2010 [1].

2. G. Burke, B. Singh, and L. Theodore, *Handbook of Environmental Management and Technology*, John Wiley & Sons, Hoboken, NJ, 1992 [2].

The reader should also note that the subject of communicating risk was not addressed, although it does receive treatment in both of the above references. As with most environmental issues of a volatile and personal nature, public perception of risk is unfortunately often based on fear rather than facts. The strategy to employ in communicating risk is therefore an important area of concern in environmental management.

14.2   Environmental Regulations [3]

Environmental regulations are not simply a collection of laws on environmental topics. They are an organized system of statutes, regulations, and guidelines that minimize, prevent, and punish the consequences of damage to the environment. This system requires each individual – whether an engineer, physicist, chemist, attorney, or consumer – to be familiar with its concepts and case-specific interpretations. Environmental regulations deal with the problems of human activities and the environment, and the uncertainties of laws associated with them.

The National Environmental Policy Act (NEPA), enacted on January 1, 1970 was considered a "political anomaly" by some. NEPA was not based on specific legislation; instead, it referred in a general manner to environment issues and quality of life concerns. The Nixon Administration at that time became preoccupied with not only trying to pass more extensive environmental legislation, but also in implementing the laws. Nixon's White House Commission on Executive Reorganization proposed in the Reorganizational Plan #3 of 1970 that a single independent agency be established separate from the Council for Environmental Quality (CEQ). The plan was sent to Congress by President Nixon on July 9, 1970, and this new U.S. Environmental Protection Agency (EPA) began operation on December 2, 1970. The EPA was officially born.

In many ways, the EPA is the most far-reaching regulatory agency in the federal government because its authority is very broad. The EPA is charged to protect the nation's land, air, and water systems. Under a mandate of national environmental laws, the EPA continues to strive to formulate and implement actions that lead to a compatible balance between human activities and the ability of natural systems to support and nurture life [1].

The EPA works with both the states and local governments to develop and implement a comprehensive environmental program. Federal laws such as the Clean Air Act (CAA), the Safe Drinking Water Act (SWDA), the Resource Conservation and Recovery Act (RCRA), and the Comprehensive Environmental Response Compensation, and Liability Act (CERCLA), etc., [3] all mandate involvement by state and local government in the details of implementation.

Pollution Prevention covers domestic and primarily industrial means of reducing pollution. This can be accomplished through (a) proper residential and commercial building design; (b) proper heating, cooling, and ventilation systems; (c) energy conservation; (d) reduction of water consumption; and, (e) attempts to reuse or reduce material before they becomes wastes. Domestic and industrial solutions to environmental problems can be addressed by considering ways to make homes and work-places more energy efficient as well as ways to reduce the amount of wastes generated within them. This topic receives attention from a regulatory per-spective under a separate section heading in this chapter.

14.3   Classification, Sources, and Effects Of Pollutants [1,2]

Not long ago, the nation's resources were exploited indiscriminately. Waterways served as industrial pollution sinks; skies dispersed smoke from factories and power plants; and, the land proved to be a cheap and convenient place to dump industrial and urban wastes. However, society is now more aware of the environment and the need to protect it. The American people have been involved in a great social movement known broadly as *environmentalism*. Society has thus been concerned with the quality of the air one breathes, the water one drinks, and the land on which one lives and works. While economic growth and prosperity are still important goals, opinion polls indicate overwhelming public support for pollution control and a pro-nounced willingness to pay for them. The next three paragraphs present the reader with information on pollutants and categorizes their effects.

Pollutants are various noxious chemicals and refuse materials that impair the purity of the atmosphere, water, and soil. In the authors' judgment, the area most affected by pollutants is the atmosphere or air. Air pollution occurs when wastes pollute the air. Artificially or synthetically created wastes are the main sources of air pollution. They can be in the form of gases or particulates which result from the burning of fuel to power motor vehicles and to heat buildings. More air pollution can be found in densely populated areas, e.g., urban areas. The air over largely populated cities often becomes so filled with pollutants that it not only harms the health of humans but also affects plants, animals, and materials of construction.

Water pollution occurs when wastes are dumped into the water. Which can cause disease. In the U.S., water supplies are disinfected to kill disease causing germs. The disinfection, in some instances, does not and often cannot remove all the chemicals and metals that may cause health problems in the distant future.

Wastes that are dumped into the soil are a form of land pollution that damages the thin layer of fertile soil, which is essential for agriculture. In nature, cycles work to keep soil fertile. Wastes, including dead plants and wastes from animals, form a substance in the soil called *humus*. Bacteria then decay the humus and convert it into nitrates, phosphates, and other nutrients that feed growing plants.

14.4   Multimedia Concerns [1,2]

The current approach to the environmental waste management requires some rethinking. A multimedia approach helps in the integration of air, water, and land pollution controls and seeks solutions that do not violate the laws of nature. Thus, it integrates air, water, and land into a single concern while seeking a solution to pollution that does not endanger society or the environment. The obvious advantage of a multimedia pollution control approach is its ability to manage the transfer of pollutants so they will not continue to cause pollution problems. Among the possible steps in the multimedia approach are understanding the cross-media nature of pollutants, modifying pollution control methods so as not to shift pollutants from one medium to another, applying available waste reduction technologies, and training environmental professionals in a total environmental concept. The challenge for the future environmental professional include:

1. Conservation of natural resources
2. Control of air-water land pollution
3. Regulation of toxics and disposal of hazardous wastes
4. Improvement of quality of life

It is now increasingly clear that some treatment technologies, while solving one pollution problem, have created others. Most contaminants, particularly toxics, present problems in more than one medium. Since nature does not recognize neat jurisdictional compartments, these same contaminants are often transferred across media. Air pollution control devices and industrial wastewater treatment plants prevent waste from going into the air and water, respectively, but the toxic ash and sludge that these systems produce can also become hazardous waste problems themselves. For example, removing trace metals from a flue gas during the pollution control phase usually transfers the products to a liquid or solid phase. Does this exchange an air quality problem for a liquid or solid waste management problem? The reader should ponder this question. Waste disposed of on land or in deep wells can contaminate ground water, and evaporation from ponds and lagoons can covert solid or liquid waste into air pollution [3]. Other examples include acid deposition, residue management, water reuse, and hazardous waste treatment and/or disposal.

Control of cross media pollutants cycling in the environment is therefore an important step in the overall management of environmental quality. Pollutants that do not remain where they are released or where they are deposited move from source to receptor by many routes, including air, water, and land. Unless information is available on how pollutants are transported, transformed, and accumulated after they enter the environment, they cannot realistically and effectively be controlled. A better understanding of the cross-media nature of pollutants and their major environmental processes – physical, chemical, and biological – is required.

14.5   ISO 14000 [1–8]

The International Organization for Standardization (ISO) is a private, nongovernmental, international standards body based in Geneva, Switzerland. Founded in 1947, ISO promotes international harmonization, and the development of manufacturing product and communications standards. It is a nongovernmental organization; however, governments are allowed to participate in the development of standards and many governments have chosen to adopt the ISO standards as their regulations. The ISO also closely interacts with the United Nations [4–5]. ISO has promulgated over 16,000 internationally accepted standards for everything from paper sizes to film speeds. Over 130 countries participate in the ISO as "Participating" members or as "Observer" members. The United States is a full-voting participating member and is officially represented in the ISO by the American National Standards Institute (ANSI).

Over the years the ISO has expanded the scope of their standards to incorporate areas such as the environment, energy conservation, service sectors, security, and managerial and organizational practice. The ISO's environmental mission is to promote the manufacturing of products in a manner that is effective, safe, and clean [6]. The ISO hopes to achieve this goal through the dedication and participation of more countries.

The ISO 14000 may be viewed as a generic environmental management standard. It can be applied to any organization and focuses on the processes and activities conducted by the company. It consists of standards and guidelines regarding environmental managed systems (EMSs). The idea for it first evolved from the United Nations conference on Environment and Development (UNCED), which took place in Rio de Janeiro in 1992. The topic of sustainable development (see last section in this chapter) was discussed there and the ISO made a commitment to support this subject [7].

The ISO 14000 standards were first written in 1996 and have subsequently been amended and updated. Their intended purpose is to assist companies and organizations to minimize their negative affect on the environment and comply with any laws, regulations, or environmental requirements that have been imposed on them. The ISO can also help to establish an organized approach to reducing any environmental impacts the company can control. Businesses that comply with these standards are eligible for certification. This certification is awarded by third-party organizations instead of the ISO [8].

14.6   The Pollution Prevention Concept [1–3,9]

The amount of waste generated in the U.S. has reached staggering proportions; according to the EPA nearly 3000 million metric tons of solid waste are generated annually. Although both the Resource Conservation and Recover Act (RCRA) and the Hazardous and Solid Waste Act (HSWA) encourage businesses to minimize the wastes they generate, the majority of current environmental protection efforts are centered around treatment and pollution cleanup.

The passage of the Pollution Prevention Act of 1990 has redirected industry's approach to environmental management: *pollution prevention* has now become the environmental option of the last decade of the 20th century and of the 21st century. Whereas typical waste management strategies concentrate on "end-of-pipe" pollution control, pollution prevention attempts to handle waste at the source (i.e., source reduction). As waste handling and disposal costs increase, the application of pollution prevention measures is becoming more attractive than ever before. Industry continues to explore the advantages of multimedia waste reduction and developing agendas to "strengthen" environmental design while "lessening" production costs.

There are significant opportunities for both the individual and industry to prevent the generation of waste; indeed, pollution prevention is today primarily stimulated by economics, legislation, liability concerns, and the enhanced environmental benefit of managing waste at the source. The EPA's Pollution Prevention Act of 1990 established pollution prevention as a national policy declaring "waste should be prevented or reduced at the source wherever feasible, while pollution that cannot be prevented should be recycled in an environmentally safe manner." [9] The EPA's policy establishes the following hierarchy of waste management:

1. Source reduction
2. Recycling/reuse
3. Treatment
4. Ultimate disposal

14.7   Green Chemistry and Green Engineering

Activities in the field of green engineering and green chemistry are increasing at a near exponential rate. This section aims to familiarize the reader with both green chemistry and green engineering by defining and setting principles to each; future trends are also discussed. Before beginning this section it is important that the term "green" should not be considered a new method or type of chemistry or engineering. Rather, it should be incorporated into the way scientists and engineers design for categories that include the environment, manufacturability, disassembly, recycle, serviceability and compliance. Today, the major element of "green" is to search for technology to reduce and/or eliminate waste from operation/processes, with an important priority being to not make it in the first place.

Green chemistry, also called *clean chemistry*, refers to that field of chemistry dealing with the synthesis, processing, and use of chemicals that reduce risks to humans and the environment [10]. It is defined as the invention, design, and application of chemical products and processes to reduce or to eliminate the use and generation of hazardous substances [11]. *Green engineering* is similar to green chemistry in many respects, as witnessed by the underlying urgency of attention to the environment seen in both sets of the principles [12].

14.8   Sustainability

Last, but not least, is the general subject of sustainability. The term *sustainability* has many different meanings to different people. To sustain is defined by some as to "support without collapse". Discussion of how sustainability should be defined was initiated by the Bruntland Commission. This group was assigned a mission to create a "global agenda for change" by the General Assembly of the United Nations in 1984. They defined sustainable very broadly: "Humanity has the ability to make development sustainable – to ensure that it meets the needs of the present without compromising the ability of future generations to meet their own needs" [13].

Sustainability involves simultaneous progress in four major areas:

1. Human
2. Economic
3. Technological, and
4. Environmental

Sustainability requires conservation of resources while minimizing depletion on nonrenewable resources, and using sustainable practices for managing renewable resources [14]. However, there can be no product development or economic activity of any kind without available resources. Other than solar and nuclear energy, the supply of resources is *finite*. Efficient designs conserve resources while also reducing impact caused by material extraction and related activities. Depletion of nonrenewable resources and overuse of otherwise renewable resources limits their availability to future generations.

Another important element of sustainability is the maintenance of the structure and function of the ecosystem. Because the health of human populations is connected to the health of the natural world, the issue of ecosystem health is a fundamental issue of concern to sustainable development. Thus, sustainability requires that the health of all diverse species as well as their interrelated ecological functions be maintained. As only one species in a complex web of ecological interactions, humans cannot separate their survivability from that of the total ecosystem.

Summarizing, sustainability is a term that is used with greater frequency in the environmental community and with sometimes varied meanings. In short, sustainability means the capacity to endure. A more specific definition of sustainability refers to the long-term maintenance of well-being, which has environmental, economic, and social dimensions, and includes the responsible management of resource use. Interest and concern in this area is increasing at a near exponential rate.

A detailed and expanded treatment of environmental management is available in the following three references.

1. M.K. Theodore and L. Theodore, *Introduction to Environmental Management*, CRC Press/ Taylor & Francis Group, Boca Raton, FL, 2010 [1].
2. G. Burke, B. Singh, and L. Theodore, *Handbook of Environmental Management and Technology*, 2nd edition, John Wiley & Sons, Hoboken, NJ, 2000 [2].
3. L. Theodore, *Chemical Engineering: The Essential Reference*, McGraw-Hill, New York City, NY, 2014 [15].

14.9   Illustrative Open-Ended Problems

This and the last section provide open-ended problems. However, solutions *are* provided for the three problems in this section in order for the reader to hopefully obtain a better understanding of these problems which differ from the traditional problems/illustrative examples. The first problem is relatively straight forward while the third (and last problem) is somewhat more difficult and/or complex. Note that solutions are not provided for the 32 open-ended problems in the next section.

Problem 1: Briefly describe the major thrust or goals or provisions of the following laws:

1. The National Environmental Policy Act of 1970 (as amended in 1989) (NEPA); (Public Law 91-190, 42 United States Code section 4321 et seq.)
2. The Resource Conservation and Recovery Act of 1976 (as amended in 1988) (RCRA); (Pubic Law 94-580, 42 United States Code section 6901 et seq.)
3. The Comprehensive Environmental Response, Compensation and Liabilities Act of 1980 (CERCLA or Superfund); Public Law 96-510, 42 United States Code section 9601 et seq.)

4. The Superfund Amendments and Reauthorization Act of 1986 (SARA) Title III – also known as The Emergency Planning and Community Right-to-Know Act; (Public Law 99-499, 42 United States Code section 11001 et seq.)
5. The Pollution Prevention Ace of 1990 (Public Law 101-508).

Comment: The texts of the laws may be found in the United States Code which many libraries possess. Discussions of these laws may also be found in many texts on the environment [1–3].

Solution: The description of the major thrust and the choice of goals and provisions given for each law is subjective and therefore one's answers may not exactly match those given here.

1. The National Environmental Policy Act (NEPA) requires any federal agency proposing a project that might affect the environment to prepare an *environmental impact statement*. The statement describes the project's adverse effect on the environment as well as its benefits. The statement also includes a discussion of alternatives.
2. The Resource Conservation and Recovery Act (RCRA) aims primarily to protect the environment from the mishandling of land disposal of municipal, industrial, commercial and hazardous wastes. A major part of the law relevant to pollution prevention is Subtitle C, regulations pertaining specifically to hazardous waste management. A waste is hazardous if it appears on a designated list or possesses certain defined characteristics (ignitability, corrosivity, reactivity, and/or extraction procedure toxicity) as determined by prescribed tests. Hazardous waste generators, transporters, and hazardous waste treatment/storage/disposal facilities are required to keep records, submit reports, meet standards set forth in regulations, and to obtain permits. RCRA attempts to create a *cradle-to-grave* hazardous waste management system.
3. The Comprehensive Environmental Response, Compensation and Liabilities Act of 1980 (CERCLA or "Superfund") attempts to provide the means by which abandoned hazardous waste sites may be remediated. A special tax is placed on certain chemicals to raise the needed funds. A list of hazardous waste sites needing clean-up (The National Priorities List) is generated through an evaluation of the relative risk to

public health and the environment posed by each site. The U.S. Environmental Protection Agency (EPA) is given the authority to seek reimbursement for the clean-up expenses from the owners/operators of the site and from the generators of the waste.

4. The Superfund Amendments and Reauthorization Act of 1986 (SARA) Title III- also known as The Emergency Planning and Community Right-to-Know Act, requires each state to establish local emergency planning districts and committees to develop, together with certain chemical facilities, contingency plans for handling and responding to the release of an "Extremely Hazardous Substance". The affected facilities are those that have on their premises any chemical found on a list of "Extremely Hazardous Substances" published by the EPA in an amount known as the "Threshold Planning Quantity" which is given on the same list. These facilities must submit an inventory of these chemicals and also report releases of reportable quantities of these chemicals to the emergency planning committee. Facilities must also account for the total quantity of substances brought in versus the amount shipped out (mass balance study). The difference could indicate a release of the chemical. The EPA is to maintain a database of release reports which is available to the public

5. The Pollution Prevention Act of 1990 seeks to make the reduction of the production of pollutants ("source reduction") the primary means for controlling pollution (as opposed to treatment, storage, and disposal). The EPA is to establish a source reduction program that: collects and disseminates information; provides financial assistance to states; establishes standard methods for measuring source reduction; provides a source reduction clearinghouse; creates an advisory panel of experts from government, industry and public groups; establishes training and award programs; promotes a multi-media approach to source reduction; and, establishes a special office to oversee the implementation of this law. (See also Overview).

Problem 2: Discuss the relative merits of transportation of goods by trucks and by railroad from the point of view of environmental management. Be as specific as possible in the comments regarding environmental pollution production and energy utilization requirements of each mode of transportation. Identify the type of data needed to provide a quantitative analysis of this problem.

Solution: Data needed for a quantitative analysis of this problem must be in the form of relative emission rates per unit mass transported. Emission data for loading and unloading activities should also be provided.

Merits of Truck Transportation:

1. Trucks can usually travel a more direct route for a delivery (fewer miles traveled) than trains. This is due to the fact that the infrastructure of roads and highways is much more extensive than that of the railroads. In many cases, a truck will carry an entire load to a single destination, taking the most direct route. Because the railroad is limited by its infrastructure, trains will most likely have to travel farther to reach the same destination.

2. Truck engines produce fewer emissions that railroad engines. Beginning in 1991, the EPA set emission standards for all classes of heavy-duty over-the-road trucks. The emission standards require heavy-duty diesel engine manufacturers to decrease engine emissions of particulates (carbon soot), hydrocarbons, NO_x, and CO. To meet these standards, engine manufacturers had to redesign many engine components and develop better combustion technologies. So far, the EPA has not set emission standards for railroad diesel engines. Railroad diesels are typically much larger than truck diesels and are built by different manufacturers. Because there are no emission standards for the railroad diesels, they continue to be built with less expensive, obsolete engine technology that produces higher emissions. To compare exhaust emissions on an equal basis, they need to be measured in grams per hour per horsepower produced, or tons of load delivered per kilogram of pollution per mile.

Merits of Railroad Transportation

1. Unlike trucks, trains can couple several hundred cars together and pull them with only a few large diesel engines. Because of this, trains accelerate very slowly but are able to reach and maintain crusing speeds (analogous to a large,

heavy car with an undersized engine). Because a train has a much lower power-to-weight ratio than a truck, it has a better fuel economy. To compare a train and truck on an equal basis, the fuel economy (mpg) should be calculated for both and divided by the total weight carried, resulting in mpg per ton values.

2. Trains travel at steadier speeds than trucks. Fuel economy and emissions are optimized during steady-state operating conditions with an internal combustion engine. Trucks are exposed to much more stop-and-go driving than trains during which the engine is repeatedly accelerated and decelerated. Engines get poor fuel economy during acceleration because more fuel must be used to prevent them from "bucking" or "stalling". Likewise, emissions are much higher during acceleration due to the overfueling that takes place. With a diesel engine, one can actually see the emissions (particulates) bellowing from the exhaust during acceleration.

Another section could be added discussing the proper balance of trucking and railroad use. Some would say that the policy has gone too far from the optimum in the direction of trucks or buses and make too little use of trains. The discussion could include an analysis of the optimal mix of trucking and trains to deliver goods in a fashion that maximizes pollution prevention.

Problem 3: Identify the following sources as pollutants, waste, or neither, and classify them as storage and handling or process, fugitive, or secondary emissions. Suggest ways to eliminate or reduce each source.

1. A truck delivers a load of coal to a plant. The coal is dumped into a storage bin at the site, generating coal dust.
2. A solvent bath has a lid that is closed during downtime when the bath is in use. The liquid is exposed to the air.
3. Spent solvent from a bath must be replaced periodically. The used solvent is stored in a drum until it is removed from the site.
4. Solvent is emitted from the stack of a chemical plant. In the atmosphere it undergoes chemical reactions and may contribute to aerosol (smog) formation. Classify both the solvent and the aerosol.

5. In a plating operation, a part is dipped in a chemical bath, and the dipped piece is transferred to a rinse bath several feet away. During the transfer, the dipped piece drips chemicals onto the plant floor ("dragout"). Most of the resulting chemical spill evaporates and the rest is washed down the drain, which eventually finds its way into the city sewer system.

6. A pump through which an organic solvent is flowing has a small, undetected leak.

7. A sulfuric acid solution is used to regenerate an ion exchange column.

8. A sulfuric acid stream is neutralized by mixing with a caustic stream from another part of the process. Classify both the reagent streams and the neutralized stream from the process.

Comment: Refer to the literature for additional information [1,2,9].

Solution: One person's solution to the eight categories is presented below.

1. The coal dust is a storage and handling emission. An accurate estimate of the amount of dust generated may be very difficult to make, as are suggestions for reducing this source. However, proper enclosure of the transfer point and exhausting the dust to a control system may minimize the emission rate. This dust contributes to the ambient particulate loading and as such is considered a pollutant. (Emission factors exist for this source).

2. The evaporating solvent is classified as a fugitive emission. An estimate of the magnitude of this emission can be made by comparing volumes of fresh and spent solvent in the bath. One possibility is for the development of an automated system that raises the lid only for purposes of inserting or removing parts. It may also be possible to reduce evaporation by temperature control of the bath (providing the bath is effective at reduced temperatures) or of the air space above the solvent liquid surface, i.e., a vapor condenser. Another option might be to install an exhaust hood to remove fumes, which can then be treated (for example, absorbed before the air is vented to the atmosphere). The evaporated solvent is considered an air pollutant.

3. The spent solvent is a process emission and may be classified as a hazardous waste, depending upon its chemical

composition. The amount of solvent waste may be reduced by recovering the solvent and reusing it (via distillation or other appropriate methods). It may be possible to replace the solvent with one that is not environmentally hazardous.

4. The solvent emitted from the stack (as a gas) is a process emission, and is classified as a pollutant. The amount of solvent lost in this way may be reduced by condensation of the solvent and separation from the stack gas. The aerosol is classified as a secondary pollutant. This term refers to pollutants that result from further reaction of primary emissions; frequently, secondary pollution occurs at different locations (downwind, for example) away from the primary emission. The production of secondary pollutants depends upon a number of ambient conditions, primarily ambient volatile organic levels, No levels, temperature, and solar irradiation. In general, smog levels can be reduced by a concerted program of emissions abatement from automobiles and industry, leading to reduced ambient concentrations.

5. The "spill" occurs as a part of the processing of the piece and can be classified as a process emission. Depending upon the chemicals used in the bath, the water used to wash the spill down the drain may become a waste or hazardous waste stream. It may be possible to reduce the problem by changing the location of the rinse bath so that the path of the piece remains over collection tanks, and by suspending the piece for a longer amount of time over the bath to allow most of the bath solution to drip back into the tank.

6. The leaked solvent is a fugitive emission. An estimate of the amount of solvent lost as fugitive emissions can be made by performing a material balance for the solvent over the entire process. Checking the integrity of the pump may reveal a repairable leak, or it may be necessary to switch to a different type of pump that is less likely to leak. It may also be possible to reduce this type of loss in process piping (at gaskets, for example) by shrink-wrapping the joint or otherwise containing the vapor. If possible, another solvent might be considered that has a lower vapor pressure or that is less corrosive to the pump interior.

7. If the acidic stream has a pH < 2, it is classified as a hazardous waste and must undergo further treatment before disposal. Similarly, if a caustic effluent has a pH > 12, it is classified as

a hazardous waste. It may be possible to replace the acidic cleaner with one that is not classified as hazardous. It may also be possible to recover spent solution and recycle it for cleaning purposes, reducing the amount of waste generated.

8. If either the acidic or caustic stream is classified as hazardous (see 7), the neutralized stream may also be considered hazardous even if its pH is greater than 2 and less than 12. This is known as the "derived-from" rule, which assigns the hazardous classification to derivatives from hazardous wastes. It may be possible to reduce the amount of neutralized solution to be processed by concentrating it; the water thus recovered may be recycled for use in other parts of the process.

14.10   Open-Ended Problems

This last section of the chapter contains open-ended problems as they relate to environmental management. No detailed and/or specific solution is provided; that task is left to the reader, noting that each problem has either a unique solution or a number of solutions or (in some cases) no solution at all. These are characteristics of open-ended problems described earlier.

There are comments associated with some, but not all, of the problems. The comments are included to assist the reader while attempting to solve the problems. However, it is recommended that the solution to each problem should initially be attempted *without* the assistance of the comments.

There are 32 open-ended problems in this section. As stated above, if difficulty is encountered in solving any particular problem, the reader should next refer to the comment, if any is provided with the problem. The reader should also note that the more difficult problems are generally located at or near the end of the section.

1. Describe the early history associated with environmental management.
2. Discuss the recent advances in the general field of environmental management.
3. Select a refereed, published technical article on the environment from the literature and provide a review.

4. Provide some normal everyday domestic applications involving the general topic of the environment.

5. Develop several original problems related to environmental management that would be suitable as illustrative examples in a book.

6. Prepare a list of the various technical books that have been written on the environment. Select the three best (be sure to select at least one written by one of the authors) and justify your answer. Also select the three weakest books and, once again, justify your answer.

7. As a well educated person in environmental management attending a public hearing on a risk-assessment study where the outcome is "no potential health or hazard to the public", what are the main concerns you would have in order to be convinced that the results are reliable and acceptable?

8. Provide a short paragraph on each of the following environmental management topics:
   - Air pollution control equipment
   - Atmospheric dispersion modeling
   - Indoor air quality
   - Industrial wastewater management
   - Wastewater treatment technologies
   - Wastewater treatment processes
   - Solid waste management
   - Superfund
   - Municipal solid waste management
   - Hospital waste management
   - Nuclear waste management

9. What does the term "multimedia analysis" refer to?

10. Acid rain results from the dissolution of nitrogen and sulfur oxides into precipitation to form acids that reach the earth's surface. In recent years studies have shown that the pH level of rain has decreased, becoming acidic as the atmospheric level of these nitrogen and sulfur oxides has increased. Areas in Scotland and the northeastern US have shown the most dramatic effects of this acidic deposition. Acid rain potentially harms forests, lakes, and even drinking water in extreme cases. How did Title IV of the Clean Air Act of 1990 address the acid rain problem in the US?

11. Define the greenhouse effect and global warming. Also discuss the relationship between the two.

12. Do you believe global warming is a problem? Explain.

13. You are a consulting engineer hired by a company to evaluate waste minimization opportunities in the organization. Outline your approach for conducting a waste minimization assessment.

14. A key to identifying and applying pollution prevention (P^2) principles is to understand the total life-cycle of a product or material. Identify and discuss potential opportunities for P^2 for a "juice-pack" of orange juice.
    Comment: Both packaging and content are components of this popular product.

15. What are the most common methods used to estimate the quantity of waste generated in a community?

16. Identify the major contaminants affecting indoor air quality, and provide a potential source and the methods of control of each contaminant.

17. Indoor air quality is a relatively recent concern in environmental management.
    - Explain why indoor air quality is a concern.
    - Explain what has caused indoor air quality to be of greater concern now than in the past.
    - Describe some of the immediate and long-term health effects of indoor air quality exposure.
    - What are some of the costs of indoor air quality problems?
    - Compare indoor air pollution with ambient air pollution. Show why indoor air quality can be of greater concern than ambient air quality.

18. Why is noise considered an environmental problem?

19. Explain the mechanism of hearing and the effect of noise exposure on individuals.

20. Define ISO in layman terms.

21. Discuss the provisions of ISO 14000 in layman terms.

22. You have been hired as the environmental officer for a manufacturing company that intends to enter the global market. As part of the business plan for the company entering the global market, the company intends to become ISO 14000 certified. What are the major components of ISO

14001 EMS that must be implemented by the company in order to pass the audit requirement for certification?

23. Refer to the previous problem. List and discuss the benefits and pitfalls of the ISO 14000 series of standards to industry from an international point of view.

24. Composting is the biological decomposition of organic waste material by microorganisms. It is one of the major treatment alternatives for municipal solid waste, especially in developing countries where solid wastes contain mostly organic carbon sources and are rich in nutrients. During the process, certain physical, chemical and biological changes take place which alter the character of the waste material. Describe these physical, chemical, and biological changes as municipal solid waste is converted to humus in the composting process.

25. Choose a small political jurisdiction that operates its own solid-waste collection system. Find out the quantity of solid waste collected, any waste processing that takes place, and the ultimate disposal option used for this waste stream. Obtain cost information for waste collection and waste disposal.

    Find out what type of recycling, if any, is done in the jurisdiction. For a class of material not yet widely recycled, develop a plan to collect, process, and sell this recycled product. Make rough engineering estimates of the cost of operating the recycling process. Determine the savings incurred because of reduced waste collection and/or disposal costs. As recycling is almost always not economically attractive, look for other values to society that this recycling represents, and incorporate these ideas as quantitatively as possible into a persuasive report selling your project.

26. Describe technical issues that must be considered in the design of a waste landfill. Also describe the issues if the waste is hazardous.

27. Hospitals are places where sick or injured people go to get cured, but themselves present health and hazard problems to patients. In examining the many different activities of a hospital, list and group as many possible health problems as you can.

28. Do you believe electromagnetic fields pose a health problem? Explain your answer.

29. The International Convention For The Prevention of Pollution From Ships, to which the U.S. is a Party, specifies the conditions under which an oil tanker may discharge oil into the sea. List and comment on these conditions.

30. The Pasquel-Gifford atmospheric dispersion coefficients (σ_x, σ_y) are provided in graphical chart form [16]. Convert these charts into any suitable equation form. Also convert the data available for instantaneous releases into equation form.

31. Sansaverino Waste Management has hired you to develop a plan to economically extract valuable resources in their landfill sites. Your plan should address both near term (this decade) and longer term action(s).

32. Whenever a difference in velocity exists between a particle and its surrounding fluid, the fluid will exert a resistive force upon the particle. Either the fluid (gas) may be at rest with the particle moving through it or the particle may be at rest with the gas flowing past it. It is generally immaterial which phase (solid or gas) is assumed to be at rest; it is the *relative* velocity between the two that is important. The resistive force exerted on the particle by the gas is called the *drag* and this effect plays a major role in the design of particulate air pollution control equipment.

    In treating fluid flow through pipes, a friction factor term is used in many engineering calculations (see Fluid Flow – Chapter 5). An analogous factor, called the drag coefficient, is employed in drag force calculations for flow past particles.

    Consider a fluid flowing past a stationary solid sphere. If F_D is the drag force and ρ is the density of the gas, the drag coefficient, C_D, is defined as

$$C_D = \frac{F_D}{A_p} \frac{2g_c}{\rho v^2} \tag{14.1}$$

    From dimensional analysis, one can then show that the drag coefficient is solely a function of the particle Reynolds number, Re, i.e.,

$$C_D = C_D(\text{Re}) \tag{14.2}$$

where

$$\text{Re} = \frac{d_p v \rho}{\mu} \tag{14.3}$$

The quantitative use of the equation of particle motion requires numerical values of the drag coefficient as a function of the Reynolds number. Graphical values are presented in Figure 14.1.

Numerous models (equations) have been developed to convert Figure 14.1 into equation form. One drag coefficient model is given by Equation (14.4) [17]:

$$\log C_D = 1.35237 - 0.60810 (\log \text{Re}) - 0.22961 (\log \text{Re})^2$$
$$+ 0.098938 (\log \text{Re})^3 + 0.041528 (\log \text{Re})^4$$
$$- 0.032717 (\log \text{Re})^5 + 0.007329 (\log \text{Re})^6$$
$$- 0.0005568 (\log \text{Re})^7 \tag{14.4}$$

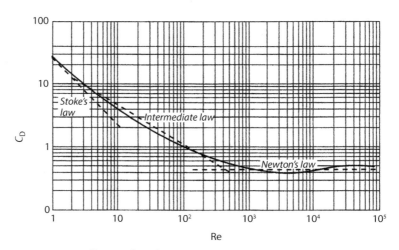

Figure 14.1 Drag coefficients for spheres.

This is an empirical equation that was obtained by the use of a statistical fitting technique. An advantage of using this correlation is that it is not partitioned for application only to a specific Reynolds number range. However, the lengthy calculation can limit its use. Still another empirical equation [18] is

$$C_D = \left[0.63 + (4.80 / \sqrt{Re}) \right]^2 \qquad (14.5)$$

This correlation is also valid over the entire spectrum of Reynolds numbers. Its agreement with literature values in the range of $30 < Re < 10,000$ is fair. This correlation lends itself easily to manual calculations.

Develop a better model to represent Figure 14.1 in equation form.

References

1. Adapted from: M.K. Theodore and L. Theodore, *Introduction to Environmental Management*, CRC Press/ Taylor & Francis Group, Boca Raton, FL, 2010.
2. Adapted from: G. Burke, B. Singh, and L. Theodore, *Handbook of Environmental Management and Technology*, 2nd edition, John Wiley & Sons, Hoboken, NJ, 2000.
3. Adapted from: L. Stander and L. Theodore, *Environmental Regulatory Calculations Handbook*, John Wiley & Sons, Hoboken, NJ, 2008.
4. ISO in Brief, August 2006. http://www.iso.org/iso/isoinbrief_2006-en.pdf
5. Introduction to ISO, www.iso.ch/infoe/intro
6. General information on ISO. http://www.iso.org/iso/support/faqs/faqs_general_information_on_iso.htm
7. Summary of ISO 14000. Lighthouse Consulting, University of Rhode Island, 2003. www.crc.uri.edu/download/12_ISO_1400_summary_ok.pdf
8. ISO 14000. http://en.wikipedia.org/wiki/ISO_1400
9. U.S. EPA, *Pollution Prevention Fact Sheet*, author unknown, Washington, DC, March 1991.
10. P. Anastas and T. Williamson, *Green Chemistry: An Overview, Green Chemistry: Designing Chemistry for the Environment*, eds. P.T. Anastas and T.C. Williamson (editor), ACS Symposium Series 626, pp. 1-17, American Chemical Society, Washington, DC, 1996.
11. P. Anastas and J. Warner, *Green Chemistry: Theory and Practice*, Oxford University Press, New York City, NY, 1998.

12. P. Anastas and J. Zimmerman, *Design Through the Twelve Principles of Green Engineering*, Environmental Science and Technology, 37, 94A-101A, Washington, DC, 2003.
13. P. Bishop, *Pollution Prevention*, Waveland Press, Inc., Prospect Heights, IL, 2000.
14. United Nations, Report of the World Commission on Environment and Development, General Assembly Resolution 42/187, December 11, 1987. Retreived October 31, 2007.
15. L. Theodore, *Chemical Engineering: The Essential Reference*, McGraw-Hill, New York City, NY, 2014.
16. L. Theodore, *Air Pollution Control Equipment Calculations*, John Wiley & Sons, Hoboken, NJ, 2008.
17. Personal notes, L. Theodore, East Williston, NY, 1974.
18. Barnea and I, Mizraki, PhD thesis, Haifa University, Haifa, Israel, 1972.

15

Environmental Health and Hazard Risk Assessment

This chapter is concerned with process environmental health and hazard risk assessment. As with all the chapters in Part II, there are several sections: Overview, several technical topics, illustrative open-ended problems, and open-ended problems. The purpose of the first section is to introduce the reader to the subject of environmental health and hazard risk assessment. As one might suppose, a comprehensive treatment is not provided although numerous references are included. The second section contains three open-ended problems; the authors' solutions (there may be other solutions) are also provided. The third (and final) section contains 45 problems; *no* solutions are provided here.

15.1   Overview

This overview section is concerned with—as can be noted from its title—environmental health and hazard risk assessment. As one might suppose, it was not possible to address all topics directly or indirectly related to this topic. However, additional details may be obtained from either the references provided at the end of this Overview and/or at the end of the chapter.

Note: Those readers already familiar with the details associated with this subject may choose to bypass this Overview.

This chapter deals not only with the dangers posed by hazardous substances but also examines the general subject of health, safety, and accident prevention. In addition, the laws and legislation passed to protect workers, the public and the environment from the effects of these chemicals and accidents are also reviewed. The chapter also discusses regulations (with particular emphasis on emergency planning) and the general subject of health and hazard risk assessment. In effect, the chapter addressed topics that one would classify as health, safety, and accident prevention. The bulk of the material has been adapted from:

1. L. Theodore, J. Reynolds, and K. Morris, *Accident and Emergency Management*, A Theodore Tutorial, Theodore Tutorials, East Williston, NY, 1994, originally published by the USEPA/APTI, RTP, NC, 1996 [1].
2. M.K. Theodore and L. Theodore, *Introduction to Environmental Management*, CRC Press/Taylor & Francis Group, Boca Raton, FL, 2010 [2].
3. L. Theodore and R. Dupont, *Environmental Health and Hazard Risk Assessment: Principles and Calculations*, CRC Press/Taylor & Francis Group, Boca Raton, FL, 2012 [3]

Two general types of potential chemical health, safety and accident exposures and/or concerns exist. These are classified as:

1. *Acute*: Exposures occur for relatively short periods of time, generally seconds to minutes to 1-2 days. The concentration of (air) contaminants is usually high relative to their protection criteria. In addition to inhalation, airborne substances might directly contact the skin, or liquids and sludges may be splashed on the skin or into the eyes, leading to toxic effects.
2. *Chronic*: Continuous exposure occurs over longer periods of time, generally several months to years. The concentrations of inhaled contaminants are usually relatively low. Direct skin contact by immersion, by splash, or by contaminated air involve contact with substances exhibiting low dermal activity.

In general, acute exposures to chemicals in air are more typical in transportation accidents, explosions, and fires, or releases at chemical manufacturing or storage facilities. High concentrations of contaminants in air usually do not persist for long periods of time. Acute skin exposure may occur when workers come in close contact with substances in order to control a release – for example, while offloading a corrosive material, uprighting a drum, or while containing and treating a spilled material.

Chronic exposures on the other hand, are usually associated with longer-term removal and remedial operations. Contaminated soil and debris from emergency operations may be involved in the around-the-clock discharges to the atmosphere. Soil and groundwater may be polluted or temporary impoundment systems may contain diluted chemicals. Abandoned waste sites typically represent chronic exposure problems. As activities start at these sites, personnel engaged in certain operations, such as sampling; handling containers; bulking compatible liquids; or, activities involving the release of vapors, gases, or particulates may be exposed to health and/or hazard problems.

The remaining Sections in this chapter include:

1. Safety and Accidents
2. Regulations
3. Emergency Planning and Response
4. Introductions to Environmental Risk Assessment
5. Health Risk Assessment
6. Hazard Risk Assessment

15.2   Safety and Accidents

There is a high risk of accidents due to the nature of the processes and the materials used in the chemical industry. Although precautions are taken to ensure that all processes run smoothly, there is always (unfortunately) room for error and accidents will occur. This is especially true for highly technical and complicated operations, as well as processes under extreme conditions such as high temperatures and pressures. In general, accidents occur due to one or more of the following 8 causes:

1. Equipment breakdown
2. Human error
3. Terrorism
4. Fire exposure and explosions

5. Control system failure
6. Natural causes
7. Utilities and ancillary system outages
8. Faulty sitting and plant layout

These causes are usually at the root of most industrial accidents. Although there is no way to guarantee that these problems will not arise, steps can be taken to minimize the number, as well as the severity, of incidents. In an effort to reduce occupational accidents, measures should be taken in the following areas [4].

1. *Training:* All personnel should be properly trained in the use of equipment and made to understand the consequences of misuse. In addition, operators should be rehearsed in the procedures to follow should something go wrong.
2. *Design:* Equipment should only be used for the purposes for which it was designed. All equipment should be periodically checked for damage or errors in the design.
3. *Human Performance:* should be closely monitored to ensure that proper procedures are followed. Also, working conditions should be such that the performance of workers is improved, thereby simultaneously reducing the chance of accidents. Periodic medical examinations should be provided to assure that workers are in good health, and that the environment of the workplace is not causing undue mental and/or physical stress. Finally, under certain conditions, it may be advisable to test for the use of alcohol or drugs— conditions that severely handicap judgment, and therefore make workers accident-prone.

15.3   Regulations

Each company must develop a health and safety program for its workers. For example, OSHA has regulations governing employee health and safety at hazardous waste operations and during emergency responses to hazardous substance releases. These regulations (29 CFR 1910.120) contain general requirements for the following 10 topics:

1. Safety and health programs
2. Training and informational programs

3. Work practices along with personal protective equipment
4. Site characterization and analysis
5. Site control and evacuation
6. Engineering controls
7. Exposure monitoring and medical surveillance
8. Material handling and decontamination
9. Emergency procedures
10. Illumination

The EPA's Standard Operating Safety Guides supplement these regulations. However, OSHA's regulations must be used for specific legal requirement in industry. For example, other OSHA regulations pertain to employees working with hazardous materials or working at hazardous waste sites. These, as well as state and local regulations, must also be considered when developing worker health and safety programs [5].

Information on chemical hazards must be dispatched from the manufacturers to employers via *material safety data sheets* (MSDSs) and container labels. This data must then be communicated to employees by means of comprehensive hazard communication programs which usually include training programs as well as the aforementioned MSDSs and container labels. Companies with multi-employer workplaces must include the MSDS methods that the employer will use for the contractors at the facility. These employers must also describe how they will inform the subcontractor (if applicable) and employees about precautions which must be followed and the specific labeling system used in the work place. This topic will be revisited in Part III, Chapter 31—Environmental Management Term Project 31.4.

15.4   Emergency Planning and Response

The extent of the need for emergency planning is significant, and continues to expand as new regulations on safety are introduced. Planning for an industrial emergency must begin at the very start, when the plant itself is still being planned. The new plant will have to pass all safety measures and OSHA standards. This is emphasized by Armenante, author of *Contingency Planning for Industrial Emergencies,* [5] "The first line of defense against industrial accidents begins at the design stage. It should be obvious that it is much easier to prevent an accident rather than to try and rectify the situation once an accident has occurred."

Successful emergency planning begins with a thorough understanding of the event or potential disaster being planned for. The impacts on public health and the environment must also be estimated. Some of the types of emergencies that should be included in the plan are:

1. Natural disasters such as earthquakes, tornadoes, hurricanes, floods, and meteorites
2. Explosions and fires
3. Acts of terrorism
4. Hazardous chemical leaks
5. Power or utility failures
6. Radiation accidents
7. Transportation accidents

In order to estimate the impact on the public or the environment, the affected area or emergency zone must be studied in depth. A hazardous gas leak, fire, or explosion may cause a toxic cloud to spread over a great distance, as it did in Bhopal, India. An estimate of the minimum affected area, and thus the area to be evacuated, should be performed based on an atmospheric dispersion model [3]. There are various models that can be used. While the more difficult models produce the most realistic results, simpler models are faster to use and usually still provide adequate data and information for planning purposes.

The main objective for any plan should be to prepare a procedure to make maximum use of the combined resources of the community in order to accomplish the following 7 steps:

1. Safeguard people during emergencies
2. Safeguard people during an act of terrorism
3. Minimize damage to livestock, property, and the environment
4. Initially contain and ultimately bring the incident under control
5. Effect the rescue and treatment of casualties
6. Provide authoritative information to the news media who will communicate the facts to the public
7. Secure the safe rehabilitation of the affect area

15.5   Introduction to Environmental Risk Assessment

Risk-based decision making and risk-based corrective action (RBCA) are decision-making processes for assessing and responding to a chemical release. The processes take into account effects on human health and the environment, in as much as chemical releases vary greatly in terms of complexity, physical, and chemical characteristics, and in the risk that they may pose. RBCA was initially designed by the American Society for Testing and Materials (ASTM) to assess petroleum releases, but the process may be tailored for use with any chemical release. For example, in the 1980s, to satisfy the need to start corrective action programs quickly, many regulatory agencies decided to uniformly apply regulatory cleanup standards developed for other purposes at underground storage tank (UST) cleanup sites. It became increasingly apparent that applying such standards without consideration of the extent of actual or potential human and environmental exposure was an inefficient means of providing adequate protection against the risks associated with UST releases. The EPA now believes that risk-based corrective-action processes are tools that can facilitate efforts to clean up sites expeditiously, as necessary, while still assuring protection of human health and the environment [6].

The EPA and several state environmental agencies have developed similar decision-making tools. The EPA refers to the process as "risk-based decision making". While the ASTM RBCA standard deals exclusively with human health risk, the EPA advises that, in some cases, ecological goals must also be considered in establishing cleanup goals.

For the purpose of this chapter, a few definitions of common terms will suffice. *Risk* is the probability that persons or the environment will suffer adverse consequences as a result of an exposure to a substance. The amount of health risk is determined by a combination of the concentration of the substance, and the toxicity of the environment to which it is exposed, the rate of intake or dose of the substance, and the toxicity of the substance. *Risk assessment* is the procedure used to attempt to quantify or estimate this risk. Risk-based decision making also distinguishes between the "point of exposure" and the "point of compliance". The *point of exposure* is the point at which the environment or the individual comes into contact with the chemical release. An individual may be exposed by methods such as inhalation of vapors, as well as physical contact with the substance. The *point of compliance* is a point in between the point of release of the chemical (i.e., the source) and the point of exposure. The *point of compliance* is selected to provide a safety buffer for effected individuals and/or environments.

15.6   Health Risk Assessment

As noted in the previous Section, there are many definitions for the word *risk*. People face all kinds of risks every day, some voluntarily and other involuntarily. Therefore, risk plays a very important role in today's world. Studies on cancer caused a turning point in the world of risk because it opened the eyes of not only risk scientists and health professionals but also chemical engineers to the world of risk assessment.

Since 1970 the field of risk assessment has received widespread attention within both the engineering, scientific, and regulatory committees. It has also attracted the attention of the public. Properly conducted risk assessments have received fairly broad acceptance, in part because they put into perspective the terms toxic, hazard, and risk. Toxicity is an inherent property of all substances. It states that all chemical and physical agents can produce adverse health effects at some dose or under specific exposure conditions. In contrast, exposure to a chemical that has the capacity to produce a particular type of adverse effect represents a health problem. As noted, risk is the probability or likelihood that an adverse outcome will occur in a person or a group that is exposed to a particular concentration or dose of the hazardous agent. Therefore, risk is generally a function of exposure *and* dose. Consequently, health risk assessment is defined as the process or procedure used to estimate the likelihood that humans or ecological systems will be adversely affected by a chemical or physical agent under a specific set of conditions [7].

More importantly, the term risk assessment is not only used to describe the likelihood of an adverse response to a chemical or physical agent, but it has also been used to describe the likelihood of any unwanted event. This subject is treated in more detail in the next section. These include risks such as: explosions or injuries in the workplace; natural catastrophes; injury or death due to various voluntary activities such as skiing, ski diving, flying, bungee jumping, diseases; death due to natural causes; and many others [8].

Health risk assessment provides an orderly, explicitly, and consistent way to deal with scientific issues in evaluating whether a health problem exists and what the magnitude of the problem might be. This evaluation typically involves large uncertainties because the available scientific data are limited, and the mechanisms for adverse health impacts or environmental damage are only imperfectly understood. When one examines risk, how does one decide how safe is safe, or how clean is clean? To begin with, the chemical engineer has to examine both sides of the risk equation, i.e., both the toxicity (and dose) of a pollutant and the extent of exposure. Information is required at both the current and potential exposures, considering all possible exposure pathways. In addition to human health risks, one needs to look at potential ecological or other environmental effects. In conducting a comprehensive risk assessment, one should remember that there are always uncertainties and these assumptions must be included in the analysis [9].

In recent years, several guidelines and handbooks have been produced to help explain the approaches for performing health risk assessments. As discussed by a special National Academy of Sciences committee convened in 1983, most human or environmental health problems can be evaluated by dissecting the analysis into four parts: health problem identification; dose-response assessment (or health problem assessment); exposure assessment; and risk characterization (see also Figure 15.1). For some perceived health problems, the risk assessment might stop with the first step, health problem identification, if no adverse effect is identified or if an agency elects to take regulatory action without further analysis [8]. Regarding identification, a health problem is defined as a toxic agent or a set of conditions that has the potential to cause adverse effects to human health or the environment. Identification involves an evaluation of various forms of information in order to identify the different problems. Dose-response or toxicity assessment is required in an overall assessment; responses/effects can vary widely since all chemicals and contaminants vary in their capacity

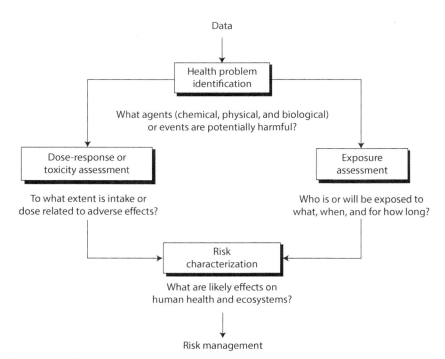

Figure 15.1 The health risk evaluation process.

to cause adverse effects. This step frequently requires that assumptions be made regarding experimental data for animals and humans. Exposure assessment is the determination of the magnitude, frequency, duration, and route of exposure on human populations and ecosystems. Finally, in risk characterization, toxicology and exposure data information are *combined* to obtain a qualitative or quantitative expression of risk.

The risk assessment involves the integration of the information and analysis associated with the above four steps to provide a complete characterization of the nature and magnitude of risk and the degree of confidence associated with this characterization. A critical component of the assessment is a full elucidation of *uncertainties* associated with each of the major steps are encompassed under this broad concept of risk assessment. It should treat uncertainty not by the application of arbitrary safety factors, but by stating them in quantitatively explicit terms, so that they are not hidden from decision makers. Risk assessment, defined in this broad way, forces an assessor to confront all the scientific uncertainties and to set forth in explicit terms the means used in specific cases to deal with these uncertainties [10].

15.7   Hazard Risk Assessment

Risk evaluation of accidents serves a dual purpose. It estimates the probability that an accident will occur and also assesses the severity of the consequences of an accident. Consequences may include damage to the surrounding environment, financial loss, or injury to life. This section is primarily concerned with the methods used to identify hazards and the causes and consequences of accidents. Issues dealing with health risks have been explored in the previous section. Risk assessment of accidents provides an effective way to help ensure either that a mishap does not occur or reduces the likelihood of an accident. The result of the risk assessment allows concerned parties to take precautions to prevent an accident before it happens.

The first thing a chemical engineer needs to understand is what exactly an accident is. An accident is defined as an unexpected event that has undesirable consequences [11]. The causes of accidents have to be identified in order to help prevent accidents from occurring. Any situation or characteristic of a system, plant, or process that has the potential to cause damage to life, property, or the environment is considered a hazard. A hazard can also be defined as any characteristic that has the potential to cause an accident. The severity of a hazard plays a large part in the potential amount of damage a hazard can cause if it occurs. The risk is the probability that human injury, damage to property, damage to the environment, or financial loss will occur.

An acceptable risk is a risk whose probability is unlikely to occur during the lifetime of the problem or plant or process. An acceptable risk can also be defined as an accident that has a high probability of occurring, with negligible consequences. Risks can be ranked qualitatively in categories of high, medium, and low. Risk can also be ranked quantitatively as annual number of fatalities per million affected individuals. This is normally denoted as a number times one millionth that is, 3×10^{-6}; this representation indicates that on the average, three individuals will die every year for every million individuals.

There are several steps in evaluating the risk of an accident (see also Figure 15.2). These are detailed below, if the system in question is a chemical plant.

1. A brief description of the equipment and chemicals used in the plant is needed.
2. Any hazard in the system has to be identified. Hazards that may occur in a chemical plant are one or a combination of the following:
    1. Corrosion
    2. Explosions
    3. Fire
    4. Rupture of a pressurized vessel
    5. Runaway reactions
    6. Slippage
    7. Unexpected leaks
    8. Temperature excursions
    9. Pressure excursions
3. The event or series of events that will initiate an accident has to be identified. An event could be a failure to follow correct safety procedures, improperly repaired equipment, or failure of a safety mechanism.
4. The probability that the accident will occur has to be determined. For example, if a nuclear power plant has a 10 year life, what is the probability that the temperature in a reactor will exceed the specified temperature range? The probability can be qualitatively ranked from low to high. A low probability means that it is unlikely for the event to occur in the life of the plant. A medium probability suggests that there is a possibility that the event will occur. A high probability

means that the event will probably occur during the life of the plant. Naturally, a quantitative estimate of the probability is preferred.

5. The severity of the consequences of the accident must be determined.

6. If the probability of the accident and the severity of its consequences are low, then the risk is usually deemed acceptable and the plant should be allowed to operate. If the probability of occurrence is too high or the damage to the surroundings is too great, then the risk is usually unacceptable and the system needs to be modified to minimize these effects.

The heart of the hazard risk assessment algorithm provided is enclosed in the dashed box in Figure 15.2. This algorithm allows for reevaluation of the process if the risk is deemed unacceptable (the process is repeated starting with either step one or two).

The reader should note that health assessment and hazard risk assessment plus accompanying calculations receives an extensive treatment by Theodore and Dupont [3]

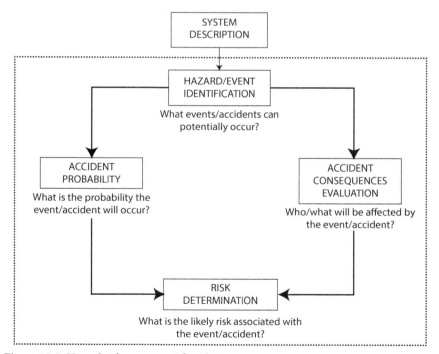

Figure 15.2 Hazard risk assessment flowchart.

A detailed and expanded treatment of environmental health and hazard risk assessment is available in the following two references.

1. L. Theodore and R. Dupont, *Environmental Health and Hazard Risk Assessment: Principles and Calculations*, CRC Press/Taylor & Francis Group, Boca Raton, FL, 2012 [3].
2. L. Theodore, *Chemical Engineering: The Essential Reference*, McGraw-Hill, New York City, NY, 2014 [12].

15.8   Illustrative Open-Ended Problems

This and the last section provide open-ended problems. However, solutions *are* provided for the three problems in this section in order for the reader to hopefully obtain a better understanding of these problems which differ from the traditional problems/illustrative examples. The first problem is relatively straightforward while the third (and last problem) is somewhat more difficult and/or complex. Note that solutions are not provided for the 45 open-ended problems in the next section.

Problem 1: The term *liability* is very often used with environmental regulations. Related terms are *strict liability*, *joint and several liability*, *retroactive liability,* and *cradle-to-grave liability.* In addition, of the terms used in connection with environmental regulations and enforcement, include their implications in environmental management. Briefly explain those terms.

Comment: Refer to the literature for assistance [13,14]

Solution: One person's definitions follow:
Liability: This means responsibility for an action. If an individual causes damage to property or other individuals, he/she are liable.
Strict Liability: This means responsibility without regard to negligence or care. A corporation could comply with all the applicable regulations in 1980 but when these regulations are changed in 1990, the corporation is liable for the new compliance.
Joint and Several Liability: In this case the responsibility is assigned (or shared) when several individuals (or corporations) do not perform properly, and it is not possible to divide the harm. If three plants contributed to a hazardous waste, each of them is liable to clean the site and mitigate the damages; this also includes the generators, the transporters, the storage facilities, and the operators. They are all collectively or individually responsible for damages.

Retroactive Liability: This is the case when a law is enacted, such as the Superfund Act; the liability goes back many years before the date enactment.

Cradle-to-Grave Liability: This implies that the generator of the waste is responsible (liable) for the waste until its ultimate destruction. Simply selling the waste to another facility does not absolve liability.

Negligence: Negligence is an act of failure to act which breaches a responsibility that one person (or company) has to another person (or company) and which unintentionally results in harm to a person (or company) or to a person's (or company's) property. Negligent behavior leads to liability. If, for example, through a failure to exercise its responsibility to take proper care, a company allows a release of a toxic gas that kills the cows on a neighboring farm, then the company would be liable for the damages caused. Violation of a regulation would be virtual proof of negligence.

Trespass: Trespass to reality is the type of trespass most often used in environmental law cases. It involves the unlawful physical invasion of another's property that interferes with the use of that property. Trespass is independent of negligence. For example, even if there was not negligence, it is a trespass if gasoline from a gas station underground storage tank leaks out and flows under a neighbor's home, filling it with toxic and flammable vapors.

Nuisance: A nuisance is the use by a person of their property in a way that causes injury or annoyance to their neighbor. Allowing bad smelling gases, dust, smoke or other annoying or harmful materials to drift over a neighbor's property would be examples of nuisance in environmental cases. These terms and liabilities imply that:

1. It is very attractive for waste generators to dispose of any waste on-site under carefully controlled conditions.
2. It places a burden on the waste generator with the threat of future costs due to the improper action of others.
3. Citizens are protected from the loss of property and/or health due to the action of others.
4. It is extremely important to select a reputable firm for waste treatment/management.

Problem 2: A dose-response relationship provides a mathematical formula or graph for estimating a person's risk of illness at each exposure level for air toxics. To estimate a dose-response relationship, measurements of health risks are needed for at least one dose level of the air toxic compared to an unexposed group. However, there is one important difference between the dose-response curve commonly used for estimating the risk of cancer and the ones used for estimating the risk of all other illnesses: the existence of a threshold dose, i.e., the highest dose at which there is no risk of illness. Because a single cancerous cell may be sufficient to cause a clinical case of cancer, EPA's and many other dose-response models for cancer assume that the threshold dose level for cancer is *zero*. In other words, people's risk of cancer is possible even at very low doses. However, it should be noted that the increased cancer risk at very low doses is likely to be very low.

1. Draw a straight line model showing the level of cancer risk increasing at a constant rate as the dose level increases. The model should illustrate increasing risk of cancer for the air toxic.
2. Also develop a straight line model to show EPA's methodology in which the EPA adjusts the observed threshold downward by dividing by uncertainty factors that range from 1 to 10,000 known as the human threshold.

Solution: It is accepted by scientists that the human body is capable of adjusting to varying amounts of cell damage without showing signs of illness. Therefore, EPA has developed models for non-cancer illnesses which include a threshold dose level that is greater than zero; this means that at low doses there may be no risk of non-cancer health effects. For non-cancer health effects, such as permanent liver or kidney damage, temporary skin rashes, or asthma attacks, information from human or animal studies is used to estimate the threshold dose levels.

1. Figure 15.3 shows the cancer dose-response curve plotted from data on a dose of 100 μg/d. this dose caused an extra change of cancer of about 1 in 100 in the study for which animals that received that dose. The straight line

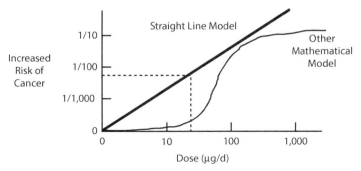

Figure 15.3 Cancer dose-response curve highlighting the straight line dose-response model.

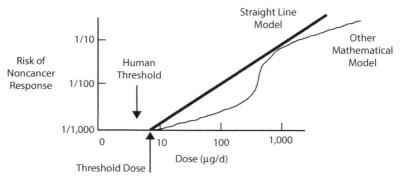

Figure 15.4 Non-cancer dose-response curve highlighting the straight line dose-response and the threshold dose.

model developed here indicates that the level of cancer risk increases at a constant rate as the dose level increases. This rate of increasing cancer risk is known as the slope factor.

2. Figure 15.4 illustrates the non-cancer dose-response curve which was drawn after converting uncertainties from animal to human data. Since individuals vary in their susceptibility to the harmful effects of toxic air pollutants, EPA adjusts the observed threshold dose downward by dividing by uncertainty factors that range from 1 to 10,000. This new adjusted value is known as the human threshold. Below the human threshold, EPA expects no appreciable risk of harmful health effects for most of the general population.

Problem 3: Storage tanks are the most common item of equipment to be involved in accidents. Although level control appears to be one of the simplest control schemes, many accidents result from overfilling storage tanks. Level indicators frequently depend on weight measurements rather than volume measurements. Level measurements can be made at the tank site by viewing a gauge glass or using a dipping device, but the operators obviously prefer to stay in the control room if it is dark or if the weather is bad. It is common practice to install a high-level alarm that actually measures volume if the level indicator measures weight. A high-level alarm may be assumed to be faulty even if it is correct, especially if the level indicator shows that the tank is not full and the operators do no understand the level measuring mechanism.

A tank that was designed to store gasoline (specific gravity 0.81) overflowed while being filled with pentane (specific gravity 0.69). The tank overflowed when the level indicator said that it was only 85 percent full. The level indicator was a differential pressure cell that measured weight not volume, but the operators did not realize that the level indicator did not measure the level directly. Prepare an event-tree diagram [3, 12] for this type of incident starting with "low-density feed". Assume that manual measurement is available, but the operators may or not check it. Assume that a high-level alarm based on volume is present, but that alarm may or may not work and the operators may or may not believe it, even if it works. Include other events that will determine whether the tank overflows.

Solution: One possible event-tree diagram for a "low density feed" tank overflow accident from a high density feed storage tank is as follows: (see Figure 15.5)

15.9 Open-Ended Problems

This last section of the chapter contains open-ended problems as they relate to environmental health and hazard risk assessment. No detailed and/or specific solution is provided; that task is left to the reader, noting that each problem has either a unique solution or a number of solutions or (in some cases) no solution at all. These are characteristics of open-ended problems described earlier.

There are comments associated with some, but not all, of the problems. The comments are included to assist the reader while attempting to solve

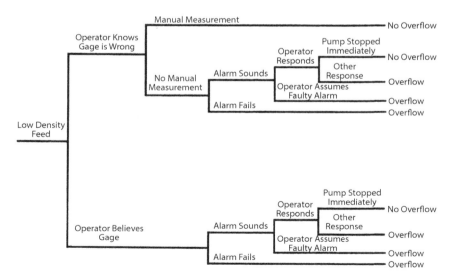

Figure 15.5 Event tree for "low density feed" tank overflow accident from high density feed storage tank.

the problems. However, it is recommended that the solution to each problem should initially be attempted *without* the assistance of the comments.

There are 45 open-ended problems in this section. As stated above, if difficulty is encountered in solving any particular problem, the reader should next refer to the comment, if any is provided with the problem. The reader should also note that the more difficult problems are generally located at or near the end of the section.

1. Describe the early history associated with health risk assessment.

2. Discuss the recent advances in the general field in health risk assessment.

3. Select a refereed, published article on health risk assessment from the literature and provide a review.

4. Provide some normal everyday domestic applications involving the general topic of health risk assessment.

5. Develop an original problem on health risk assessment that would be suitable as an illustrative example in a book.

6. Prepare a list of the various technical books that have been written on health risk assessment. Select the three best (try to include a book written by one of the authors) and justify your answer. Also select the three weakest books and, once again, justify your answer.

7. Select three chemical compounds from the list of "Extremely Hazardous Substances" as given in the Federal Register, April 22, 1987. For each of the three compounds find the:
   - Chemical formula
   - Molecular weight
   - Vapor pressure at 25°C
   - Boiling point
   - Freezing point
   - Flash point
   - Occupational threshold limit value

8. The National Air Toxics Information Clearinghouse (NATICH) data base contains information on selected EPA risk analysis results calculated using the Human Exposure Model (HEM). Explain the steps that regulatory bodies use for quantifying the number of people exposed to air pollutants emitted by stationary sources.

9. Suggest a method for estimating the uncertainty associated with health risk assessment calculations.

10. A possible health risk policy for a company could take the following form

    1. The average individual risk level for the public should be less than _____.
    2. The maximum individual risk for employees should be less than _____.
    3. The probability of one or more public deaths should be less than _____.
    4. The probability of 100 or more public deaths should be less than _____.
    5. The probability of one or more public illnesses should be less than _____.
    6. The probability of 100 or more public illnesses should be less than _____.

    The above form can be applied on either an annual or lifetime basis.

    Your company has requested that you improve on the present/proposed company policy. Submit your recommendations.

11. What is the technical definition of carcinogenic "unit risk"? How can this term be best described to a layman?

12. Describe and illustrate the process of setting a reference dose (RfD) using a schematic dose-response curve. Correctly label the axes and all other important information on your illustration. Develop another method of setting an RfD.

13. Discuss in general terms the means available for protecting humans from the *health* effects associated with radiation.

14. With reference to the previous problem can you suggest a better means of protecting the public from this health effect?

15. Can health-risk and hazard-risk communication aid the general population in ranking the importance of risk? Comment: Does the general population rank risk the same way that the "experts" do?

16. Communications about hazards and health risk present difficult problems in information presentation. What are some of these problems and how would you deal with them? Comment: Refer to the literature [1–3] for more details.

17. Describe the early history associated with hazard risk assessment.

18. Discuss the recent advances in the general field in hazard risk assessment.

19. The Occupational Safety and Health Act (OSHA) enforces basic duties which must be carried out by employers. Discuss these basic duties as they apply to an industry of your choice.

20. State the major roles of the National Institutes of Safety and Health (NIOSH) and the Occupational Safety and Health Administration (OSHA).

21. If one informs workers about the hazard risk associated with their jobs, what steps and/or guidelines should be followed to best get the message (information) across to the worker(s)?

22. Provide a list of the various safety hazard regulatory groups. Comment: Examples can include:
    - American Petroleum Institute (API)
    - American Society for Testing Materials (ASTM)
    - Associated Factory Mutual Insurance Companies, Boston.
    - Manufacturing Chemists' Association.
    - National Board of Fire Underwriters.

23. Discuss in general terms the means available for protecting humans from the *hazard* effects associated with radiation. Can you improve on this?

24. Select a refereed, published article on hazard risk assessment from the literature and provide a review.

25. Provide some normal everyday domestic applications involving the general topic of hazard risk assessment.

26. Develop an original problem on hazard risk assessment that would be suitable as an illustrative example in a book.

27. Prepare a list of the various technical books which have been written on hazard risk assessment. Select the three best (try to include a book written by one of the authors) and justify your answer. Also select the three weakest books and, once again, justify your answer.

28. Education and training of personnel are critical components of efforts to reduce hazards in a chemical processing plant or a chemical laboratory. Identify the major topics that should be included in an effective education and training program.

29. The effectiveness of accident and emergency management plans can be enhanced by informed and receptive citizens. What factor or factors have been identified as most important in addressing public opposition to siting industrial facilities involving toxic and/or hazardous chemicals? With this in mind, what efforts must be undertaken to counter this opposition?

30. Suggest a method for estimating the uncertainties associated with hazard risk assessment calculations.

31. Discuss the many pieces of equipment and protective clothing available for routine safety practices and emergencies in a well organized chemical laboratory.

32. Locate newspaper and/or news magazine articles in the library about a recent accident that involved evacuation of a population due to the risk to their health and/or safety. From these sources, write a brief essay describing what happened, where the incident took place, the number of people killed and/or injured, the immediate impact on the community, etc.

33. Very often, the actions of the first person on the scene of an accident can have a significant impact on the final outcome of the incident. However, it is crucial that this person does not subject himself or herself to personal injury. The actions of an untrained person and trained person may differ greatly.
    - What sequence of actions would you recommend for the *untrained* individual who observes a large spill from a tank truck accident?
    - How would you, the *trained* leader of an emergency response team, behave in a similar circumstance?

34. One of the authors has employed the Delphi Panel approach in consulting activities. Provide your definition of this approach as it applies to environmental risk assessment.
    Comment: Refer to the literature [1–3] for background information.

35. You are the manager of a 25,000 gallon tank facility storing oil and oil-based products. This facility is located on a hillside 200 ft above and 2,000 ft away from a navigable river. The river intake for a water treatment plant serving 30,000 people is located 3 miles downstream from your location. This is a relatively new facility with no history of past chemical spills. You are responsible for the preparation of

the Spill Prevention Control and Countermeasures (SPCC) plan for the emergency response to a tank failure or other spill in this facility.

Prepare an outline listing the items or topics that the plan must address under the Clean Water Act (CWA) Amendments, the Resource Conservation and Recovery Act (RCRA), and the Comprehensive Environmental Response, Compensation, and Liability Act (CERCLA). This is to be a "first step" outline addressing what is required and who (job title or job description) is involved. You may assume that the details of implementation are not required at this stage of the response-planning process.

36. An out-of-control forklift has ruptured a water line and a benzene storage tank in a small building. The tank contains 1000 kg of benzene. The water line discharged 800 m³ of water before a cutoff valve could be activated. Both fluids accumulated in the basement of the building, which is not vented. The remaining airspace in the basement is 1000 m³.
    - Determine if a separate phase of benzene exists on the top of the water.
    - Does a flammability hazard exist in this basement?
    - Suggest several methods of cleanup that do not exacerbate the flammability problem.

    Assume the following information applies at 25°C and 1 atm:
    Water vapor pressure = 23.8 mmHg
    Benzene aqueous solubility = 1800 mg/L
    Benzene LFL = 1.4%v/v (by volume)
    Benzene UFL = 8%v/v (by volume)
    Henry's Law constant for benzene = $10^{-2.25}$ atm·m³/gmol

37. Consider two plants that process and store large quantities of hazardous materials. Plant A is located in a narrow valley approximately 200 ft from the nearest major stream. The climate can be described as humid. A thick clay extends from the land surface in the area to a depth of 100 ft. Unfractured shale at least 200 ft thick underlies the clay. Plant B is located on a hillside in an area where the climate is arid. At the surface there is a 20 ft thick layer of till (an unsorted glacial deposit of differing grain sizes). The till is underlain by a fractured basalt that is several hundred feet thick. Discuss how the features of these two sites would influence dispersion of hazardous materials at the two sites.

38. You are a member of the Local Emergency Planning Committee (LEPC) for your community. You have been assigned the task of planning for evacuation of the community in the event of a major accident at the nearby railroad yard. Propose a scheme or schemes for notifying all people in the community, as quickly and efficiently as possible, that they should immediately leave the area and assemble at the regional high school four miles away. Your community covers an area of 0.6 square miles with a total population of 5,000. Of this total, 2,200 live in 850 single-family housing units; 1,600 live in 900 two-story apartments; and 1,200 live in four high-rise apartment buildings. List the advantages and disadvantages of each notification scheme that addresses cost, manpower requirements, effectiveness, and other concerns you believe appropriate.
    Comment: Since this is an open-ended problem involving a great deal of individual judgment, considerable variation in answers can therefore be expected.

39. You have been hired by the World Health Organization (WHO) to *quantify* the risk(s) associated with global *warming* over the next 25 year period. Repeat the exercise for this century.

40. You have been hired by the World Health Organization (WHO) to *quantify* the risk(s) associated with global *cooling* over the next 25 year period. Repeat the exercise for this century.

41. The Pentagon has hired you as an outside consultant to develop a procedure that could be employed to estimate the probability and associated risk that a foreign power will launch a nuclear attack at another nation.

42. With reference to the previous problem, the Pentagon has asked that you *quantify* (your best estimate) the probability and risk that North Korea will successfully deliver a nuclear attack on the U.S mainland.

43. With reference to the previous problem, provide your best estimate of the probability and associated risk of Iran launching a nuclear attack on Israel.

44. It is 2:00 am, and you have just received a call from your boss stating that a boxcar used for transporting a hygroscopic (moisture adsorbing) dry solid chemical, manufactured in your department, has been involved in a train derailment just outside the quiet little town of

Smallville. The boxcar is in a section of the train that has not derailed, but fumes issuing from the car are at high enough concentrations to cause irritation to personnel in the immediate vicinity of the boxcar. Local EPA personnel are headed to the site to take control of the situation, and have already notified the press that your company will pay for any and all costs in cleaning up the derailment. You must develop a plan of action and issue direction to personnel at the site.

- What information should you obtain from the personnel at the site related to the derailment?
- What information is necessary from personnel who are not at the site related to the chemical?
- What are your instructions to personnel at the site after you obtain all of the following data:

  There is a brisk 20 km/h wind blowing normal to the railroad tracks and away from Smallville.

  There is a high probability the boxcar involved is actually returning from a customer carrying empty shipping containers. This means that a thing layer of chemical dust is deposited on the inner walls of each container.

  It is not unusual for the customers to return containers that are only partially closed.

  The earliest you or any other personnel familiar with this chemical can arrive at the site is eight hours.

Comment: Fumes resulting from the reaction of the chemical with moisture are characterized as:

- noticeable in concentrations of 0.1 ppmv,
- hazardous for eight hour exposures in excess of 200 ppmv,
- can cause loss of consciousness in concentrations above 400 ppmv, and
  - can be fatal if inhaled when concentrations exceed 700 ppmv.

The diffusivity of the hazardous gas is similar to hydrogen. The rate of reaction between water vapor and the chemical is slow under ambient conditions.

45. In 1963, two 20-year-old (or, possibly older) bottles labeled "diisopropyl ether ($C_3H_7OC_3H_7$) – student preparation" were discovered in a basement storeroom at a major New

England university. Diisopropyl ether is a clear liquid. Both bottles were approximately one-third full with a solid water-insoluble material, presumably the corresponding ether peroxide ($C_3H_7OOC_3H_7$). Each bottle had a total capacity of approximately 2.5 L. The bottles were decanted and the solids were thrown (while still inside the bottles) into a dump at the edge of the nearest town. When the bottles did not break, stones were cast at them until, at last, they suddenly exploded. The density of solid dialkyl ether peroxides is approximately 0.72 g/mL. The total heat released may be estimated from the heat of combustion, which can be obtained by applying Hess' Law using the bond energies supplied below in Table 15.1. Assume sufficient oxygen was available to allow complete combustion to CO_2 and H_2O. The energy equivalent of 1.0 lb of TNT = 2,000 Btu.

Estimate the lb of TNT equivalent released in this explosion and comment on the choice of a disposal method used in this case.

Table 15.1			
Bond Energies (Btu/lbmol)			
C-H	178,000	C-O	144,000
C-C	149,000	O_2	212,400
O-O	59,300	C=O	304,000
O-H	199,000		

References

1. L. Theodore, J. Reynolds, and K. Morris, *Accident and Emergency Management*, A Theodore Tutorial, Theodore Tutorials, East Williston, NY, 1994, originally published by the USEPA/APTI, RTP, NC, 1996.
2. M.K. Theodore and L. Theodore, *Introduction to Environmental Management*, CRC Press/Taylor & Francis Group, Boca Raton, FL, 2010.
3. L. Theodore and R. Dupont, *Environmental Health and Hazard Risk Assessment: Principles and Calculations*, CRC Press/Taylor & Francis Group, Boca Raton, FL, 2012.
4. Adapted from: A.M. Flynn and L. Theodore, *Accident and Emergency Management in the Chemical Process Industries*, CRC Press/Taylor & Francis Group, Boca Raton, FL, 2004.

5. P. Armenante, *Contingency Planning for Industrial Emergencies*, Van Nostrand Reinhold, New York City, NY, 1991.

6. EPA Office of Emergency and Remedial Division, *Standard Operating Safety Guides*, location unknown July 1988.

7. USEPA, *Use of Risk-Based Decision Making*, OSWER Directive 9610.17, US Environmental Protection Agency, Washington DC, March 1995.

8. D. Paustenbach, *The Risk Assessment of Environmental and Human Health Hazards: A Textbook of Case Studies*, John Wiley & Sons, Hoboken, NJ, 1989.

9. Adapted from: G. Burke, B. Singh, and L. Theodore, *Handbook of Environmental Management and Technology*, 2nd edition, John Wiley & Sons, Hoboken, NJ, 2000.

10. J. Rodricks and R. Tardiff, *Assessment and Management of Chemical Risks*, American Chemical Society, Washington, DC, 1984.

11. AIChE, *Guidelines for Hazard Evaluation Procedures*, Batelle Columbus Division for the Center for Chemical Process Safety of the American Institute of Chemical Engineers, 1985.

12. L. Theodore, *Chemical Engineering: The Essential Reference*, McGraw-Hill, New York City, NY, 2014.

13. *Environmental Law Handbook*, Government Institutes, Rockland, MD, 1991.

14. L. Stander and L. Theodore, *Environmental Regulatory Calculations Handbook*, John Wiley & Sons, Hoboken, NJ, 2010.

16

Energy Management

This chapter is concerned with process energy management. As with all the chapters in Part II, there are several sections: Overview, several technical topics, illustrative open-ended problems, and open-ended problems. The purpose of the first section is to introduce the reader to the subject of energy management. As one might suppose, a comprehensive treatment is not provided, although several technical topics are also included. The next section contains three open-ended problems; the authors' solutions (there may be other solutions) are also provided. The final section contains 37 problems; *no* solutions are provided here.

16.1 Overview

This overview section is concerned with—as can be noted from its title—energy management. As one might suppose, it was not possible to address all topics directly or indirectly related to energy management. However, additional details may be obtained from either the references provided at the end of this Overview section and/or at the end of the chapter.

Note: Those readers already familiar with the details associated with this subject may choose to bypass this Overview.

Over the past four decades, an acute awareness of energy as a problem of impending critical magnitude on the national scene has arisen among informed leaders of industry, government, and the environmental movement. The energy crisis, or problem, or shortage, or dilemma, as it has been called, is created by the continually increasing demand for energy and a lack of management policy. This situation has resulted in two issues that are fast becoming pervasive concerns of this nation. One is the need for an adequate, reliable supply of all forms of energy, and the other, the growing public concern with the environment and social consequences of producing this much energy and further, of the environmental and social ramifications of its expenditure.

The solution to the energy problem amazingly *may* simply involve the application of meaningful conservation measures and the development of new, less problematic energy forms. Energy conservation may sharply reduce the terrible waste of resources that many have argued has also been at the very heart of the energy problem. Moreover, an extensive conservation program can be implemented in a very short period of time. Such an effort can play a major role in slowing the growth in the demand for energy and in causing energy to be used more efficiency and effectively. However, at the same time, new sources of energy must be developed to ensure the availability of adequate, long-term energy supplies. The *feasibility* of solar power, wind, tidal, geothermal, fusion, etc., and other unconventional sources of energy should continue to be investigated and developed.

In the final analysis, both society and the technical community can either accept or reject the grim projections for the future obtained by extending the energy consumption patterns and trends of the past that establish the basis for defining "energy demand". Once it has been determined that the demand exists, the choice among the various sources of energy and the means of energy conversion systems, either available at present or in some stage of development, can be made. This requires an evaluation for each means of power generation of the available fuel resources, including the environmental implications and their relation to relevant economic, political and social issues. However, all of these considerations are themselves influenced by assumptions regarding future demands for power; these, too, must be re-examined. For example, by analyzing the various components that presently constitute energy demand, resources, and transmission/transportation options, various alternatives can be devised to *maximize* the long-term social return per unit of energy consumed. In turn, such alternatives may have important implications for the economic system, social processes, and lifestyles. Topics (where applicable), such as resource quantity, resource availability, economics, energy quality, conservation requirements, transportation requirements, delivery requirements, operating and manufacturing, regulatory issues, political issues, environmental concerns, cost consequences, advantages and disadvantages, and public acceptance, need to be reviewed. A verifiable quantitative detailed review and practical evaluation of all viable energy options, categories, availability, cost, etc., has yet to be accomplished. These considerations define the energy issues and provide a means of solving and managing energy problems that exist today and defining the optimal course for future generations.

Regarding the future, are conservation and fossil fuels the answer? One of the authors believes that fossil fuels will dominate the energy landscape for at least the next two generations (50 years), if not longer. It is for this reason, that much of material in this chapter deal with fossil fuels.

Topics to be reviewed in this chapter include:

1. Energy Resources
2. Energy Quantity/Availability
3. General Conservation Practices in Industry
4. General Domestic Conservation Applications
5. General Commercial Real Estate Property Conservation Applications
6. Architecture and the Role of Urban Planning
7. The U.S. Energy Policy/Independence

16.2   Energy Resources

All the major energy resources may be classified into the following categories:

1. Natural gas
2. Liquid fuels (oil)
3. Coal
4. Shale oil
5. Tar sands
6. Solar
7. Nuclear (fission)
8. Hydroelectric
9. Wind
10. Geothermic

11. Hydrogen
12. Bioenergy
13. Waste

Extensive details on each of these energy resources are available on the Department of Energy (DOE) website and the Internet, as well as the work of Skipka and Theodore. [1]

16.3   Energy Quantity/Availability

An evaluation methodology was established by one of the authors [2] for a comparative analysis of energy resources. Its purpose was to provide an answer to the question, "Can a procedure be developed that can realistically and practically quantify the overall advantages and disadvantages of the various energy resource options?" A list of 12 categories/parameters that affect the answer to the question for each energy category was prepared by the author and are listed below. This analysis was applied to different sectors of the world, including the U.S. [1]

1. Resource Quantity
2. Resource Availability
3. Energy Quality
4. Economic Considerations
5. Conversion Requirements
6. Transportation Requirements
7. Delivery Requirements
8. Operation and Maintenance
9. Regulatory Issues
10. Environmental Concerns
11. Consumer Experience
12. Public Acceptance

16.4   General Conservation Practices in Industry

There are numerous general energy conservation practices that can be instituted at plants. Ten of the simpler practices are detailed below.

1. Lubricate fans
2. Lubricate pumps
3. Lubricate compressors

4. Repair steam and compressed air leaks
5. Insulate bare steam lines
6. Inspect and repair steam traps
7. Increase condensate return
8. Minimize boiler blowdown
9. Maintain and inspect temperature measuring devices
10. Maintain and inspect pressure measuring devices

Providing details on fans, pumps, compressors, and steam lines is beyond the scope of this chapter. Descriptive information [2,3] and calculation procedures [4] are available in the literature.

Eight energy conservation practices applicable to specific chemical operations are also provided below.

1. Recover energy from hot gases and/or liquids
2. Cover tanks of heated liquid to reduce heat loss
3. Reduce reflux ratios in distillation columns
4. Reuse hot wash water
5. Add effects to existing evaporators
6. Use liquefiers/gases as refrigerants
7. Recompress vapors or low-pressure steam
8. Generate low-pressure steam from flash operations

Providing details on distillation columns, evaporators, refrigerators is also beyond the scope of this chapter. Descriptive information and calculation procedures [4,5] are available in the literature. Additional details are available in several chapters in Part II.

16.5   General Domestic Conservation Applications

Domestic applications involving energy conservation are divided into 6 topic areas. These include:

1. Cooling
2. Heating
3. Hot water
4. Cooking
5. Lighting
6. New appliances

Specific details are provided in the literature [6].

Action by Congress and state legislatures, rulings by courts pronouncements by important people, or wishing alone cannot solve the energy problem. Individual efforts by everyone can make things happen and can help to win the battle against wasting energy. Each individual is important in that battle. An individual working alone or cooperating with neighbors, working with schools and colleges, with industry, with government, and with nonprofit organizations can make a difference. Here are specific suggestions that one can implement to help reduce energy waste.

1. Purchase energy-efficient automobiles
2. Purchase energy-efficient recreation vehicles
3. Purchase efficient appliances
4. Purchase energy-efficient toys
5. A well-tuned internal combustion engine makes a car, boat, lawnmower, or tractor more efficient and safer to both the individual and the environment
6. Carpooling, biking, walking, and using mass transit results in less pollution and energy savings
7. Use natural ventilation in the automobile whenever possible
8. Use natural ventilation in the home whenever possible
9. Avoid travel/trips that are not necessary
10. Do not waste food
11. Do not overeat
12. Make a conscious effort to operate on an energy efficient basis

16.6   General Commercial Real Estate Conservation Applications

Perhaps the major conservation measures in the future will occur in the commercial property arena since practitioners have come to realize the enormous financial investment possibilities that exist today. As a result, interest in this area has increased at a near exponential rate. The initial energy efficiency investments made in the commercial real estate (CRE) market have been associated with lower cost improvements having relatively short payback periods (less than 2-3 years) and involving low technology risk. As a result, the CRE industry now has the opportunity to move from this initial phase of low cost, short payback energy efficiency improvements to the multifaceted second phase of implementing deep energy retrofits where the capital need is much more intensive and the payback period is often longer. These technology changes range from minor changes that can be implemented quickly and at low cost, to major changes involving replacement of process equipment or processes at a very high cost. The challenge associated with these deeper, more capital-intensive energy efficiency retrofit improvements is complicated when internal financing is limited or not available. Fortunately, this is changing, and the market-ready, commercially-attractive financing mechanisms have arisen to meet these needs [7]. As noted above, energy conservation projects involving commercial properties are certain to increase in the future. This activity will be enhanced by:

1. convincing CRE properties to operate in a more efficient manner, and
2. providing incentives, e.g., underwriting loans, etc., to implement attractive energy efficient projects

This topic is reviewed in the next section, Problem 2.

16.7   Architecture and the Role of Urban Planning [6,7]

As energy concerns present some of the most pressing issues to the world, both professional and academic architects have begun to address how planning and *built form* affect society. Although the term *built environment* has come to mean different things to different people, one may state in general terms that it is the result of human activities that impact society. It essentially includes everything that is constructed or built, i.e., all types of buildings, chemical plants, utilities, arenas, roads, railways, parks, farms, gardens, bridges, etc. Thus, the built environment includes everything that one can describe as a structure or "green" space. Generally the built environment is organized into 6 interrelated components:

1. Products
2. Interiors
3. Structures
4. Landscapes
5. Cities
6. Regions

While "architecture" may appear to be one of the many contributors to the current environmental state, in reality, energy consumption and pollution affiliated with the materials, the construction, and the use of building contributes to most of the major environmental problems. In fact, architectural planning, design, and building can significantly contribute to the destruction of the rain forest, the extinction of plant and animal species, the depletion of nonrenewable energy sources, the reduction of the ozone layer, the proliferation of chloroflourocarbons (CFCs), exposure to carcinogens and other hazardous materials, and potential global warming problems. Where one chooses to build, which construction materials are selected, how a comfortable temperature is maintained or what type of transportation is needed to reach it—each issue decided by both architect and user—significantly impacts these overall conditions. Sadly, despite these opportunities to shape a healthier future, an analysis of American planning and building has in the past represented an assault on the existing energy ecological conditions.

Most architects today are committed to build "green". Most new buildings will incorporate a range of green elements including: radiant ceiling panels that heat and cool—saving energy and improving occupant comfort; a cogeneration plant that utilizes waste heat; a green roof that is irrigated exclusively with rainwater in order to mitigate the heat island effect; materials that are rapidly renewable and regionally manufactured, etc. Additionally, buildings are being designed to maximize weather patterns, day-lighting, and air circulation. For example a birds nest design has been emulated that is efficient, withstanding wind loads and wind shear, while simultaneously enabling light and air to move through it. Throughout the building process, construction, and demolition, energy is conserved and waste is recycled. Measurement and verification plans are also being employed to track utility usage for sustainability purposes.

Urban planners, who used to be called architects, are employing designs that operate like a wall of morning glories—adjusting to sunlight throughout the day, both regulating light and gathering solar energy. In effect, the design can often create an energy surplus that can be employed elsewhere in the system and/or process.

16.8   The U.S. Energy Policy/Independence [1,8]

Energy is central to all current and future human activities. That being said, this nation is approaching a crisis stage. Historically, the primary impetus for the emergence of the U.S. as a global leader has been its ability to satisfy energy needs independently. Without a strategy for determining the most beneficial path to achieve energy independence, the U.S. is positioning itself to be at the mercy of those that will control energy.

The current energy policy being pursued by the U.S. is disjointed, random, and fraught with vested interests. Achieving *energy independence* by pursuing current approaches will be costly, inefficient and disruptive, especially as resources diminish. In addition, environmental impacts will become greater concerns, and the consequences of continuing on the current course may be surprisingly difficult to correct.

Skipka [8] has suggested one approach to developing an energy policy that would achieve energy independence. He has proposed investigations and solutions that will be performed under a three-phase study program. The first phase will be devoted to finding sponsorship for: organizing research and analytical teams; defining team work scopes and objectives; and conducting literature reviews, outreach, matrix formulation, plan refinements, industrial/commercial/governmental coordination, and other tasks.

The second phase will involve: finalization of teams, work scopes and objectives; setting schedules; finalizing matrix criteria; coordinating teams; initiating work scopes, periodic reviews and realignments, cost/benefit and fatal/flaw analyses; draft findings/documents; industrial/commercial/governmental coordination; and other tasks.

The final phase will be implementation of the optimal energy strategy defined during the first two phases. The study group will be supplemented with industrial, academic, and business consultants.

The project will attempt to maximize long-term social return per unit of energy consumed. As such, other alternatives may also have important implications for the economic system, for social processes, and for lifestyles. Topics (where applicable), such as resource quantity, resource availability, economics, energy quality, conservation requirements, transportation requirements, delivery requirements, operation and manufacturing, regulatory issues, political issues, environmental concerns, cost consequences, advantages, disadvantages, and public acceptance, will be studied. Another feature of the project will include an analysis that provides a rated, quantitative detailed review and practical evaluation of all viable energy options, categories (see earlier sections), and corresponding weighting factors that are contained in this analysis. These considerations presently define the energy issues and can provide a means of solving and managing energy problems that exists today while at the same time defining the optimal course for future generations.

A detailed and expanded treatment of energy management is available in the following two references:

1. K. Skipka and L. Theodore, *Energy Resources: Past, Present, and Future Management*, CRC Press/Taylor & Francis Group, Boca Raton, FL, 2014 [1].
2. L. Theodore, *Chemical Engineering: The Essential Reference*, McGraw-Hill, New York City, NY, 2014 [9].

16.9   Illustrative Open-Ended Problems

This and the last section provide open-ended problems. However, solutions *are* provided for the three problems in this section in order for the reader to hopefully obtain a better understanding of these problems, which differ from the traditional problems/illustrative examples. The first problem is relatively straightforward while the third (and last problem) is somewhat more difficult and/or complex. Note that solutions are not provided for the 37 open-ended problems in the next section.

Problem 1: Discuss the general subject of energy conservation from a layman's perspective.

Solution: Many still remember the fuel shortages of the 1970s with long lines at filling stations, and buildings under-heated or closed in the winter. Others remember, more nostalgically, gasoline prices of 25 cents per gallon and spending the evening driving around on a dollars worth of gas. Gasoline at that price is long gone, probably never to return. And a return to the days of fuel shortages may be lurking just around the corner. In a highly industrialized society, Americans are by far the largest energy users in the world and must seek ways to conserve energy. Small improvements in the efficient use of energy translate into the release of large amounts of coal, fuel oil, and gasoline to be used for other purposes or that may be banked for future generations. Such increased efficiency also reduces the production of air pollutants, such as the greenhouse gas, carbon dioxide, which may contribute to global warming. Emissions of sulfur and nitrogen oxides, which are involved in the production of acid rain, are also reduced when energy is conserved.

Whatever the reasons, energy conservation should be an important part of any energy-planning strategies. There are many things, both small and inexpensive and large and costly, which can be done to improve the manner in which one uses this valuable resource.

Problem 2: Discuss why the commercial real estate (CRE) market is "ripe" for energy conservation activity in the future.

Solution: The commercial real estate (CRE) market in the U.S. consists of approximately 4.8 million office, retail, service, lodging, multifamily, warehouse, and storage buildings, representing a significant opportunity for building owners to reduce energy use and monetize their energy savings. Moreover, it is now evident to CRE owners and lenders that a building's energy performance can impact property value. As a result, less energy-efficient buildings are at a growing competitive disadvantage and in danger of accelerated obsolescence.

The above market developments have stimulated a number of retrofit projects designed to increase energy efficiency on any projects involving properties that still rely on original mechanical and electrical equipment often near the end of their useful life; this has resulted in a substantial demand for equipment upgrades and replacement. Assuming this to be true, the floodgates holding back the demand for equipment replacement and upgrading may finally be at the cusp of opening. This dynamic will represent a significant opportunity for replacing or upgrading dated energy-consuming equipment with more efficient units. The end result of this powerful business driving force will likely result in rapid acceleration of the deep energy efficiency retrofit market.

As noted earlier, the initial energy efficiency investments made in the CRE market have been associated with lower-cost improvements having relatively short payback periods (less than 2-3 years) and involving low technology risk. As a result, the CRE industry now has the opportunity to move from this initial phase of low-cost, short-payback energy efficiency improvements to the multifaceted second phase of implementing energy retrofits (defined as resulting in at least a 30% reduction in whole-building energy use) where the capital need is much more intensive and the payback period often longer.

The execution challenge associated with deeper, more capital-intensive energy efficiency retrofit improvements is complicated when internal financing is limited or not available. While some financing for energy efficiency upgrades has been available to CRE owners, the availability of "commercially-attractive" financing often has not. Fortunately, this is changing, and market-ready, commercially attractive financing mechanisms have arisen to meet the need.

Problem 3: It would normally seem that the thicker the insulation, the less the heat loss; i.e., increasing the insulation should reduce heat loss to the surroundings. But, this is not always the case. There is a "critical insulation thickness" above which the system will experience a greater cost to prevent heat loss due to an increase in insulation thickness. This situation arises for "small" diameter pipes when the increase in area increases more rapidly than the resistance opposed by the thicker insulation. Provide a technical analysis of this phenomenon.

Solution: The reader is first referred to Chapter 6 for a solution to this problem. Consider the system shown in Figure 16.1. The area terms for the heat transfer equations in rectangular coordinates are no longer valid and cylindrical coordinates must be employed. Applying the energy equation to a pipe or cylinder leads to

$$q = \frac{T_i - T_o}{\dfrac{1}{2\pi r_i L}\left(\dfrac{1}{h_i}\right) + \dfrac{\Delta x_w}{k_w 2\pi L r_{\text{lm},w}} + \dfrac{\Delta x_i}{k_i 2\pi L r_{\text{lm},i}} + \dfrac{1}{2\pi r L}\left(\dfrac{1}{h_o}\right)}$$

$$= \frac{2\pi L(T_i - T_o)}{\dfrac{1}{r_i h_i} + \dfrac{\ln(r_o / r_i)}{k_w} + \dfrac{\ln(r / r_o)}{k_i} + \dfrac{1}{r h_o}} = \frac{2\pi L(T_i - T_o)}{f(r)} \qquad (16.1)$$

where

$$f(r) = \frac{1}{r_i h_i} + \frac{\ln(r_o / r_i)}{k_w} + \frac{\ln(r / r_o)}{k_i} + \frac{1}{r h_o} \qquad (16.2)$$

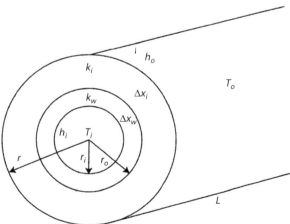

Figure 16.1 Critical insulation thickness for a pipe

The development presented applies when r is less than r_c. If r_o is larger than r_c, the above analysis again applies, but only to the results presented for $r > r_c$, i.e., r_c will decrease indefinitely as r increases. Note that there is *no* maximum/minimum (inflection) for this case since values of $\ln(r/r_o)$ are indeterminate for $r < r_o$. Once again, g approaches zero in the limit as r approaches infinity.

As noted earlier in Chapter 6, as the thickness of the insulation is increased, the cost associated with heat lost decreases, but the insulation cost increases. The optimum thickness is determined by the minimum of the total costs. Thus, as the thickness of the insulation is increased, the heat loss reaches a maximum value and then decreases with further increases in insulation. Reducing this effect can also be accomplished by using an insulation with low thermal conductivity.

16.10   Open-Ended Problems

This last section of the chapter contains open-ended problems as they relate to energy management. No detailed and/or specific solution is provided; that task is left to the reader, noting that each problem has either a unique solution or a number of solutions or (in some cases) no solution at all. These are characteristics of open-ended problems described earlier.

There are comments associated with some, but not all, of the problems. The comments are included to assist the reader while attempting to solve the problems. However, it is recommended that the solution to each problem should initially be attempted *without* the assistance of the comments.

There are 37 open-ended problems in this section. As stated above, if difficulty is encountered in solving any particular problem, the reader should next refer to the comment, if any is provided with the problem. The reader should also note that the more difficult problems are generally located at or near the end of the section.

1. Describe the early history associated with energy management.
2. Discuss the recent advances in the general field in energy management.
3. Select a refereed, published article on energy management from the literature and provide a review.
4. Provide some normal everyday domestic applications involving the general topic of energy management.
5. Develop an original problem on energy management that would be suitable as an illustrative example in a book.

6. Prepare a list of the various technical books which have been written on energy management. Select the three best (try to include a book written by one of the authors) and justify your answer. Also select the three weakest books and, once again, justify your answer.

7. List and discuss the various life expectancies for the following fuels:
   - Coal
   - Oil
   - Natural gas
   - Solar
   - Fission
   - Fusion

   Comment: Refer to the literature [1] for some assistance.

8. Provide at least a dozen energy conservation measures that can be implemented at home.
   Comment: Refer to the literature [1,6] for some assistance.

9. Provide at least a dozen energy conservation measures that can be implemented at the office.
   Comment: Refer to the literature [1,6] for some assistance.

10. Provide at least a dozen energy conservation measures that can be implemented at a chemical plant.
    Comment: Refer to the literature[1,6] for some assistance.

11. Discuss the role insulation plays in energy management.
    Comment: See Problem 3 in the previous section

12. List and discuss the various types of insulation currently available. Outline how one might go about developing a new insulation.

13. Discuss the benefits of including an entropy/exergy analysis in energy conservation analyses.

14. For the purposes of implementing an energy conservation strategy, process changes and/or design can be divided into four phases, each presenting different opportunities for implementing energy conservation measures. These include product conception, laboratory research, process development, and mechanical design. Discuss each of these measures.

15. Discuss the present state of the nuclear power industry in the U.S. Extend the discussion to that of the world.

16. Provide a layman with a definition of fission and fusion, and detail the difference.

17. Provide a technical definition of fission and fusion, and detail the difference.

18. Ivory Tower University runs its own coal-fired power plant, consuming Utah bituminous coal with an energy content (in the combustion literature, energy content is defined as the lower heating value, LHV) of 25,000 kJ/kg. The coal contains, on average, 1.0 wt% sulfur and 1.2 wt% ash (based on the total mass of the coal). The power plant is 35% efficient (indicating that 35% of the energy in the coal is actually converted to electrical energy), and is operated at a 2.0 megawatt average daily electrical load (ADL). Assume that the coal is completely burned during combustion, and also that the power plant captures 99% of the ash and 70% of the sulfur dioxide produced during combustion. The current coal price is $120/ton delivered to the university. Ash-hauling charges to the regional landfill are $40/metric tonne (1 metric tonne = 1,000 kg = 2,205 pounds = 1.1025 US tons).

After a US EPA Green Lights energy audit, Ivory Tower finds that it can install energy-efficient lighting and reduce its average daily electrical generation. The materials and labor costs for the energy-efficient lighting upgrades are $350,000, which Ivory Tower will pay from cash on hand.

Using the information given above, calculate the average reduction in electrical load, and the new average daily load for the power plant. Base the calculations on reductions (due to the justification of energy-efficient lighting) in the 10-50% range. Comment on the results.

19. Refer to the previous problem. Using the efficiency of the power plant, the heating value of the coal, and the results from Problem 18, calculate the daily reduction in the quantity of coal (kg/d) consumed by the university's power plant for each percentage range value. Again, comment on the results.

20. Using the results from the two previous problems, calculate the daily reduction in the quantity of ash (kg/d) produced when the university implements this energy-saving lighting program. Comment on the results.

21. Using the results from the previous problems, calculate the annual reduction in cost of coal supplied to the power

plant when the university implements this energy-saving lighting program. Comment on the results.

22. Using the results from the previous problems, calculate the annual reduction in cost of ash hauled to the landfill when the university implements this energy-saving lighting program. Comment on the results.

23. Refer to the previous problems. Sulfur dioxide (SO_2) is produced from the coal's sulfur during combustion. Using this information, calculate the annual reduction (metric tonnes/year) of SO_2 emissions that will result from the various reduction possibilities in average daily electrical generating needs. Comment on the results.

24. It took about 20,000 Btu fuel input to produce 1kW·h of electricity in 1900. Estimate the efficiency of conversion and compare it with a typical value for today's power industry.

    Comment: Refer the to the utility industry literature for conversion efficiency values.

25. A steam turbine in a power plant operates between the temperatures of 300 and 900 K. The efficiency has already been increased to the maximum by mechanical changes, and the plant is now trying to improve the efficiency thermodynamically. One can increase the operating temperature of the heat source by 30 K or decrease the temperature of the heat sink (the cool reservoir) by 30 K. Which is most effective? Do you think it would be a significant improvement to use water from the bottom of a lake rather than the top for cooling water? Repeat the above calculating for temperature increases/decreases in the 20-40 K range.

26. Even with an aggressive energy conservation program, the Earth's growing population will continue to demand increasing amounts of electricity. Identify and describe the environmental impacts, both positive and negative, of the following alternative means of power generation: coal-fired steam boilers; nuclear power; photovoltaic solar panels; and hydroelectric dams.

27. The opportunity for energy conservation in the field of transportation is enormous. Forty-five million gallons of gasoline could be saved for future generations *each day in the U.S. alone* with an increase in fuel efficiency of only 2 mpg. And, as has been stressed by many, this is a "win-win"

situation, since a reduction in pollution is proportional to the reduction in fuel use. One partial solution to the production of combustion pollutants is the addition of gasoline additives to increase the amount of oxygen in the fuel. Methyl-*tert*-butyl ether (MTBE) is currently used as an oxygenating fuel additive, providing more power per gallon as well as more complete gasoline combustion in internal combustion engines. This is achieved at a cost of about 1 cent per gallon of fuel.

Assuming that the average fuel usage rating for automobiles in the U.S. is 18 mpg and that the increase in efficiency with the MTBE additive is 5%, calculate the savings in gallons of fuel in the United States per year if MTBE were used in every car. Repeat the calculation for various mpg and additive values. Also analyze the results.

28. An important consideration for the use of the MTBE fuel additive described in the previous problem is the "break-even" point. This occurs when the cost of the additive equals the savings in gasoline. Given that the use of MTBE costs 1 cent per gallon and the cost of gasoline averages at $3.20 per gallon nationwide, calculate the "break-even" point for its use. Use any relevant data from the previous problem in the solution. Comment on the results.

29. Discuss the use of the MTBE fuel additive in light of the results of the two previous problems.

Comment: Two additional factors have been recently added to the arguments against MTBE addition. First, MTBE is thought to be a possible health threat to consumers as they pump their own gas, and secondly, MTBE is highly mobile in groundwater, has generated large groundwater plumes below a number of leaking underground storage tank sites, and poses a potential health threat due to contamination of drinking water supplies.

30. A good plan for energy conservation can include a variety of strategies. One of these should be recycling of materials. Most recycling efforts have multiple positive benefits, including reducing energy demands and landfill space, as well as reducing the pollution of land, air and water.

Society continues in its transition from a "throw-away" society to one that is giving more thought to each consumer product and its containers. This change has been brought about by a variety of pressures, including dwindling landfill space and the increased cost of solid waste disposal. Incentives in the form of bottle deposits and recycling efforts of communities have helped in this transition. The engineers at the Steincke Quality Container Corporation have designed three new containers and are making a presentation to the Vice President of Marketing. She will make the choice of which container to use to package a new product. Differences in appearance of the three options are negligible and none has an effect on the product in any way.

Using the data below, Table 16.1 help the Vice President make the energy-conscious decision with regard to this container.

Table 16.1 Container Information, Problem 30

	Container A	Container B	Container C
E_1	200 J/unit	450 J/unit	650 J/unit
E_2	0 J/unit	250 J/unit	150 J/unit

Note: E_1 = energy to produce the container, and E_2 = energy to manufacture the container from recycled materials.

Assume that the raw materials cost to produce the item are the same for each container. The only difference is in the amount of energy needed to improve the structure of the container to make it more durable. Container A cannot be recycled, while containers B and C can be recycled. However, Container B can only be recycled five times, on average; Container C can be recycled up to 20 times without deterioration of performance.

- Create a table for each container showing the number of units required for each recycle and the associated costs, E_1, E_2, and Total E. Calculate the energy costs for one million unit-uses at a recycle rate of 25%.
- Which container would you advise the Vice President to select, based only upon these energy considerations? Justify your answer.

31. Using the data from the previous problem, calculate the number of times Containers B and C would have to

be recycled to make them competitive with Container A, assuming a recycle rate (RR) of various values in the 75-95% range. Comment on the results.

32. The authors have often stated that most technical individuals believe that the world's energy problems will be solved by either the nuclear or solar route. Express your position on this issue.

33. The authors of this book have voiced a concern that employing windmills to recover the energy of the wind for useful purposes may adversely affect the rotation of the Earth. Comment on this concern.

34. Develop an energy policy for the U.S.
    Comment: Refer to the literature [1] for some assistance.

35. Develop an energy policy for a third world country.
    Comment: Refer to the literature [1] for some assistance.

36. Develop an energy policy for China.
    Comment: Refer to the literature [1] for some assistance.

37. In terms of future activities, discuss whether energy management or water management will be more important. (See also Chapter 17).

References

1. Adapted from: K. Skipka and L. Theodore, *Energy Resources: Past, Present, and Future Management*, CRC Press/Taylor & Francis Group, Boca Raton, FL, 2014.

2. J. Santoleri, J. Reynolds, and L. Theodore, *Introduction to Hazardous Waste Incineration*, 2nd edition, John Wiley & Sons, Hoboken, NJ, 2002.

3. J. Reynolds, J. Jeris, and L. Theodore, *Handbook of Chemical and Environmental Engineering Calculations*, John Wiley & Sons, Hoboken, NJ, 2004.

4. J.P. Abulencia and L. Theodore, *Fluid Flow for the Practicing Chemical Engineer*, John Wiley & Sons, Hoboken, NJ, 2009.

5. L. Theodore and F. Ricci, *Mass Transfer Operations for the Practicing Engineer*, John Wiley & Sons, Hoboken, NJ, 2011.

6. M.K. Theodore and L. Theodore, *Introduction to Environmental Management*, CRC Press/ Taylor & Francis Group, Boca Raton, FL, 2010.

7. L. Theodore, unpublished article, *The Case for Energy Conservation: Commercial Property, Industrial Sector and Domestic Level*, East Williston, NY, 2013.

8. Personal notes: K. Skipka, Carle Place, NY, 2013.

9. L. Theodore, *Chemical Engineering: The Essential Reference*, McGraw-Hill, New York City, NY, 2014.

17

Water Management

This chapter is concerned with process water management. As with all the chapters in Part II, there are several sections: Overview, several technical topics, illustrative open-ended problems open-ended problems. The purpose of the first section is to introduce the reader to the subject of water management. As one might suppose, a comprehensive treatment is not provided although several technical topics are also included. The second section contains three open-ended problems; the authors' solutions (there may be other solutions) are also provided. The third (and final) section contains 46 problems; *no* solutions are provided here.

17.1   Overview

This overview section is concerned with—as can be noted from its title—water management. As one might suppose, it was not possible to address all topics directly or indirectly related to water management. However, additional details may be obtained from either the references provided at the end of this Overview section and/or at the end of the chapter.

Note: Those readers already familiar with the details associated with this subject may choose to bypass this Overview.

Traditionally, water management emphasized the importance of this natural resource for navigation, commerce, farming and agriculture, consumption, and recreation. More recently, water policy discussions have expanded to encompass the sustainable use of water for both current and future generations. From a broad perspective, water management incorporates aspects of addressing water as both a natural resource *and* a commodity. More scrutiny is being placed on water management priorities competition for readily available water increases.

The authors believe that the number one global *environmental* problem is the lack of potable water. Perhaps it is or will soon become the number one global problem. At a minimum, it will achieve greater significance in the years ahead [1]. Nations go to war over oil (or other natural resources) but there are resources that can replace oil. Water? There really is no substitute. And therein lies the problem, particularly with many of the undeveloped nations.

The EPA, in partnership with state and local governments is responsible for improving and maintaining water quality. These efforts are centered around one theme: maintaining the quality of drinking water. This is addressed by monitoring and treating drinking water prior to consumption and by minimizing the contamination of surface waters and protecting against contamination of ground water needed for human consumption.

The most severe and acute public health effects from contaminated drinking water, such as cholera and typhoid, have been essentially eliminated in the U.S. However, some less acute and immediate hazards remain in the nations tap water. These hazards are associated with a number of specific contaminants in drinking water. Contaminants of special concern to the EPA are lead, radionuclides, microbiological contaminants, and disinfection byproducts. These are detailed below.

The primary source of lead in drinking water is corrosion of plumbing materials such as lead service lines and lead solders in water distribution systems, as well as in houses and larger buildings. Virtually all public water systems serve households with lead solders of varying ages, and most faucets are made of materials that can contribute some lead to drinking water.

Radionuclides are radioactive isotopes that emit radiation as they decay. The most significant radionuclides in drinking water are radium, uranium, and radon, all of which occur naturally in nature. While radium and uranium enter the body by ingestion, radon is usually inhaled after being released in the air during showers, baths, and other activities such as washing clothes or dishes. Radionuclides in drinking water occur primarily in those systems that use ground water. Naturally occurring radionuclides are seldom found in surface waters (such as rivers, lakes, and streams) [1].

Water contains many microbes – bacteria, viruses, and protozoa. Although some organisms are harmless, others can cause disease. Contamination continues to be a national concern because contaminated drinking water systems can rapidly spread disease.

Disinfection byproducts are produced during water treatment by the chemical reactions of disinfectants with naturally occurring or synthetic organic materials present in untreated water. Since these disinfectants are essential to ensure safe drinking water, the EPA is presently looking at ways to minimize the risks from byproducts.

Drinking-water safety cannot be taken for granted. There are many chemical and physical threats to drinking certain water supplies. Chemical threats include contamination from improper chemical handling and disposal, animal wastes, pesticides, human waste, and naturally occurring substances. Drinking water that is not properly treated or disinfected, or that travels through an improperly maintained distribution system, can also become contaminated and subsequently pose a health risk. Physical threats include failing water supply infrastructure and threats posed by tampering or terrorist activity [2]. These topics receive treatment in this chapter. The remainder of the chapter addresses the following areas:

1. Water as a Commodity and Human Right
2. The Hydrologic Cycle
3. Water Usage
4. Regulatory Status
5. Acid Rain
6. Treatment Processes
7. Future Concerns

It should be noted that much of the material has been adapted from the work of Carbonaro [2].

17.2   Water as a Commodity and as a Human Right

From a commodity perspective, arguments are made that assessing a price for water promotes investment in developing reliable water sources and encourages conservation. Placing a value on water will have the positive affect of encouraging better water management policies. Conversely, the lack of an economic driver to penalize water use will invariably work against conservation and reuse.

The term "full-cost pricing of water" recently has worked its way into the debate on water management. The recognition of the full cost of accessing and supplying water highlights the value and economic impact of providing water in sufficient quantity and quality to meet the desired needs of society.

There is another water policy perspective that is gaining more widespread attention. Should water be managed as a human right? In effect, all people should have access to clean and adequate water resources for basic personal and domestic needs. One can argue that the human right aspect of water management is an integral part of sound water management practices. However, the human right perspective of water supply and management is an evolving principal with likely increasing impacts to sustainable water management in the coming years. Conflicts exist between all the various water management strategies. Yet, all play a role in providing a sustainable water supply. The importance of water management in protecting the environment and providing a sustainable and affordable water supply is evolving, and needs to be more fully addressed in the future.

17.3 The Hydrologic Cycle [2]

Water is the original renewable resource. Although the total amount of water on the surface of the Earth remains fairly constant over time, individual water molecules carry with them what the authors describe as a rich history. The water molecules contained in the water one drank yesterday may have fallen as rain last year from a distant place or could have been used decades, centuries, or even millennia ago by one's ancestors.

Water may be assumed to be always in motion, and the hydrologic cycle describes this movement from place to place. The vast majority (96.5%) of water on the surface of the Earth is contained in the oceans. With respect to the cycle, solar energy heats the water at the ocean surface and some of it evaporates to form water vapor. Air currents take the water vapor up into the atmosphere along with water transpired from plants and evaporated from the soil. The cooler temperatures in the atmosphere cause the vapor to condense into clouds. Clouds transverse the world until the moisture capacity of the could is exceeded (supersaturated) and the water falls as precipitation. Most precipitation in warm climates is displaced into the oceans or onto land, where the water flows over the ground as surface *run-off*. Runoff can enter rivers and streams which in turn transport the water to the oceans, accumulate, and be stored as freshwater for long periods of time. In cold climates, precipitation often falls as snow and can accumulate as ice caps and glaciers which can store water for thousands of years [3,4]

Throughout this cycle, water picks up contaminants originating from both naturally occurring and anthropogenic sources. Depending upon the type and amount of contaminant present, water present in river, lakes and streams or beneath the ground may become unsafe for use.

17.4   Water Usage

The term *natural waters* consist of surface waters and ground waters. The term *surface water* refers to the freshwater in rivers, streams, creeks, lakes, and reservoirs, and the saline's water present in inland seas, estuaries and the oceans. The source of freshwater is vitally important to everyday life. The main uses of surface water include drinking water and other public uses, irrigation uses, and for use by the thermoelectric power industry to cool the electricity-generating equipment.

17.5   Regulatory Status [5]

The development to follow in this section will solely address the Safe Drink Water Act. A more comprehensive treatment is available in the literature [5].

17.5.1   The Safe Drinking Water Act (SDWA)

The first legislation enacted in the U.S. to protect the quality of drinking water was the Public Health Service Act (PHSA) of 1912. The PHSA brought together the various federal health authorities and programs at that time, such as the Public Health Service and the Marine Hospital Service, under one statute. The PHSA authorized scientific studies on the impact of water pollution and human health, and introduced the concept of water quality standards. True national drinking water standards were not established, however, until 60 years later with the Safe Drinking Water Act.

The SDWA, originally passed by Congress in 1974, was authorized by the EPA to set national health-based standards for drinking water to protect against both naturally occurring and man-made contaminants that may be found in drinking water. Since its enactment, there have been over 10 major revisions and additions, the most substantial changes occurring in earlier amendments in 1986 and 1996.

The SDWA applies to every public water supply systems (PWS) in the U.S., and approximately 87% of all water used in the U.S. is drawn from PWSs [6]. There are currently more than 160,000 PWS systems currently in the U.S. PWS include: municipal water companies, homeowner associations, schools, businesses, campgrounds, and shopping malls. The EPA works with these PWS systems, along with state and city agencies, to assure that these standards are met. Originally, the SDWA focused primarily on treatment as the means of providing safe drinking water. The 1996 amendments greatly expanded the existing law which now includes source water protection, protection of wells and collection systems making certain water is treated by qualified operators, funding for water system improvements, and making information available to the public on the quality of their local drinking water.

Drinking water standards are regulations that EPA has established to control the concentration of contaminants in the drinking water supply. In most cases, EPA delegates responsibility for implementing drinking water standards to states (and tribes). Drinking water standards apply to PWSs, which provide water for human consumption through at least 15 service connections, or regularly serve at least 25 individuals.

The SDWA 1996 Amendments also required the EPA to identify potential drinking water problems, establish a prioritized list of chemical of concern, and establish a set of standards where appropriate. Peer-reviewed science and data are required to support an intensive technological evaluation which includes many factors such as: the occurrence of the chemicals in the environment; human exposure and risks of adverse health effects in the general population and sensitive subpopulations; analytical methods of detection; technical feasibility; and, impacts of any regulations on water systems, the economy, and public health.

National Primary Drinking Water Regulations (NPDWRs or primary standards) are the legally enforceable Maximum Contaminant Levels (MCLs) and treatment techniques that apply to public water supply. The contaminants are divided up into the following groups, according to the type of contaminant: inorganic chemicals; organic chemicals; microorganisms; disinfectants; disinfection byproducts; and, radionuclides. A list of these contaminants and their respective standard is available in the literature [2,5].

National Secondary Drinking Water Regulations (NSDWRs or secondary standards) are nonenforceable guidelines regulating contaminants that may cause cosmetic effects (such as skin or tooth discoloration) or aesthetic effects (such as taste, odor, or color) in drinking water. The EPA recommends secondary standards for water systems but does not require systems to comply; state and local agencies may choose to adopt these as enforceable standards. The SDWA also includes a process where new contaminants are identified that may require regulation in the future with a primary standard. EPA is required to periodically release a Contaminant Candidate List (CCL) which is used to prioritize research and data collection efforts to help determine whether a specific contaminant should be regulated.

17.5.2   The Clean Water Act (CWA)

Along with the SDWA, the CWA has played an important role in assuring and maintaining the safety of both the sources and quality of drinking water. Growing public awareness and concern for controlling water pollution led to the enactment of the Federal Water Pollution Control Act Amendments of 1972. As amended in 1977, this law became commonly known as the CWA. The CWA established the basic structure for regulating discharges of pollutants into the waters of the U.S. It gave EPA the authority to implement pollution control programs such as setting wastewater standards. The CWA also extended earlier requirements to set water quality standards for all contaminants in surface waters. The CWA made it unlawful for any person or organization to discharge any pollutant from a source into navigable waters unless a permit was obtained that dictated the terms of the release. It also funded the construction of wastewater treatment plants under the construction grants program. Pollutants regulated under the CWA include biochemical oxygen demand (BOD), total suspended solids (TSS), fecal coliform, oil and grease, toxic chemicals (priority pollutants); and, various contaminants not identified as either conventional or priority (nonconventional pollutants).

The CWA introduced a *permit system* for regulating *point sources* of pollution. A point source is defined as a single identifiable and localized source of a contaminant. Point-source pollution can usually be traced back to a single origin or source. Examples of point source include industrial facilities (e.g., manufacturing, mining, oil and gas extraction, etc.), municipal and some agricultural facilities (e.g., animal feedlots), etc. Point-sources are not allowed to be discharged into surface waters without a permit from the National Pollutant Discharge Elimination System (NPDES). This system is managed by the EPA in partnership with the pertinent state environmental agencies. EPA has authorized 45 states to issue permits directly to the discharging facilities. The EPA regional office directly issues permits in the remaining water quality-based standards states and territories [7].

Water quality standards (WQS) are risk-based requirements which set site-specific allowable pollutant levels for individual water bodies such as rivers, lakes, streams, and wetlands. A water quality standard defines the water quality goals of a water body by designating the use(s) to be made of the water (e.g., recreation, water supply, aquatic life, and agriculture), by setting criteria necessary to protect the users, and by preventing degradation of water quality through anti-degradation provisions. The criteria are numeric pollutant concentrations similar to an MCL for drinking water. States adopt water quality standards to protect public health or welfare, enhance the quality of water, and serve the purposes of the CWA.

A *total maximum daily load* (TMDL) is defined as the maximum amount of a pollutant that a water body can receive and still meet WQS. It is the collective sum of the allowable loads of a single pollutant from all contributing point and non point sources. The calculation includes a *margin of safety* to ensure that the water body can be used for the purposes the state has designated [7].

17.6   Acid Rain

Acid rain with a pH below 5.6 is formed when certain anthropogenic air pollutants travel into the atmosphere and react with moisture and sunlight to produce acidic compounds. Sulfur and nitrogen compounds released into the atmosphere from different industrial sources are now believed to play the biggest role in the formation of acid rain. The natural processes which contribute to acid rain include lightning, ocean spray, decaying plants and bacterial activity in the soil and volcanic eruptions. Anthropogenic sources include those utilities, industries, businesses, and homes that burn fossil fuels, plus motor vehicle emissions. Sulfuric acid is the type of acid most commonly formed in areas that burn coal for electricity, while nitric acid is more common in areas that have a high density of automobiles and other internal combustion engines.

There are several ways that acid rain affects the environment [7].

1. Contact with plants can harm plants by damaging outer leaf surfaces and by changing the root environment.
2. Contact with soil and water resources can harm fish and leach away nutrients in the soil. Due to the acid in the rain, fish-kills in ponds, lakes and oceans, as well as effects on aquatic organisms, are common occurrences. Acid rain can cause

minerals in the soil to dissolve and be leached away. Many of these minerals are nutrients for both plants and animals.

3. Acid rain mobilizes trace metals, such as lead and mercury. When significant levels of these metals dissolve from surface soils they may accumulate elsewhere, leading to poisoning.

4. Acid rain had been known to damage structures and automobiles due to accelerated corrosion rates.

The environmental effects of acid rain are usually classified into four general categories.

1. Aquatic
2. Terrestrial
3. Materials
4. Human

Although there is evidence that acid rain can cause certain effects in each category, the extent of those effects is uncertain. The risks that these effects may pose to public health and welfare are also unclear and difficult to quantify.

17.7 Treatment Processes

Municipal wastewater is composed of a mixture of dissolved, colloidal, and particulate organic and inorganic materials. However, municipal wastewater contains 99.9% water. The total amount of the substances accumulated in a body of waste water is referred to as the *mass loading*. The concentration of any individual component is constantly changing as a result of sedimentation, hydrolysis, and microbial transformation and degradation of organic compounds.

Treatment technologies can essentially be divided into three broad categories [8]:

1. Physical
2. Chemical
3. Biological

Many treatment processes combine two or all three categories to provide the most economical treatment. There are a multitude of treatment technologies for each of these categories.

Finally, the biochemical oxygen demand (BOD) test was developed in an attempt to reflect the depletion of oxygen that would occur in a stream due to utilization by living organisms as they metabolize organic matter. BOD is often used as the sole basis for determining the efficiency of the treatment plant in stabilizing organic matter. Effluent ammonia-nitrogen poses an analytical problem in measuring BOD. At 20°C (68°F), the nitrifying bacteria in raw domestic wastewater usually are significant in number and normally will not grow sufficiently during the 5-day BOD test to exert a measurable oxygen demand. Thus, the BOD test may require correction for nitrification to obtain a true measure of the treatment plant performance in removing organic matter.

The *chemical oxygen demand (COD)* analysis is more reproducible and less time-consuming. The COD test and the BOD test can be correlated, but the correlation ultimately gives *qualitative* value. The COD test measures the non-biodegradable as well as the ultimate biodegradable organics. A change in the ratio of biodegradable to non-biodegradable organics affects the correlation between the COD and BOD. Such a correlation is specific for a particular waste, but may vary considerable between treatment plant influent and effluent. Additional information is available in the literature [2,7,8].

17.8   Future Concerns

Is water conservation the answer? Water conservation has become an important topic in today's environmentally conscious society. Numerous urban centers are experiencing water shortages and are building dams to help curb the ever-increasing demand for water. These dams often have an undeniable impact on the natural environment, killing numerous species of animals and plants. In any event, conservation of this resource is very much needed and can be accomplished through personal conservation efforts, as well as conscientious building design. (See previous Chapter).

A detailed and expanded treatment of water management is available in the following three references.

1. R. Carbonaro, *Introduction to Environmental Management*, M.K. Theodore and L. Theodore, CRC Press/ Taylor & Francis Group, Boca Raton, FL, 2010 [2].
2. L. Theodore, *Chemical Engineering: The Essential Reference*, McGraw-Hill, New York City, NY, 2014 [9]

3. R. Thomann and J. Mueller, *Principles of Surface Water Quality – Modeling and Control*, Harper and Row, New York City, NY, 1987 [10].

17.9 Illustrative Open-Ended Problems

This and the last section provides open-ended problems. However, solutions *are* provided for the three problems in this section in order for the reader to hopefully obtain a better understanding of these problems which differ from the traditional problems/illustrative examples. The first problem is relatively straightforward while the third (and last problem) is somewhat more difficult and/or complex. Note that solutions are not provided for the 46 open-ended problems in the next section.

Problem 1: Microbial regrowth can be defined as an increase in viable microorganism concentrations in drinking water downstream of the point of disinfection after treatment. These microorganisms may be coliform bacteria, bacteria enumerated by the heterotrophic plate count (HPC bacteria), other bacteria, fungi, or yeasts. Regrowth of bacteria in drinking water can lead to numerous associated problems including multiplication of pathogenic bacteria such as *Legionella pneumophila*, deterioration of taste, odor, and color of treated water, and intensified degradation of the water mains, particularly cast iron, by creating anaerobic conditions and reducing pH in a limited area. In order to obtain stable drinking water (i.e., to control regrowth), one needs to understand the sources of biological instability. List 5 factors that affect regrowth in a water distribution system.

Solution:
Five factors that affect regrowth in a water distribution system are:

1. Water quality (e.g., concentration of organic carbon and nutrients, temperature, disinfectant residual, etc.)
2. Pipe materials and conditions.
3. Flow conditions in the distribution systems (e.g., detention time, flow velocity, wall shear stress, flow reversals, etc.)
4. Water treatment processes
5. Presence, type, concentration, and physiological state of the bacteria.

Problem 2: Land treatment of industrial wastewater is a process in which wastewater is applied directly to the land. This type of treatment is most common for food processing wastewater including meat, poultry, dairy, brewery, and winery wastes. The principal rationale of this practice is that the soil is a highly efficient biological treatment reactor, and food process-ing wastewater is highly degradable. This treatment practice is usually car-ried out by distributing the wastewater through spray nozzles onto the land or letting the water run through irrigation channels.

Suppose that the rate of the wastewater flowing to a land application site is 178 gal/acre·min and the irrigated land area is 5.63 acres. The entire irrigation process lasts 7.5 h/d. Calculate the mass of BOD_5 remaining in the soil after the land treatment process is complete (assume a BOD_5 removal efficiency by the land treatment process is 95%)? Perform the calculations for various wastewater BOD_5 concentrations, e.g., 50 mg/L. Comment on the results.

Solution: Perform the first calculation for a BOD_5 concentration of 50 mg/L. The amount of the wastewater flowing to the land is:

$$(178 \text{ gal/acre} \cdot \text{min})(60 \text{ min/h})(7.5 \text{ h/d})(5.63 \text{ acre})$$
$$= 450,963 \text{ gal/d}$$

The amount of BOD_5 remaining, $BOD_5(R)$, is

$$BOD_5(R) = (450,963 \text{ gal/d})(1.00 - 0.95)(50 \text{ mg/L})(3.785 \text{ L/gal})$$
$$= 4,267,688 \text{ mg/d}$$

$$BOD_5(R) = 4,267,688 \text{ mg/d } (1 \text{ g}/1,000 \text{ mg})(454 \text{ g/lb})$$
$$= 9.4 \text{ lb/d}$$

The reader is left the exercise of repeating the calculation for other BOD_5 concentrations and commenting on the results.

Problem 3: In 1947, two ships docked in Texas City, Texas, with tons of ammonium nitrate fertilizer and other cargo aboard. These ships caught fire, burned and exploded over a period of more than 16 hours. The explo-sions were so powerful that almost 600 people were killed and more than 3,500 were injured. The port area and much of the city was destroyed. One of the ship's anchors was thrown approximately 2 miles inland where it still lies today as a memorial to the incident.

In 1986, Congress approved Title III of the Superfund Amendments and Reauthorization Act (SARA), also known as the Emergency Planning and Community Right-to-Know Act (EPCRA) of 1986. Among other things, this law requires any facility that produces, uses or stores any chemical on a published list in excess of the "Threshold Planning Quantity" to notify local emergency response entities (such as the fire department, police department, hospitals, etc.) of the quantity, identity, and nature of these chemicals; to cooperate with a Local Emergency Planning Committee (LEPC); and, to develop an emergency plan to be used in the event of a release.

While the regulatory definition of a "facility" includes transportation vessels and port authorities for release reporting, these entities are exempted from notification and emergency planning requirements. As a result, emergency response planning against another Texas City disaster is not a requirement of the EPCRA legislation.

Prepare a list of areas of concerns that would have to be addressed if the Title III notification and emergency planning requirements were applied to port areas; particularly addressing the water management aspect of this problem.

Comment: Among other things, the reader may wish to address matters such as the short residence time of in-transit materials and the political (as opposed to legal) ramifications of applying regulation of this kind to foreign flag carriers. See also Chapter 15.

Solution: The following baker's dozen areas should be addressed in the notification and emergency planning for port areas:

1. Who shall be responsible for notification and/or emergency planning – the shipper, transporter, or port operator?
2. How shall the inventory of the materials flowing in and out of the area to be maintained?
3. Should there be a minimum storage time that triggers notification and emergency planning?
4. Should an emergency plan be developed for the release of every chemical that ever flowed through the port, even though some of those chemicals may never be present in the area again?
5. What notification and emergency planning criteria should be adopted for large quantities of listed materials that frequently flow through the port but are present for only short periods of time?
6. Should limits be placed on the quantities of some materials being stored in the port area at a given time?
7. Should ports be classified as to what materials are allowed to enter them?
8. Should segregation of cargo by compatibility groups be required for materials waiting to be loaded or trans-shipped?

9. Should port areas be rezoned to reduce the surrounding population?
10. Are evacuation plans possible for port areas in large cities?
11. If there sufficient authority under current law to accomplish this task or is new legislation required?
12. What would be the political consequences of requiring foreign ships to adhere to these regulations?
13. What would be the cost of applying these regulations to port areas?

Any other suggestions?

17.10   Open-Ended Problems

This last section of the chapter contains open-ended problems as they relate to water management. No detailed and/or specific solution is provided; that task is left to the reader, noting that each problem has either a unique solution or a number of solutions or (in some cases) no solution at all. These are characteristics of open-ended problems described earlier.

There are comments associated with some, but not all, of the problems. The comments are included to assist the reader while attempting to solve the problems. However, it is recommended that the solution to each problem should initially be attempted *without* the assistance of the comments.

There are 46 open-ended problems in this section. As stated above, if difficulty is encountered in solving any particular problem, the reader should next refer to the comment, if any is provided with the problem. The reader should also note that the more difficult problems are generally located at or near the end of the section.

1. Describe the early history associated with water management.
2. Discuss the recent advances in the water management.
3. Select a refereed, published article on water management from the literature and provide a review.
4. Provide some normal everyday domestic applications involving the general topic of water management.
5. Develop an original problem on water management that would be suitable as an illustrative example in a book.

6. Prepare a list of the various technical books that have been written on water management. Select the three best and justify your answer. Also select the three weakest books and, once again, justify your answer.
7. Prepare a list of the various physical properties of water.
8. Prepare a list of the various chemical properties of water.
9. Prepare a list of the average chemical composition of natural waters.
10. Describe some of the dissolved minerals in natural waters.
11. Describe some of the dissolved gases in natural waters.
12. Describe some of the heavy metals in natural waters.
13. Describe some of the organic constituents in natural waters.
14. Describe some of the nutrients in natural waters.
15. Provide your description of the hydrologic cycle.
16. Provide a one paragraph description of the Safe Drinking Water Act (SDWA) and the Clean Water Act (CWA).
17. Discuss both the national primary drinking water regulations and the national secondary standards for drinking water.
18. Describe the regulations associated with municipal wastewater treatment.
19. Describe the regulations associated with industrial wastewater treatment.
20. Explain why the water quality in the U.S. has improved in the last 40+ years.
21. What are the characteristics of municipal wastewater? What are the characteristics of industrial wastewater?
22. What are some of the sources of industrial wastewater pollution?
23. List and describe the various wastewater treatment processes.
24. Describe sludge characteristics.
25. Provide a layman's description of eutrophication.
26. Discuss the advantages of modeling water systems.
   Comment: Refer to the classic work of R. Thomann and J. Mueller, "Principles of Surface Water Quality – Modeling and Control," Harper and Row, New York City, NY, 1987 [10].
27. Define acid rain and offer your thoughts on controlling/reducing/eliminating acid rain.

28. List the possible sources of highly toxic hexavalent chromium (Cr^{6+}) and methods to remove it from a wastewater stream.

29. Provide your definition of advanced wastewater treatment.

30. Provide a layman definition of nonpoint source (NPS) water pollution.

    Comment: Consider the role agriculture urban runoff, abandoned mines, construction, land disposal, etc... can play in addressing this issue. Also refer to the literature for additional details [2,10]

31. Describe the complexity of factors that arise in determining water quality.

32. What are some of the water quality issues that affect the operating of water treatment plants for treating domestic water supplies?

33. What are the three major process components used to treat water sources to be used for potable water supplies. Explain why are they utilized.

34. One important aspect of water quality management is the assignment of allowable point source wastewater discharges to a receiving water body such that the water quality objectives for that water body can be maintained. This management process is called Waste Load Allocation (WLA).

    • What are some of the important steps that should be included in the WLA process?

    • What are the three major parameters regulated under the NPDES program for municipal wastewater discharges and what are the maximum concentrations allowed for each of these parameter in NPDES permits? Comment on these regulations.

35. Sludge generated in a water or waste treatment plant usually contains substantial amounts of water and therefore needs to be processed to reduce the water content (the process is called sludge dewatering) for ultimate disposal or landfilling. A municipal water treatment plant in Kansas produces 1.05 tons of sludge everyday. The wet sludge (before watering) has a density of 1.05 g/cm³, which increases after being treated. How much additional space would be needed in a landfill site annually if the wet sludge (before dewatering) were treated. How much additional space would be needed in a landfill site annually if the wet

sludge were dumped directly into the landfill site without dewatering?

Comment: This practice is no longer permitted by law because of leachate and gas production concerns in sanitary landfills.

36. Flocculation is a physical process used to encourage small particles to aggregate into larger particles, or floc. It is an essential component of most water treatment plants in which flocculation, sedimentation, and filtration processes are integrated to effectively remove suspended particles from water. Chemicals (such as alum, polyelectrolytes, etc.) are usually added to achieve agglomeration among small particles in water.

Jar tests are used to estimate the optimum amount of chemicals needed to ensure proper flocculation. In a jar test conducted for a given water treatment plant, 4L samples were poured into a series of jars. After the test, the jar that has been dosed with 20 mL of alum solution containing 5 mg Al(III)/mL showed optimal results. Calculate the pounds per day of alum [$Al_2(SO_4)_3 \bullet 14H_2O$, MW = 594.4 g/gmol] that should be added to the raw water for various water treatment flow rates, e.g., 90 Mgal/d.

37. Cyanide-bearing waste is to be treated by a batch process using alkaline chlorination. In this process, cyanide is reacted with chlorine under alkaline conditions to produce carbon dioxide and nitrogen as end products. The cyanide holding tank contains a cyanide concentration of 18 mg/L. Assuming that the reaction proceeds according to its stoichiometry, answer the following questions.

- How many pounds of chlorine are needed?
- How long will the hypochlorinator have to operate if the hypochlorinator can deliver 900 L/d of chlorine?
- How long should the caustic soda feed pump operate if the pump delivers 900 L/d of 10 wt% caustic soda solution?

Perform the above calculations for various holding tank volumes, e.g., 28 m³.

38. Natural water bodies possess a capacity to stabilize organic matter without seriously affecting their general quality and aesthetics. The process is called self-purification, which involves various microorganisms (e.g., bacteria, algae,

etc.). Draw a diagram incorporating a processes that could occur in a lake or pond when wastewater is discharged intermittently into it. Describe the carbon and nitrogen cycles related to these processes and locate where they should occur in the water body.

39. The state of Florida has contracted Theodore Partners (TP) to determine whether the present water requirements will be adequate in the future. You have been hired by TP to conduct the study. Submit a report detailing the results of your analysis.

40. Refer to Problem 2 in the previous Section. Perform the calculations for various BOD removal efficiencies at given wastewater BOD concentrations. Comment on the results.

41. Refer to the previous problem. Perform the calculations for various combnations of wastewater BOD_5 concentrations and BOD_5 removal efficiencies. Comment on the results. Also, attempt to correlate the results.

42. The following 5-day biochemical oxygen demand (BOD_5) and total suspended solids (TSS) data were collected at a local municipal wastewater treatment plant over a 7-day period. (See Table 17.1). The NPDES permit limitation for BOD_5 and TSS effluent concentrations from this wastewater treatment plant are 45 mg/L on a 7-day average. Based on this information, is the treatment plant within its NPDES

Table 17.1 Daily BOD_5 and TSS Effluent Concentration Data Collected Over a 7-Day Period at a Municipal Wastewater Treatment Plant

Day	BOD (mg/L)	TSS (mg/L)
1	45	20
2	79	100
3	64	50
4	50	42
5	30	33
6	25	25
7	21	15

permit limits? Based on your calculations, comment on how to bring the plant into compliance if it is out of compliance.

43. Estimate your daily water requirements. Then, actually calculate/record your daily water requirements for each day over a period of month. Also, calculate the mean and standard deviation of those data, and compare the calculated results with your output estimate. (See also Chapter 19).

44. Research the literature and develop present-day water availability and requirements for third-world nations.

45. Research the literature and develop water availability and requirements at the turn of the century for third-world nations.

46. One of the authors regularly visits Florida for sun-and-fun activities that include paramutual wagering. Many Floridians are concerned about a "water problem". Discuss this problem and outline what steps can be taken to alleviate future concerns in this area.

References

1. L. Theodore, *On Water*, The Williston Times, East Williston, NY, September 15, 2006.

2. R. Carbonaro, *Introduction to Environmental Management*, M.K. Theodore and L. Theodore, CRC Press/ Taylor & Francis Group, Boca Raton, FL, 2010.

3. P. Gleick, *Water in Crisis: A Guide to the World's Fresh Water Resources*, Oxford University Press, New York City, NY, 1993.

4. S. Hutson, N. Barber, J. Kenny, K. Linsey, D. Lumia, and M. Maupin, *Estimated Use of Water in the United States in 2000*, U.S. Department of the Interior US Geological Survey, Reston, VA, 2004.

5. Adapted from: L. Stander and L. Theodore, *Environmental Regulatory Calculations Handbook*, CRC Press/ Taylor & Francis Group, Boca Raton, FL, 2012.

6. MWH, *Water Treatment: Principles and Design*, John Wiley & Sons, Hoboken, NJ, 2005.

7. G. Burke, B. Singh, and L. Theodore, *Handbook of Environmental Management and Technology*, 2nd edition, John Wiley & Sons, Hoboken, NJ, 2000.

8. Personal communication, J. Jeris, lecture notes (with permission), Manhattan College, Bronx, NY, 1992.

9. L. Theodore, *Chemical Engineering: The Essential Reference*, McGraw-Hill, New York City, NY, 2014.

10. R. Thomann and J. Mueller, *Principles of Surface Water Quality – Modeling and Control*, Harper and Row, New York City, NY, 1987.

11. L. Theodore and R. Dupont, *Environmental Health and Hazard Risk Assessment: Principles and Calculations*, CRC Press/Taylor & Francis Group, Boca Raton, FL, 2012.

18

Biochemical Engineering

This chapter is concerned with biochemical engineering. As with all the chapters in Part II, there are sereval sections: Overview, several technical topics, illustrative open-ended problems, and open-ended problems. The purpose of the first section is to introduce the reader to the subject of biochemical engineering. As one might suppose, a comprehensive treatment is not provided although numerous references are included. The second section contains three open-ended problems; the authors' solutions (there may be other solutions) are also provided. The third (and final) section contains 40 problems; *no* solutions are provided here.

18.1 Overview

This overview section is concerned with—as can be noted from the chapter title—biochemical engineering. As one might suppose, it was not possible to address all topics directly or indirectly related to biochemical engineering. However, additional details may be obtained from either the references provided at the end of this Overview and/or at the end of the chapter.

Note: Those readers already familiar with the details associated with biochemical engineering may choose to bypass this Overview.

Biochemical Engineering (BChE) is a relatively new discipline in the chemical engineering profession and, as one might suspect, it has come to mean different things to different people. Terms such as biomedical engineering, bioengineering, biotechnology, biological engineering, genetic engineering, etc., have been used interchangeable by many in the technical community. To date, standard definitions have not been created to distinguish between these terms. Consequently, the authors have lumped them all together using the term BChE for the sake of simplicity. What one may conclude from all the above is that BChE involves applying the concepts, knowledge, basic fundamentals, and approaches of virtually all engineering disciplines (not only chemical engineering) to solve specific health and health-care related problems in the biochemical field; the opportunities for interaction between chemical engineers and health-care professionals are therefore many and varied.

Because of the broad nature of this subject, this chapter can only serve as an introduction. The reader is referred to the three excellent references in the literature for an extensive comprehensive treatment of this discipline [1–3]. However, the bulk of the material in the chapter has been drawn from P. Vasudevan, "Biochemical Engineering," A Theodore Tutorial, Theodore Tutorials, East Williston, NY, a text originally published by the USEPA/APTI, RTP, NC, 1994 [4].

This subject has served as the title for numerous books. Condensing this subject matter into one chapter was a particularly difficult task. In the end, the authors decided to provide a superficial treatment that would introduce the reader to the subject. As noted, Vasudevan [4] provides an excellent and detailed review of key related topics.

Following this introductory section, the chapter consists of four sections:

1. Enzyme and Microbial Kinetics
2. Enzyme Reaction Mechanisms
3. Effectiveness Factors
4. Design Procedures

Additional information is available in the work of Vasudevan for the interested reader [4]. Also note that the notation adopted by Vasudevan is employed throughout this chapter.

18.2   Enzyme and Microbial Kinetics [4]

Enzyme and microbial kinetics involve the study of reaction rates and the variables that affect these rates. It is a topic that is critical for the analysis of enzyme and microbial reacting systems. The rate of a biochemical reaction can be described in many different ways. The most commonly used definition is similar to that employed for traditional reactors (see also Chapter 8). It involves the time change in the amount of one of the components participating in the reaction or in one of the products of the reaction; this rate is also based on some arbitrary factor related to the system size or geometry, such as volume, mass, and interfacial area. In the case of immobilized enzyme catalyzed reactions, it is common to express the rate per unit mass or volume of the catalyst.

The Michaelis-Menten rate equation is as follows

$$v = \frac{v_{max}[S]}{K_m + [S]} \tag{18.1}$$

where v = reaction rate = $-\dfrac{d[S]}{dt}$

v_{max} = constant defined as the maximum reaction rate
K_m Michaelis constant (dissociation constant)
[S] = substrate or reactan concentration
t = time

Note that when K_m >> [S], the equation reduces to the following first order rate of reaction.

$$v = \frac{v_{max}}{K_m}[S] \tag{18.2}$$

For an enzyme reaction in which [S] << K_m

$$\ln\left(\frac{[S]}{[S]_o}\right) = \frac{v_{max}}{K_m}t \tag{18.4}$$

With respect to microbial cell growth, the rate equation can be written as:

$$\frac{d[X]}{dt} = \mu[X] \tag{18.4}$$

Integrating:

$$\ln\left(\frac{[X]}{[X]_o}\right) = \mu t \tag{18.5}$$

Where $[X]_o$ represents the initial cell concentration at $t = 0$.

Unlike chemical reactions both enzyme and microbial reactions are generally complex. The mechanism of enzyme catalyzed reactions is discussed in the next secion.

18.3 Enzyme Reaction Mechanisms

The rate of a chemical or biochemical reaction is similar to that defined earlier, i.e., the time rate or change in the quantity of a particular species participating in a reaction divided by a factor that characterizes the reacting system's geometry. The choice of this factor is also a matter of convenience. For homogeneous media, the factor is almost always the volume of the reacting system. For most fluid-solid reaction systems, the factor is often the mass of the solid. For example, in immobilized enzyme reactions, the factor is the mass of the immobilized or supported enzyme catalyst.

Consider the following scenario for a simple enzyme catalyzed reaction; the reaction scheme is as follows:

$$E + S \underset{k_{-1}}{\overset{k_1}{\rightleftharpoons}} ES$$

$$ES \xrightarrow{k} E + P \tag{18.6}$$

where E = free enzyme
S = substrate or reactant
ES = primary enzyme-substrate complex
P = product
k = reaction rate constant

The decomposition of the primary complex ES to the free enzyme E and the product P is assumed to be the rate-determining (slow) step. The expression below is valid for both homogeneous (where the enzyme is used in the native or soluble form) and for immobilized enzyme reactions. The reaction rate v is given by:

$$v = -\frac{d[S]}{dt} = \frac{d[P]}{dt}$$

(18.7)

where [S] and [P] are once again the concentrations of substrate and product, respectively.

There are two approaches in deriving an expression for the reaction rate. In the *Michaelis-Menten approach* the first reaction in Equation (18.6) is assumed to be in equilibrium. The decomposition of the enzyme-substrate complex ES to form E and P is, as noted above, the rate-determining step. In the second approach, it is assumed that after an initial period, the rate of change of the concentration of the enzyme-substrate complex is essentially zero. Mathematically, this can be expressed as

$$\frac{d([ES])}{dt} \approx 0$$

(18.8)

This is known as the *quasi steady-state approximation*, and is valid for enzyme catalyzed reactions if the initial total enzyme concentration is much less than the initial substrate concentration, i.e., $[E]_o << [S]_o$.

The maximum reaction rate v_{max} is equal to $k[E]_o$. When $K_m = [S]$, $v = v_{max}/2$, one can derive an expression for the reaction rate. The total enzyme balance can be written as:

$$[E]_o = [E] + [ES]$$

(18.9)

where $[E]_o$ = the total enzyme concentration
$[E]$ = concentration of free enzyme
$[ES]$ = concentration of the enzyme-substrate complex

Since Michaelis-Menten kinetics is valid, the reaction between the free enzyme and substrate to from ES in Equation (18.6) may be assumed to be in equilibrium, so that

$$K_m = -\frac{k_{-1}}{k_1} = \frac{[E][S]}{[ES]} \tag{18.10}$$

where K_m is once again the Michaelis-Menten constant. The preceding equation can be combined with the total enzyme balance to provide a relationship between [ES] and the total enzyme concentration $[E]_o$.

$$[ES] = \frac{[E]_o[S]}{K_m + [S]} \tag{18.11}$$

The reaction rate v then equals:

$$v = k[ES] = \frac{k[E]_o[S]}{K_m + [S]} \tag{18.12}$$

In the quasi steady-state approximation, the constant K_m is known as the dissociation constant. Assuming quasi steady-state, the rate of disappearance of the enzyme-substrate complex, ES is:

$$\frac{d[ES]}{dt} = k_1[E][S] - k_1[ES] - k[ES] = 0 \tag{18.13}$$

Eliminating [E] by combining with the equation for the total enzyme balance, and solving for [ES],

$$[ES] = \frac{[E]_o[S]}{\left(\dfrac{k + k_{-1}}{k_1}\right) + [S]} \tag{18.14}$$

The reaction rate v then equals

$$v = k[ES] = \frac{k[E]_o[S]}{K_m + [S]} \tag{18.15}$$

which is identical to the expression obtained earlier; the only difference lies in the definition of K_m which is equal to $\dfrac{k+k_{-1}}{k_1}$ instead of $\dfrac{k_{-1}}{k_1}$ (equilibrium assumption).

18.4   Effectiveness Factor [4]

The effectiveness factor is defined as the ratio of the reaction rate in the presence of internal or pore diffusion to the reaction rate in the absence of pore diffusion. The value of the effectiveness factor is thus a measure of the extent of diffusion limitation. For isothermal reactions (generally true of most biochemical reactions), diffusional limitations are negligible when the effectiveness factor (η) is close to unity. If $\eta<1$, the reaction is diffusion limited.

The problem of pore diffusion is only limited to immobilized enzyme catalysts, and not enzyme catalyzed reactions in which the enzyme is used in the native or soluble form. The shape of the immobilized enzyme pellet may be spherical, cylindrical, or rectangular (as in a slab). If the reaction follows Michaelis-Menten kinetics as discussed above, a shell balance around a spherical enzyme pellet can be shown to result in the following second order differential equation:

$$D_e\left(\frac{d^2[S]}{dr^2}+\frac{2}{r}\frac{d[S]}{dr}\right)=\frac{V_{max}[S]}{K_m+[S]} \tag{18.16}$$

where D_e = the effective diffusivity, cm²/s
The boundary conditions (BC) are:

1. $[S]=[S]_o$ at r=R where R is the radius of the spherical catalyst pellet and $[S]_o$ is the substrate concentration in the bulk liquids, and
2. $D[S]/dr=0$ at r=0 (due to the symmetry of the pellet, the concentration gradient is zero at the center)

The preceding differential equation can be solved analytically or numerically to determine the concentration profile inside the pellet.

From the definition of the effectiveness factor, the actual or observed reaction rate, v_s (in the presence of pore diffusion), is equal to:

$$v_s = \eta v = \eta \left[\frac{V_{max}[S]_o}{K_m + [S]_o} \right] \tag{18.17}$$

When the substrate concentration is low, i.e., when $[S] << K_m$, the reaction rate becomes first order. In this case, the preceding differential equation can be solved analytically to obtain the concentration profile inside the catalyst pellet. By defining the following dimensionless parameters, the differential equation can be written in a dimensionless form as follows:

$$\overline{S} = \frac{[S]}{[S]_o}, \quad \overline{r} = \frac{r}{R} \tag{18.18}$$

and

$$\left(\frac{d^2\overline{S}}{d\overline{r}^2} + \frac{2}{\overline{r}} \frac{d\overline{S}}{d\overline{r}} \right) = 9\phi^2 \overline{S} \tag{18.19}$$

where

$$\phi = \frac{R}{3} \sqrt{\frac{V_{max} / K_m}{D_e}} = \text{the Thiele modulus} \tag{18.20}$$

Once the concentration profile is known, the effectiveness factor can be expressed as a function of the Thiele modulus by the following relationship:

$$\eta = \frac{1}{\phi} \left[\frac{1}{\tanh 3\phi} - \frac{1}{3\phi} \right] \tag{18.21}$$

The observed reaction rate can then be determined from Equation (18.17).

For a rectangular catalyst slab or a rectangular membrane in which both sides are exposed to the substrate, the effectiveness factor is related to the Thiele modulus as follows:

$$\eta = \frac{\tanh \phi}{\phi} \tag{18.22}$$

where

$$\phi = \frac{L}{2} \sqrt{\frac{V_{max} / K_m}{D_e}}$$ (18.23)

and L is the thickness of the membrane or catalyst slab.

18.5   Design Procedures

There are a host of design topics associated with biochemical reactors [5]. Discussing these in any detail is also beyond the scope of this book. Topics of interest to the practicing chemical engineer are primarily concerned with sterilization. Four specific areas include:

1. Design of a batch sterilization unit
2. Design of a continuous sterilization unit
3. Design of an air sterilizer
4. Scale-up of a fermentation unit

Each of these briefly receives qualitative treatment below with Vasudevan [4] providing quantitative details and analyses plus illustrative examples.

18.5.1   Design of a Batch Sterilization Unit

Sterilization is the process of inactivation or removal of viable organisms. Sterilization can be accomplished by the steaming of equipment and/or medium, plus the additives; it is an important operation in the fermentation industry. The main objective of media sterilization is to kill all living organisms present before inoculation and to eliminate any possible competition or interference with the growth and metabolism of the desired organism. This objective should be accomplished with minimal damage to the media ingredients.

One of the drawbacks of batch sterilization is that Del factors [4], which is a measure of the extent of sterilization, are scale dependent. Large fermenters require longer heating and cooling times. The consequences of longer times can be severe if the medium is thermolabile since the destruction due to heat is dependent on the value of the thermal rate constant. A better alternative for heat sensitive material is to use continuous sterilization, a topic discussed in the next subsection.

18.5.2   Design of a Continuous Sterilization Unit

As noted above, the consequences of long heat-up and cool-down times in batch sterilization can be severe if the medium components are heat-sensitive. Since the destruction due to heat is dependent on the value of the *thermal death constant* [4] it can be shown that the spore B. *stearothermophilus* is only significantly inactivated above 110°C due to its high activation energy of 67.7 kcal/mol. On the other hand, many organic nutrients, which follow the Arrhenius relationship for thermal degradation, have a much lower activation energy of approximately 25-30 kcal/gmol. This analysis implies that longer exposure to lower sterilization temperatures, due to slow heat-up or cool-down, can cause more damager to nutrients. The best alternative for heat sensitive materials is to use continuous sterilization.

In continuous sterilization, the raw medium is mixed with water, and then continuously pumped through the sterilizer to a sterilized fermenter. In the sterilizer, the media is instantaneously heated by either direct or indirect contact with steam [6], and held at a very high temperature (about 140°C) for a relatively short time. The residence time or holding time of the medium is fixed by adjusting the flow rate and length of the insulated holding pipe. The hot steam from the sterilizer is rapidly cooled by a heat exchanger (with or without heat recovery) and/or by flash cooling before it enters the fermenter.

The design of continuous sterilizers must also allow some flexibility in operating conditions to adapt the system to a different medium. The system must incorporate automatic recycle to recirculate the medium if the temperature falls below the design value. The design should include:

1. The ability to fill the fermenter within 2-3 hours (this might be considered down time)
2. The recovery of 60-70% of the heat
3. Plug flow in the holding section
4. The option of either direct of indirect heating

The continuous sterilization process has several advantages.

1. The temperature profile of the medium is almost one of instant heating and cooling, allowing an easy estimate of the Del factor required [4].
2. Scale-up is very simple since the medium is exposed to high temperatures for very short times, thereby minimizing nutrient degradation.

3. The energy requirements of the sterilization process can be dramatically reduced by using the incoming raw medium to cool the hot sterile medium.

The difficulties with continuous sterilizers are typically due to exchanger fouling and control instability. In general, any medium containing starches requires special attention.

18.5.3   Design of an Air Sterilizer [4]

In aerobic fermentations, it is necessary to sterilize air. Since the volume of air required in aerobic fermentations is usually large, conventional techniques of heat sterilization are uneconomical. Effective and viable alternatives include the use of membranes or fibrous filters. An important consideration of the filter medium is that it should not be wetted since this can lead to deposition contamination. Materials such as glass fibers are often used to avoid this problem.

The mechanisms by which particles suspended in a flowing stream of air are removed include *impaction, diffusion, and interception* [7]. Impaction occurs when particles in the air collide with the fibrous filter due to their higher momentum as compare to air. Smaller particles, on the other hand, travel towards the fiber as a result of molecular diffusion caused by Brownian motion; particles less than 1μm are collected by this mechanism [7,8]. Particles that are not small or heavy but large in size may be intercepted by the fiber. The efficiency of air filtration is therefore a combination of the three mechanisms.

The mathematical equations for designing air sterilizers may be developed by considering the effect of each of the above mechanisms separately, and then developing a combined expression. In the case of collision or impaction, the efficiency of the process is a function of the Reynolds (Re) and Stokes (St) numbers with the Reynolds number being based on the diameter of the filter or collection device.

The effect of the air velocity on efficiency is considerable. Removal of particles by collision and interception is enhanced as the air velocity is increased, whereas the efficiency of particulate removal by diffusion is lowered because of the residence time. This is of course dependent on the particle size [7]. In general, it should be remembered that most equations for the removal efficiency are empirical.

18.5.4   Scale-Up of a Fermentation Unit

Scale-up is a fundamental problem in the fermentation industry because of the need to perform microbial operations in different size equipment. The scale-up problem arises from the difference in transport phenomena when the scale and geometry of the equipment is changed. Describing transport and kinetic phenomena in fermenters requires knowledge of both the kinetics and flow patterns. Since the balances are very complex and the flow patterns are largely unknown, the use of fundamental principles for design is limited. Instead, various empirical procedures are usually employed.

The first step in scaling up a fermentation process is to use the production requirements to determine the size and number of fermenters that will be required. In sizing a new unit, or evaluating an existing fermenter for a biological process, it is necessary to establish the desired product rate R on an annual or daily basis. Then the productivity of the individual fermenter – defined as the weight of product that the fermenter can produce per unit volume per unit time –may be determined.

In addition to the fermentation phase, the overall productivity must account for the time spent in activities such as cleaning, filling and sterilizing; this is normally referred to as *down-time*. This is important if the process is to be operated batch-wise. The total installed fermenter capacity, V, required can then be calculated by dividing the desired product rate by the productivity. The number of fermenters can then be easily calculated. The next step is to size the number of inoculum stages. Fermenter pre-cultures must be made in order to have sufficient inoculum for a large fermenter. The production fermenter volume and the optimal inoculum levels will determine how many pre-culture stages are necessary. Typical inoculum concentrations are:

Bacteria (0.1-3%)
Actinomycetes (5-10%)
Fungi (5-10%)

A detailed and expanded treatment of this topic is available in the following two references.

1. P. Vasudevan, *Biochemical Engineering*, A Theodore Tutorial, Theodore Tutorials, East Williston, NY, a text originally published by the USEPA/APTI, RTP, NC, 1994 [1].
2. Adapted from, L. Theodore, *Chemical Engineering: The Essential Reference*, McGraw-Hill, New York City, NY, 2014 [1].

18.6 Illustrative Open-Ended Problems

This and the last section provide open-ended problems. However, solutions *are* provided for the three problems in this section in order for the reader to hopefully obtain a better understanding of these problems which differ from the traditional problems/illustrative examples. The first problem is relatively straightforward while the third (and last problem) is somewhat more difficult and/or complex. Note that solutions are not provided for the 40 open-ended problems in the next section.

Problem 1: The kinetics of mixed cell cultures are very important in the biological treatment of wastewater. Mixed cell populations exist in a number of natural environments, and it is very important to understand how these cells interact with one another. Define and discuss the following terms:

1. Neutralism
2. Mutalism
3. Symbiosis
4. Methanobacillus omelianskii
5. Competition
6. Commensalism
7. Ammensalism

Also discuss the predator-prey problem.

Solution: *Neutralism* refers to mixed growth of two microbial populations in which the growth rate of either microorganism is not affected by the presence of the other. On the other hand, *mutualism* refers to a case where both populations grow faster together. This may be due to the exchange of growth factors or nutrients. If the partnership between the two species is necessary for the survival of either species, then the mutualistic relationship is referred to as *symbiosis*. In anaerobic sludges, *methanobacillus amelianskii*, has been found to be a mixture of two species. The first species of the bacterium converts ethanol to hydrogen and acetate, but is inhibited by the product hydrogen. The second species consumes hydrogen and produces methane. *Competition* occurs when each species exerts a negative influence on the other. If the second microbial species alone enjoys the benefits of interacting with the first one, this relationship is referred to as *commensalism*. If the growth of the second species is affected by the first population, the process is known as *amensalism*.

Multiple interacting species can give rise to complex behavior. For example, in predation and parasitism, one species benefits at the expense of the other. The consumption of bacteria by protozoa is a classic example of a *predator-prey* relationship. The populations of predator and prey do not reach steady state values but oscillate for the protozoan-bacterium system. If the concentration of prey is high and the concentration of predator is low, the number of predators increases while the prey population simultaneously decreases. When the population of prey is sufficiently low, the predator population declines and the prey population starts to increase. This cycle will then repeat again and again.

Problem 2: Discuss why recombinant DNA has revolutionized the biotechnology industry.

Solution: This technique involves the production of a hybrid gene by joining pieces of DNA from different organisms *in vitro*, and then inserting this hybrid material into a host cell. As a result of developments in recombinant DNA (rDNA), there exists vast potential for directing cells to synthesize new products in the agriculture, chemical, environment, food and pharmaceutical industries. Some of the valuable products that have resulted from rDNA include insulin, interferon, human growth hormone and human serum albumin.

There are four basic steps common to genetic engineering. First, break up the DNA into short stretches or sequences. Second, join each bit into a suitable vector. Third, have the vector invade the host cells one at a time, and grow into large numbers. Finally, find the cells with the right transplanted genes in them.

In the first step, the DNA to be engineered is broken into small pieces by enzymes (known as restriction endonucleases) that recognize a particular sequence, and cut the DNA double helix only where such a sequence appears. The resultant shorter sequences of DNA often have "sticky ends". In the second step, the DNA is inserted into a vector, which is usually a bacterial plasmid (a circular double-stranded DNA molecule). The plasmid is cleaved by the same enzyme as used for the DNA to be transplanted, leaving sticky ends complementary to those of the DNA. Bits of DNA pair up and create a recombined plasmid. The third step involves the growth of bacteria carrying recombinant DNA. The plasmids are mixed with the host bacteria (such as *E. coli*), and the bacteria carrying recombinant plasmids are grown on the surface of a culture medium. Each bacteria divides until a colony of millions of bacteria are formed. A library of bacteria containing recombinant plasmids has, somewhere within it, the gene of interest. This can be identified and isolated.

Several types of instability may limit the productivity of recombinant DNA. First, a certain fraction of new cells may be born without plasmids. The plasmid-free cells may outgrow the plasmid-bearing cells. This situation arises because the plasmid-bearing cells are at a disadvantage when competing with plasmid-free cells for essential nutrients. This is known as segregational instability. On the other hand structural instability results in the inability of cells to synthesize the active product. Many methods are being pursued to genetically eliminate plasmid instability. For example, the application of *selective pressure*, such as the addition of an antibiotic supplement to the medium will kill plasmid-free cells since plasmids contain a marker for antibiotic resistance. Cells containing the plasmid will not be affected as a result. Growth-rate dominated instability or growth-rate differential occurs as a result of the redirection of cellular (catabolic and anabolic) activity in the recombinant cells when they synthesize the desired product. Consequently, plasmid-bearing cells do not have full use of their own resources and grow more slowly than those which are plasmid-free. The plasmid-free cells born by plasmid segregation are thus able to rapidly overtake the population of plasmid-bearing cells. To alleviate this problem, a two stage fermentation process can be used in which the first stage is optimized to produce viable plasmid containing cells, and the second stage is used to produce the target protein by turning on an inducing promoter.

Problem 3: Bovine liver catalase is used to decompose hydrogen peroxide to water and oxygen. The concentration of hydrogen peroxide is given in Table 18.1 as a function of time for a reaction mixture at 25°C with a pH of 7.0.

Table 18.1 Time-Concentration Data for Problem 3.

Time, min	0	10	20	50	100
H_2O_2, molar concentration	0.02	0.018	0.016	0.011	0.005

Determine the Michaelis-Menten parameters by any suitable means. If the total concentration of the enzyme changes, what will the substrate concentration be after various periods of time?

Comment: Refer to the work of P.T. Vasudevan [4] for additional details.

Solution: The rate (v) equation for the quasi steady-state approximation is given by:

$$v = -\frac{d[S]}{dt} = \frac{v_{max}[S]}{K_m + [S]} \qquad (18.1)$$

This equation can be integrated between the limits $[S]_o$ (initial substrate concentration and $[S]$ (substrate concentration at time t):

$$\int_{[S]_o}^{[S]} \frac{K_m + [S]}{[S]} = -\int_0^t v_{max} dt \tag{18.24}$$

One may linearize the integrated form of the equation. The resulting equation is implicit in the substrate (reactant) concentration. A numerical linear regression analysis provide estimates of the two parameters [10]. Or, a graphical technique can be used and the parameters estimated from the slope and intercept.

$$\frac{[S]_o - [S]}{\ln\dfrac{[S]_o}{[S]}} = \frac{v_{max} t}{\ln\dfrac{[S]_o}{[S]}} - K_m \tag{18.25}$$

A plot of $\dfrac{[S]_o - [S]}{\ln\dfrac{[S]_o}{[S]}}$ versus $\dfrac{t}{\ln\dfrac{[S]_o}{[S]}}$ should be linear with an intercept $= -K_m$

and a slope $= v_{max}$. Employing a numerical linear regression analysis, $v_{max} = 3.605 \times 10^{-4}$ mol/L·min and $K_m = 0.015$ M.

Assume the enzyme concentrations is now doubled so that v_{max} will be twice the value estimated above. The reaction rate constant k is an intrinsic parameter that does not change. Thus, $v_{max} = 2 \times 3.605 \times 10^{-4} = 7.21 \times 10^{-4}$ mol/Lmin.

Substituting into the integrated equation above and calculate the substrate concentration for the conditions specified. This requires an iterative procedure. The substrate concentration is given by

$$[S] = [S]_o - v_{max} t + K_m \ln\frac{[S]_o}{[S]} \tag{18.25}$$

Substituting t = 30 minutes, $[S]_o = 0.02$ M, v_{max} 7.21 $\times 10^{-4}$ mol/L·min and $K_m = 0.015$ M into the above equation and solving iteratively leads to

$$[S] = 0.0095 \text{ M}$$

The above calculation can be repeated for other periods of time and other total enzyme concentrations.

Finally, the reader should note that integral analysis of reaction data in enzyme catalyzed reactions does not yield very accurate values of the reaction parameters. It is important to measure initial reaction rates since the quasi steady-state approximation is not valid as $[S]$ approaches $[E]_0$.

18.7   Open-Ended Problems

This last section of the chapter contains open-ended problems as they relate to biochemical engineering. No detailed and/or specific solution is provided; that task is left to the reader, noting that each problem has either a unique solution or a number of solutions or (in some cases) no solution at all. These are characteristics of open-ended problems described earlier.

There are comments associated with some, but not all, of the problems. The comments are included to assist the reader while attempting to solve the problems. However, it is recommended that the solution to each problem should initially be attempted *without* the assistance of the comments.

There are 40 open-ended problems in this section. As stated above, if difficulty is encountered in solving any particular problem, the reader should next refer to the comment, if any is provided with the problem. The reader should also note that the more difficult problems are generally located at or near the end of the section.

1. Describe the early history associated with biochemical engineering.
2. Discuss the recent advances in biochemical engineering.
3. Select a refereed, published article on biochemical engineering from the literature and provide a review.
4. Provide some normal everyday domestic applications involving the general topic of biochemical engineering.
5. Develop an original problem on biochemical engineering that would be suitable as an illustrative example in a book.
6. Prepare a list of the various technical books which have been written on biochemical engineering. Select the three best and justify your answer. Also select the three weakest books and, once again, justify your answer.
7. Discuss both enzyme and microbial kinetics.
8. Discuss the general subject of enzyme reaction mechanisms in layman terms.

9. Provide the equations employing traditional chemical engineering notation for the Michaelis-Menten approach to describing expressions for the rate of reaction.

10. Provide the equations employing traditional chemical engineering notation for describing expressions for the quasi steady-state approach.

11. Define and discuss the effectiveness factor.

12. Provide a layman's definition of the Thiele modulus.

13. Provide a layman's definition of the biological oxygen demand (BOD).

14. Provide a layman's definition of the chemical oxygen demand (COD).

15. Provide a layman's definition of the total organic carbon (TOC).

16. Describe the various biomedical engineering opportunities that exist for chemical engineers.
    Comment: Refer to the work of Abulencia and Theodore [11] for additional details.

17. Define and discuss the power number as it applies in the design of agitated tanks.

18. Three dimensionless numbers involved in particulate behavior are the Peclet, Stokes, and Schmidt numbers [7]. Define these numbers and discuss the relationship between the three.

19. Describe the Cunningham Correction Factor (CCF) [7]. Attempt to improve on this equation.
    Comment: Refer to the literature [7,8] for additional details.

20. Describe the batch sterilization process.

21. Discuss the various types of heat transfer mechanisms employed in batch sterilization.

22. Define and describe both competitive inhibition and substrate inhibition.

23. Discuss the general subject of immobilized enzymes and enzyme deactivation. Are there any major differences?

24. Discuss metabolism, catabolism, and anabolism. What are the differences?

25. Describe Monod growth kinetics in layman terms.

26. Discuss the role membrane separators [12] plays in biochemical engineering processes.

27. In terms of biochemical concepts, provide a brief definition and description of

- bacteria
- viruses
- cells
- fungi
- mutation
6. genetic concepts
28. Describe the different phases of bacteria cell growth.
29. Provide a layman's definition of chemostats.
30. Discuss the difference between biochemical and biomedical engineering.
31. Provide fluid flow analogies with the following biomedical terms:
    - Blood
    - Blood vessels
    - Heart
    - Plasma cell flow
    Comment: Refer to the literature [11] for additional details.
32. List and briefly describe the various classes of biochemical reactors.
    Comment: Refer to the literature for additional details [5].
33. Describe trickling biological filters.
34. Instead of a Lineweaver-Burke plot, can you think of any other way of linearizing the describing equation and deducing the nature of inhibition? Can the nature of inhibition and the inhibition parameter be determined from a Dixon plot? Discuss the significance of a high inhibition constant.
    Comment: Refer to the literature [4]
35. Describe fluidized bed bioreactors and discuss the differences with the traditional fluid bed reactors.
36. Describe how the immobilization of cells on an inert support is carried out.
37. Explain why scale-up is a fundamental problem in the fermentation industry.
38. Consider the growth of an organism that follows Monod growth kinetics in a batch reactor. If the yield $Y_{x/s} = 0.5$, $\mu_{max} = 0.5$ h^{-1}, $K_s = 0.004$ g/L, the initial cell concentration $[X]_o = 1.0$ g/L, and the intial substrate concentration $[S]_o = 20$ g/L, provide equations on how $\ln[X]$, $[X]$, $[S]$, $d[X]/dt$ vary with respect to time.
    Comment: Refer to the literature [4] for additional details.

39. A simple batch fermentation gave the results provided in Table 18.2.
    Calculate the following:
    - Yield
    - Maximum growth rate
    - Doubling time
    - Lag time
    Specific growth rate of various times.
    Provide a general analysis of the results.
    Comment: Refer to the work of Vasudevan [4] for additional details.
40. Data for the enzymatic breakdown of a phosphate chemical is provided in Table 18.3.

Table 18.2 Time-Concentration Data for Problem 39.

Time, min	$[X]$, g/L	$[s]$, g/L
0	0.19	9.2
120	0.21	9.2
240	0.31	9.1
480	1.0	8.0
600	1.8	6.8
720	3.2	4.6
840	5.6	0.9
960	6.2	0.08
1080	6.3	0

Table 18.3 Rate-Concentration Data for Problem 40.

Substrate concentration, μmol/L	Initial rate, μmol/L·min at initial inhibitor concentration, μmol/L	
	$[i] = 0.0$	$[i] = 146$
6.7	0.300	0.1075
3.5	0.238	0.0800
1.7	0.160	0.0562

What can one deduce about the nature of the inhibition?

Comment: Refer to the literature [4] for additional details.

References

1. J. Enderle, S. Blanchard, and J. Bronzing, *Introduction to Biomedical Engineering*, 2nd ed, Elseview/Academic Press, New York City, NY, 2000.
2. J. Bronzing (editor), *Biomedical Engineering Fundamentals*, 3rd ed, CRC/Taylor & Francis Group, Boca Raton, FL, 2000.
3. S. Vogel, *Life in Moving Fluids*, 2nd ed, Princeton University Press, Princeton, NJ, 1994.
4. P. Vasudevan, *Biochemical Engineering*, A Theodore Tutorial, Theodore Tutorials, East Williston, originally published by the USEPA/APTI, RTP, NC, 1994.
5. Adapted from: L. Theodore, *Chemical Reactor Design and Analysis for the Practicing Engineer*, John Wiley & Sons, Hoboken, NJ, 2012.
6. L. Theodore, *Heat Transfer for the Practicing Engineer*, John Wiley & Sons, Hoboken, NJ, 2011.
7. L. Theodore, *Air Pollution Control Equipment Calculations*, John Wiley & Sons, Hoboken, NJ, 2008.
8. C.E. Cunningham, *Proc. Roy. Soc. London*, Ser. A, 83, 357, 1910.
9. L. Theodore, *Chemical Engineering: The Essential Reference*, McGraw-Hill, New York City, NY, 2014.
10. S. shaefer and L. Theodore, *Probability and Statistics Applications in Environmental Science*, CRC Press/ Taylor & Francis Group, Boca Raton, FL, 2007.
11. P. Abulencia and L. Theodore, *Fluid Flow for the Practicing Chemical Engineer*, John Wiley & Sons, Hoboken, NJ, 2009.
12. L. Theodore and F. Ricci, *Mass Transfer Operations for the Practicing Engineer*, John Wiley & Sons, Hoboken, NJ, 2010.

19

Probability and Statistics

This chapter is concerned with process probability and statistics. As with all the chapters in Part II, there are several sections: Overview, several technical topics, illustrative open-ended problems, and open-ended problems. The purpose of the first section is to introduce the reader to the subject of probability and statistics. As one might suppose, a comprehensive treatment is not provided although several technical topics are also included. The next section contains three open-ended problems; the authors' solution (there may be other solutions) are also provided. The final section contains 42 problems; *no* solutions are provided here.

19.1 Overview

This overview section is concerned with—as can be noted from the chapter title—probability and statistics. As one might suppose, it was not possible to address all topics directly or indirectly related to probability and statistics. However, additional details may be obtained from either the references provided at the end of this Overview and/or at the end of the chapter.

Note: Those readers already familiar with the details associated with this topic may choose to bypass this Overview.

The title of this chapter is Probability and Statistics. Webster [1] defines *probability* as "the quality or state of being probable; likelihood; something probable; the number of times something will probably occur over the range of possible occurrences, expressed as a ratio" and the term *statistics* as "facts or data of a numerical kind, assembled, classified, and tabulated so as to present significant information about a given subject; the science of assembling, classifying, tabulating, and analyzing such facts or data". There are obviously many other definitions.

The key area of interest to chemical engineers is however, statistics. The problem often encountered is usually related to interpreting limited data and/or information. This can entail any one of several topics.

1. Obtaining additional data and/or information
2. Deciding which data and/or information to use
3. Generating a mathematical model (generally an equation) to represent the data and/or information
4. Generating information about unknowns, a process often referred to as *inference*

The material in this chapter is divided into 6 sections. Titles are provided below.

1. Probability Definitions and Interpretations
2. Introduction to Probability Distributions
3. Discrete and Continuous Probability Distributions
4. Contemporary Statistics
5. Regression Analysis
6. Analysis of Variance

Topic (5) receives the bulk of the treatment in the presentation to follow.

Finally, the reader should note that much of the material in this chapter was adopted from L. Theodore and F. Taylor, "Probability and Statistics," Theodore Tutorials, East Williston, NY, originally published by USEPA/APTI, RTP, NC, 1993 [2].

19.2   Probability Definitions and Interpretations [2,3]

Probabilities are nonnegative numbers associated with the outcomes of so-called *random* experiments. A random experiment is an experiment whose outcome is uncertain. Examples include throwing a pair of dice, tossing a coin, counting the number of defectives in a sample from a lot of manufactured items, and observing the time to failure of a tube in a heat exchanger or a seal in a pump or a bus section in an electrostatic precipitator. The set of possible outcomes of a random experiment is called the *sample space* and is usually designated by S. Then P(A), the probability of an event A, is the sum of the probabilities assigned to the outcomes constituting the subset A of the space S. A *population* is a collection of *objects* having observable or measureable characteristics defined as a *variate* while a *sample* is a group of "objects drawn from a population (usually random) where each is likely to be drawn."

Consider, for example, tossing a coin twice. The sample space can be described as

$$S = [HH, HT, TH, TT] \tag{19.1}$$

If probability ¼ is assigned to each element of S and A is the event of at least one head, then

$$A = [HH, HT, TH] \tag{19.2}$$

The sum of the probabilities assigned to the elements of A is ¾. Therefore, P(A) = ¾. The description of the sample space is not unique. The sample space S in the case of tossing a coin twice could be describe in terms of the number of heads obtained. Then

$$S = [0, 1, 2] \tag{19.3}$$

Suppose probabilities ¼, ½, and ¼ are assigned to the outcomes 0, 1, and 2, respectively. Then A, the event of at least one head, would have for its probability,

$$P(A) = P(1, 2) = 3 / 4 \tag{19.4}$$

Probability P(A) can also be interpreted *subjectively* as a measure of degree of belief, on a scale from 0 to 1, that the event A occurs. This interpretation is frequently used in ordinary conversation. For example, if someone says, "The probability that I will go the race track tonight is 90%," then 90% is a measure of the person's belief that he or she will go to a race track (a site regularly visited by one of the authors). This interpretation is also used when, in the absence of concrete data needed to estimate an unknown probability on the basis of observed relative frequency, the personal opinion of an expert is sought. For example, a chemical engineer might be asked to estimate the probability that the seals in a newly designed pump will leak at high pressures. The estimate would be based on the expert's familiarity with the history of pumps of similar design.

19.3   Introduction to Probability Distributions [2,3]

The probability distribution of a random variable concerns the distribution of probability over the range of the random variable. The distribution of probability, i.e., the values of random variables together with their associated probabilities, is specified by the *probability distribution function* (pdf). This section is devoted to providing general properties of the pdf for the case of discrete and continuous random variables as well as an introduction to the *cumulative distribution function* (cfd). Special pdfs finding extensive application in chemical engineering analysis are presented in the next section.

The pdf of a *discrete* random variable X is specified by f(x), where f(x) has the following essential properties

1. $F(x) = P(X)=x)$ = probability assigned to the outcome corresponding to the number x in the range of X; i.e., X is a *specifically designated* value of x.   (19.5)
2. $F(x) \geq 0$ (19.6)
3. $\sum_x f(x) = 1$ (19.7)

Property 1 indicates that the pdf of a discrete random variable generates probability by substitution. Property 2 and Property 3 restrict the values of $f(x)$ to nonnegative real numbers and numbers whose sum is 1, respectively.

The pdf of a *continuous* random variable X has the following properties:

1. $\int_a^b f(x)dx = P(a < X < b)$ (19.8)

2. $F(x) \geq 0$ (19.9)

3. $\int_{-\infty}^{\infty} f(x) dx = 1$ (19.10)

Equation (19.8) indicated that the pdf of a continuous random variable generates probability by integration of the pdf over the interval whose probability is required. When this interval contracts or reduces to a single value, the integral over the interval becomes *zero*. Therefore, the probability associated with any particular value of a continuous random variable is therefore zero. Consequently, X is continuous

$$P(a < X \leq b) = P(a < X \leq b)$$ (19.11)
$$= P(a < X < b)$$
$$= P(a \leq X < b)$$

Equation (19.9) restricts the values of f(x) to nonnegative numbers. Equation (19.10) follows from the fact that

$$P(-\infty < X < \infty) = 1$$ (19.12)

The expression $P(a < X < b)$ can be interpreted geometrically as the area under the pdf curve over the interval (a, b). Integration of the pdf over the interval yields the probability assigned to the interval. For example, the probability that the time in hours between successive failures of an aircraft air conditioning system X is greater than 6 but less than 10 is $P(6 < X < 10)$.

Another function used to describe the probability distribution of a random variable X is the *cumulative distribution function* (cdf). If f(x) specifies the pdf of a random variable X, then F(x) is used to specify the cdf. The cdf of X is defined by for both discrete and continuous random variables

$$F(x) = P(X \geq x); -\infty < x < \infty$$ (19.13)

Note that the cdf is defined for all real numbers, not just the values assumed by the random variable. It is helpful to think of F(x) as an accumulator of probability as x increases through all real numbers. In the case of a discrete random variable, the cdf is a step function increasing by finite jumps at the values of x in the range of X. In the case of a continuous random variable, the cdf is a continuous function.

The following properties of the cdf of a random variable X can be deduced directly from the definition of F(x).

1. $F(b) - F(a) = P(a < X \le b)$ (19.14)
2. $F(+\infty) = 1$
3. $F(-\infty) = 0$
4. Also note that $F(x)$ is a nondecreasing function of X

As noted above, these properties apply to the cases of both discrete and continuous random variable.

Shaefer and Theodore [3] provided numerous illustrative examples dealing with probability distribution.

19.4   Discrete and Continuous Probability Distributions [2,3]

19.4.1   Discrete Probability Distributions

There are numerous discrete probability distributions. However, this section simply lists the three distributions the chemical engineer is most likely to encounter in his/her career.

1. The Binomial distribution
2. The Hypergeometric distribution
3. The Poisson distribution

Each receives treatment in Shaefer and Theodore [3], including several other distributions.

There are numerous continuous probability distributions. The four distributions the chemical engineer is most likely to encounter in his/her career are listed below

1. The exponential distribution
2. The Weibull distribution
3. The normal distribution
4. The log-normal distribution

The literature [3] provides details on these as well as other continuous probability distribution topics [4].

19.5   Contemporary Statistics

This section examines topics that Shaefer and Theodore [3] have classified as contemporary statistics; their development includes numerous illustrative examples. The following subject areas are reviewed by the authors.

1. Confidence Intervals for Means
2. Confidence Intervals for Proportions
3. Hypothesis Testing
4. Hypothesis Test for Means and Proportions
5. Chi-Square Distribution
6. The F Distribution
7. Nonparametric Tests

19.6   Regression Analysis (3)

It is no secret that many statistical calculations are now performed with spreadsheets or packaged programs; this statement is particularly true with *regression analysis*, a topic of interest to all chemical engineers. The result of this shortsighted approach has been to reduce or eliminate one's fundamental understanding of this subject. This section attempts to correct this shortcoming.

Chemical engineers and scientists in numerous disciplines often encounter applications that require the need to develop a mathematical relationship between data for two or more variables. For example, if Y (a dependent variable) is a function of or is dependent of X (an independent variable), that is

$$Y = f(X) \qquad (19.15)$$

one may be required to express this (X, Y) data in equation form. This process is referred to as *regression analysis*, and the regression method most often employed is the method of *least squares*.

An important step in this procedure – which is often omitted – is to prepare a plot of Y vs. X. The result, referred to as a *scatter* diagram, could take on any form. This topic received treatment in Chapter 2, but is provided again in Figure 19.1 (a)-(c). The first plot (A) suggests a linear relationship between X and Y, i.e.,

$$Y = a_0 + a_1 X \qquad (19.16)$$

The second graph (B) appears to be best represented by a second order (or parabolic) relationship, i.e.,

$$Y = a_0 + a_1 X + a_2 X^2 \qquad (19.17)$$

The third plot (C) suggests a linear model that applies over two different ranges, i.e., it should represent the data as

$$Y = a_0 + a_1 X; \quad X_0 < X < X_M \tag{19.18}$$

and

$$Y = a'_0 + a'_1 X; \quad X_M < X < X_L \tag{19.19}$$

This multiequation model finds application in representing adsorption equilibria, multiparticle size distributions, and quantum energy relationship. In any event, a scatter diagram and individual judgment can suggest an appropriate model at an early stage in the analysis.

Some of the models often employed by chemical engineers are as follows

1.	$a_0 + a_1 X$	Linear	(19.20)
2.	$a_0 + a_1 X + a_2 X^2$	Parabolic	(19.21)
3.	$a_0 + a_1 X + a_2 X^2 + a_3 X^3$	Cubic	(19.22)
4.	$a_0 + a_1 X + a_2 X^2 + a_3 X^3 + a_4 X^4 + \ldots$	Higher Order	(19.23)

Procedures to evaluate the regression coefficients a_0, a_1, a_2, etc., are available in the literature. The least squares technique provides numerical values for the regression coefficients a_i such that the sum of the square of the difference (error) between the actual Y and Y_e predicted by the equation or model is minimized.

The correlation coefficient provides information on how well the model, or line of regression, fits the data; it is denoted by r. The procedure to calculate r and its properties are also available [2]. It should be noted that the correlation coefficient *only* provides information on how well the model fits the data. It is emphasized that r provides no information on how good the model is or, to reword this, whether this is the correct or best model to describe the functional relationship of the data. This topic is briefly discussed in the next section.

19.7 Analysis of Variance

Analysis of variance is a special statistical technique that chemical engineers may become faced with during their career. It features the splitting of the total variation of data into components measuring variation attributable to one or more factors or combinations of factors. The simplest application of analysis of variance involves data classified in categories (levels) of one factor. This topic receives extensive treatment in Shaefer and Theodore [3]. Interested readers should review the other literature resources of a statistics nature.

A detailed and expanded treated of probability and statistics is available in the following three references.

1. L. Theodore and F. Taylor, *Probability and Statistics*, A Theodore Tutorial, Theodore Tutorials, East Williston, NY, originally published by USEPA/APTI, RTP, NC, 1993 [2].
2. S. Schaefer and L. Theodore, *Probability and Statistics Applications in Environmental Science,* CRC Press/ Taylor & Francis Group, Boca Raton, FL, 2007 [3].
3. L. Theodore, =*Chemical Engineering: The Essential Reference*, McGraw-Hill, New York City, NY, 2014 [6].

19.8   Illustrative Open-Ended Problems

This and the last section provide open-ended problems. However, solutions *are* provided for the three problems in this section in order for the reader to hopefully obtain a better understanding of these problems which differ from the traditional problems/illustrative examples. The first problem is relatively straightforward while the third (and last problem) is somewhat more difficult and/or complex. Note that solutions are not provided for the 42 open-ended problems in the next section.

Problem 1: The problem of calculating probabilities of *objects* or *events* in a finite group – defined earlier as the *sample space* – in which equal probabilities are assigned to the elements in the sample space requires counting the elements which make up the events. The counting of such events if often greatly simplified by employing the rules for permutations and combinations. Define and describe these two terms.

Solution: Permutations and combinations deal with the grouping and arrangement of objects or events. By definition, each different ordering in a given manner or arrangement *with* regard to order of all or part of the objects is called a *permutation*. Alternately, each of the sets which can be made by using all or part of a given collection of objects *without* regard to order of the objexts in the set is called a *combination*. Although,

TABLE 19.1 Subsets of Permutations and Combinations

Permutations (With Regard to Order)	Combinations (Without Regard to Order)
Without replacement	Without replacement
With replacement	With replacement

permutations or combination can be obtained with *replacement* or *without replacement*, most analyses of permutations and combinations are based on a sample that is performed without replacement; i.e., each object or element can be used only once.

For each of the two *with/without* pairs (with/without regard to order and with/without replacement), four subsets of two may be drawn. These four are provided in Table 19.1. To personalize this, the reader could consider the options (games of change) one of the authors faces while on a one-day visit to a casino. The only three options normally considered are dice (often referred to as craps), blackjack (occasionally referred to as 21), and pari-mutual (horses, trotters, dogs, and jai alai) simulcasting betting. All three of these may be played during a visit, although playing two or only one is also an option. In addition the order may vary and the option may be repeated. Some possibilities include the following:

1. Dice, blackjack, and then simulcast wagering
2. Blackjack, wagering, and dice
3. Wagering, dice, and wagering
4. Wagering and dice (the author's usual sequence)
5. Blackjack, blackjack (following a break), and dice

Problem 2: Provide a general introduction to the subject of design of experiments (DOE).

Solution: One of the more difficult decisions facing the authors during the writing of this text was how to handle/treat the more advanced subject of "design of experiments" (DOE), not to be confused with Department of Energy. The material presented in the introduction discussed statistical techniques employed to analyze data obtained from "experiments", with little or no consideration given to the experiment itself. For most individuals desiring an understanding of statistics, the planning and design of experiments is not only outside their control, but also beyond the required background of that individual. It is only on rare occasions that the procedures and details of this topic are mandated. Ultimately, it was decided primarily to provide a very short introduction to DOE because of the enormous depth of the topic. That introduction follows.

The terminology employed in DOE is different to some degree from that employed earlier. Traditionally, experimental variables are usually referred to as *factors*. The factor may be continuous or discrete. Continuous factors include pollutant concentration, temperature, and drug dosage; discrete factors include year, type of process, time, machine operator, and drug classification. The particular value of the variable is defined as the *level* of the factor. For example, if one were interested in studying a property (physical or chemical) of a nanoparticle at two different sizes, there are two levels of the factor (in this case, the nanoparticle's size). If a combination of factors is employed, they are called *treatments* with the result defined as the *effect*. If the study is to include the effect of the shape of the nanoparticle, these different shapes are called *blocks*. Repeating an experiment at the same conditions is called a *replication*.

In a very general sense, the overall subject of DOE can be thought of as consisting of four steps:

1. Determining the importance of (sets of) observations that can be made to the variable or variables one is interested in
2. The procedure by which the observations will be made, i.e., how the data is gathered
3. The review of the observations, i.e., how the data are treated/examined,
4. The analysis of the final results

The reader should also note that obtaining data can be:

1. Difficult
2. Time consuming
3. Dangerous
4. Expensive
5. Affected by limited resources available for experimentation

For these reasons there are obviously significant advantages to expending time on the design of an experiment *before* the start of the experiment. The variables can have a significant effect on the final results; careful selection can prove invaluable. In summary, the purpose of experimental design is to obtain as much information concerning a "response" as possible with the minimum amount of experimental work possible.

Problem 3: A technical definition of both a probability distribution function (pdf) and a cumulative distribution function (cdf) was presented in the Overview. Provide some specific examples of both pdfs and cdfs.

Solution: Consider, for example, a box of 100 transistors containing 5 defectives. Suppose that a transistor selected at random is to be classified as defective or nondefective. Then X is a discrete random variable with pdf specified by

$$f(x) = 0.05; \quad x = 1$$
$$= 0.95; \quad x = 0$$

For another example of the pdf or a discrete random variable, let X denote the number of the throw on which the first failure of an electrical switch occurs. Suppose the probability that a switch fails on any throw is 0.001 and that successive throws are independent with respect to failure. If the switch fails for the first time on throw x, it must have been successful on each of the preceding $x - 1$ trials. In effect, the switch survives up to $x - 1$ trials and fails at trial x. Therefore, the pdf of X is given by

$$f(x) = (0.999)^{x-1}(0.001); \quad x = 1, 2, 3, \ldots, n, \ldots$$

Note that the range of X consists of a countable infinitude of values.

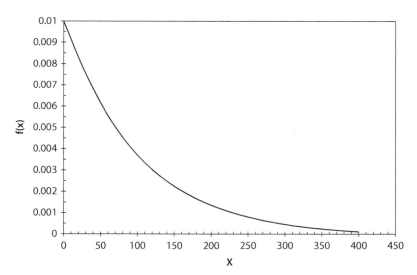

Figure 19.1 The pdf of time in hours between successive failures of an aircraft air conditioning system.

As an example of the pdf of a continuous random variable, consider the pdf of the time x in hours between successive failures of an aircraft air conditioning system. Suppose the pdf of x is specified by

$$f(x) = 0.01e^{-0.01x}; \quad x > 0$$
$$= 0; \text{ elsewhere}$$

A plot of $f(x)$ vs. x for positive values of x is provided in Figure 19.2. Inspection of the graph indicates that intervals in the lower part of the range of x are assigned greater probabilities than intervals of the same length in the upper part of the range of x because the areas over the former are greater than the areas over the latter.

To illustrate the derivation of the cdf from the pdf, consider the case of a random variable X whose pdf is specified by

$$f(x) = 0.2; \quad x = 2$$
$$= 0.3; \quad x = 5$$
$$= 0.5; \quad x = 7$$

Applying the definition of the cdf provided earlier, one obtains for the cdf of X (see also Figure 19.3).

$$f(x) = 0; \quad x < 2$$
$$= 0.2; \quad 2 \le x < 5$$
$$= 0.5; \quad 5 \le x < 7$$
$$= 7; \quad x \ge 7$$

It is helpful to think of $f(x)$ as an accumulator of probability as x increases through all real numbers. In the case of a discrete random variable, the cdf is a step function increasing by finite jumps at the values of x in the range of x. In the aforementioned example, these jumps occur at the values 2, 5, and 7. The magnitude of each jump is equal to the probability assigned to the value at which the jump occurs. This is depicted in Figure 19.2.

Finally, in the case of a continuous random variable, the cdf is a continuous function. Suppose, for example, that x is a continuous random variable with pdf specified by

$$f(x) = 2x; \quad 0 \le x < 1$$
$$= 0; \quad \text{elsewhere}$$

Applying the definition once again, one obtains

$$F(x) = 0; \qquad x < 0$$

$$\int_0^x 2x\,dx = x^2; \qquad 0 \le x < 1$$

$$= 1; \qquad x \ge 1$$

Figure 19.3 displays the graph of this cdf, which is simply a plot of $f(x)$ vs. x. Differentiating cdf and setting pdf equal to zero where the derivative of cdf does not exist can provide the pdf of a continuous random variable. For example, differentiating the cdf of x^2 yields the specified pdf of $2x$. In this case, the derivative of cdf does not exist for $x = 1$.

19.9 Open-Ended Problems

This last section of the chapter contains open-ended problems as they relate to probability and statistics. No detailed and/or specific solution is provided; that task is left to the reader, noting that each problem has either a unique solution or a number of solutions or (in some cases) no solution at all. These are characteristics of open-ended problems described earlier.

There are comments associated with some, but not all, of the problems. The comments are included to assist the reader while attempting to solve the problems. However, it is recommended that the solution to each problem should initially be attempted *without* the assistance of the comments.

There are 42 open-ended problems in this section. As stated above, if difficulty is encountered in solving any particular problem, the reader

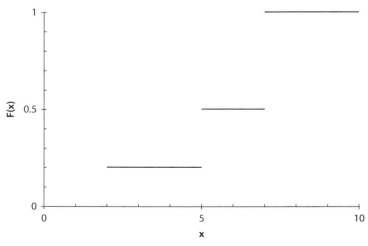

Figure 19.2 Graph of the cdf of a discrete random variable x.

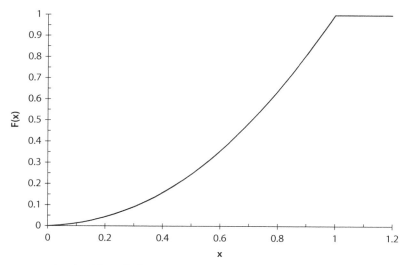

Figure 19.3 Graph of the cdf of a continuous random variable *x*.

should next refer to the comment, if any is provided with the problem. The reader should also note that the more difficult problems are generally located at or near the end of the section.

1. Describe the early history associated with probability.
2. Describe the early history associated with statistics.
3. Discuss the recent advances in statistics.
4. Select a refereed, published article on probability and statistics from the literature and provide a review.
5. Provide some normal everyday domestic applications involving the general topic of probability and statistics.
6. Develop an original problem on either probability or statistics that would be suitable as an illustrative example in a book.
7. Prepare a list of the various technical books which have been written on probability and statistics. Select the three best (hopefully including the text written by one of the authors) and justify your answer. Also select the three weakest books and, once again, justify your answer.
8. Why is the general subject of probability and statistics important to the chemical engineer?
9. Define set notation in layman terms.
10. Define conditional probability in layman terms.
11. What is a random variable?

12. What are random numbers?

13. Define Bayes' Theorem in layman terms.

14. Describe how the pdf and cdf for a random discrete variable differ from that of a random continuous variable.

15. Describe the hypergeometric distribution.

16. Discuss the relationships between the hypergeometric and binomial distributions.

    Comment: Refer to the literature [3] for additional details.

17. The average number of breakdowns of personal computers during 1000 hours of operating of a computer center is 2. What is the probability of no breakdowns during a work period in the 1 – 12 hour range? Comment on the results.

18. One of the authors recently bet on ten basketball games at the Mirage simulcasting center in Las Vegas. Assume the odds of winning the bet are 0.5. The probability of breaking even, i.e., winning 5 of the bets, can be calculated via application of the binomial theorem with $n = 10$, $p = 0.5$, and $x = 5$.

$$P(X = 5) = \frac{10!}{5!5!}(0.5)^5(0.5)^5$$
$$= (252)(0.03125)(0.03125)$$
$$= 0.246 = 24.6\%$$

Revise and rewrite the above Problem statement so that it relates to a technical application of your choice.

19. Revise and rewrite Problem 18 so that it applies to a domestic (everyday) application.

20. The probability that U.S. citizens of age 72 to 73 (once the age of one of the authors) will die within the year was 0.0417. With a group of 1,000,000 such individuals, what is the probability that exactly X will die within the year? Perform the calculations for values of X in the 1×10^4 – 20×10^4 range. Analyze the results.

    Comment: As one might surmise, one of the authors survived that year. The author is currently 80 and the probability of dying has unfortunately increased significantly.

21. Provide a layman's definition of the Weibull distribution.

22. One of the authors [4] is currently attempting to develop a failure model to replace the Weibull distribution. You have been assigned a similar task. Note that the Weibull distribution contains 6 coefficients—two for each of three

different ranges. The author(4) has developed a of 4-coefficient equation to replace the Weibull equation. You have been hired to develop "another" failure rate equation that consists of *less* than 4 coefficients.

23. Define the log-normal distribution and discuss its relationship to the normal distribution.

24. Attempt to develop another probability distribution.

25. Provide technical definitions for the following two terms.
    • Confidence interval
    • Confidence coefficient

26. Describe both the confidence limit and confidence coefficient in layman terms.

27. Provide definitions for the following 9 terms:
    • Statistical hypothesis
    • Null hypothesis
    • Alternative hypothesis
    • Test of a statistical hypothesis
    • Type I error
    • Type II error
    • Level of significance
    • Test statistic
    • Critical region

28. Describe both fault trees and event trees. What are the differences between the two trees?

29. A distillation column explosion can occur if the overhead cooler fails (*OC*) and condenser fails (*CO*) or there is a problem with the reboiler (*RB*). The overhead unit fails (*OUC*) only if both the coolers fail (*OC*) and the condenser fails (*CO*). Reboiler problems develop if there is a power failure (*PF*) or there is a failed tube (*FT*). A power failure occurs only if there are both operator error (*OE*) and instrument failure (*IF*). Add any other failure possibilities of your choice and then construct a fault tree.
    Comment: Refer to the literature [2] for possible assistance.

30. Describe hypothesis testing in layman terms.

31. One of the authors [5] once outlined a general procedure that may be employed for testing a hypothesis, as described below
    • Choose a probability distribution and a random variable associated with it. This choice may be based on previous experience, intuition, or the literature.

- Set H_0 and H_1. These must be carefully formulated to permit a meaningful conclusion.
- Specify the test statistic and choose a level of significance α for the test.
- Determine the distribution of the test statistic and the "critical region" for the test statistic.
- Calculate the value of the test statistic from sample data. Accept of reject H_0 by comparing the calculated value of the test statistic with the critical region.

Apply this procedure to a technical application of your choice.

32. Provide a layman's definition of nonparametric tests.
33. Provide a layman's definition of analysis of variance.
34. Describe the role design of experiments can play on the career of a chemical engineer.
35. A redundant system consisting of three operating pumps can survive two pump failures. Assume that the pumps are independent with respect to failure and each has a probability of failure of x. Obtain the equations describing the reliability of the system in terms of x. Calculate the reliability of the system for various x values in the 0.05 – 0.25 range. Analyze the results.
36. Consider a standby redundancy system with one operating unit and one on standby, and a system that can survive one failure. If the failure rate is 2 units per year, what is the reliability of the system over a 2 – 24 month range? Comment on the results.
37. Consider the following numbers:

$$8, 14, 23, 34, 42, 49, 57, 66$$

Comment on whether these are random numbers, a result of a set progression, or the results of an analytical expression or whether there is some other explanation for this sequence. Comment: This is a tough one. Care should be exercised in interpreting numbers. The numbers given in the Problem statement do *not* represent random numbers. Refer to the literature [3] for the solution.
38. One of the major gambling options during the professional football championship game (Super Bowl) is to "buy a box" in a uniquely arranged square, usually referred to as the

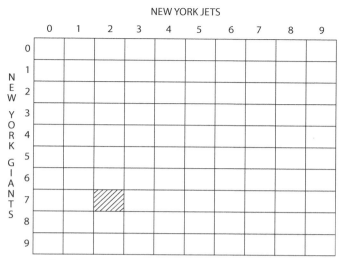

Figure 19.4 Football pool

pool. An example of a pool is shown in Figure 19.4. As can be seen, there are 100 boxes. If each box costs $1,000, the total cash pool is $100,000. The individual, who select the box with the last digit of the final score for each team takes home the bacon, i.e., wins the $100,000. If the final score of the Super Bowl is Jets 22/Giants 7, the owner of the shaded box is the winner. Scores such as Jets 12/Giants 27 or Jets 22/Giants 37 would also be winners. Note that this format does not provide each person buying a box an equal chance of winning (unless the numbers are selected at random).

Revise and rewrite the above Problem statement so that it relates to a technical application.

39. A series system consists of two electrical components, A and B. Component A has a time to failure, T_A, assumed to be normally distributed with mean 100 hours and standard deviation 20 hours. Component B has a time to failure, T_B, assumed to be normally distributed with mean 90 hours and standard deviation 10 hours. The system fails whenever either Component A or Component B fails. Therefore, T_S, the time to failure of the system is the minimum of the times to failure of components A and B. Estimate the average value of T_s on the basis of the simulated values of 10 simulated values of T_A and 10 simulated values of T_B using any convenient method of performing the calculation [2,4].

Comment: Can applying a Monte Carlo method of solution assist?

40. The life (time to failure) of a machine component has a Weibull distribution. Determine the probability that the component lasts at least 25,000 hours if

   1. the failure rate is constant and equal to 0.01 per thousand hours where t is measured in thousand of hours, and

   2. the failure rate is $t^{-1/2}$ where t is measured in thousand of hours. Also perform the calculation for different failure rate values. Comment on the results.

41. Table 19.2 below shows 8 pairs of observations on X and Y where Y is the observed percent yield of a chemical reaction at various centigrade temperatures, X. Select a model of your choice and obtain the least squares line of regression of Y and X and use it to estimate the average percent yield at 235° degrees C° [2]

   Comment: First plot the observed values of Y against the associated values of X and observer whether the scatter diagrams exhibit a linear pattern.

42. Refer to Problems 21 and 22. Attempt to develop a *two* coefficient failure rate model.

TABLE 19.2 Temperature – Yield for Problem 4

Temperature, °C X	% Yield Y
150	75.4
175	79.4
200	82.1
225	86.6
250	90.9
275	93.3
300	95.9
325	96.1

References

1. Webster's New World Dictionary, Second College Edition, Prentice-Hall, Upper Saddle River, NJ, 1971.
2. L. Theodore and F. Taylor, *Probability and Statistics*, A Theodore Tutorial, Theodore Tutorials, East Williston, NY, originally published by USEPA/APTI, RTP, NC, 1993.
3. Adapted from, S. Schaefer and L. Theodore, *Probability and Statistics Applications in Environmental Science,* CRC Press/ Taylor & Francis Group, Boca Raton, FL, 2007.
4. L. Theodore and R. Dupont, *Environmental Health and Hazard Risk Assessment: Principles and Calculations*, CRC Press/Taylor & Francis Group, Boca Raton, FL, 2012.
5. Personal notes: L. Theodore, East Williston, NY, 2013.
6. L. Theodore, *Chemical Engineering: The Essential Reference*, McGraw-Hill, New York City, NY, 2014.

20

Nanotechnology

This chapter is concerned with nanotechnology. As with all the chapters in Part II, there are sereval sections: Overview, several technical topics, illustrative open-ended problems, and open-ended problems. The purpose of the first section is to introduce the reader to the subject of nanotechnology. As one might suppose, a comprehensive treatment is not provided although several technical topics, are also included. The next section contains three open-ended problems; the authors' solution (there may be other solutions) are also provided. The final section contains 26 problems; *no* solutions are provided here.

20.1   Overview

This overview section is concerned with—as can be noted from the chapter title—nanotechnology. As one might suppose, it was not possible to address all topics directly or indirectly related to nanotechnology. However, additional details may be obtained from either the references provided at the end of this Overview and/or at the end of the chapter.

Note: Those readers already familiar with the details associated with this topic may choose to bypass this Overview.

Nanotechnology. Everybody's talking about it. It's the new kid on the block. The financial markets and some readers don't quite know what to make of it. Although not to be found in Webster's Dictionary, it is concerned with the world of invisible miniscule particles that are dominated by forces of physics and chemistry that cannot be applied at the macro— or human—scale level. These particles, when combined, have come to be defined by some as nanomaterials, and these materials possess unusual properties not present in traditional and/or ordinary material.

Regarding the word nanotechnology, it is derived from the Greek word *nano* (for dwarf) and technology. Nano, typically employed as a prefix, is defined as one billionth of a quantity or term and is represented mathematically as 1×10^{-9} or simply as 10^{-9}. *Technology* generally refers to "the system by which a society provides its members with those things needed or desired". The term nanotechnology has come to be defined as those systems or processes that provide goods and/or services that are obtained from matter at the nanometer level, i.e., from sizes at or below one billionth of a meter. The new technology thus allows the engineering of matter by systems and/or processes that deal with atoms. One of the major problems that remain is the development of nanomachines that can produce other nanomachines in a manner similar to what many routinely describe as mass production.

Interesting, the fundamental laws of chemistry and physics work differently when particles reach the nanoscale as the powers of hydrogen bonding, quantum energy, and van der Waals' forces endow some nanomaterials with some very unusual properties. Carbon nanotubes for instance, discovered in the sooty residue of vaporized carbon rods, defy the standard laws of physics. Stronger and more flexible than steel, yet measuring about 10,000 times smaller than the diameter or a human hair, these cylindrical sheets of carbon atoms are also useful as coatings on computers and other electrical devices. Nanoparticles, another manifestation of nanotechnology, are known to foster stubborn reactions because they have an enormous surface area relative to their volume [2].

As noted above, the classical laws of science are different at the nanoscale. Nanoparticles possess large surface areas and essentially no inner mass, i.e., their surface to mass ratio is extremely high. This new "science" is based on the knowledge that particles in the nanometer range, and nanostructures or nanomachines that are developed from these nanoparticles, possess special properties, and, in conjunction with their unique behavior, can significantly impact physical, chemical, electrical, biological, mechanical, etc., functional properties. These new characteristics can be harnessed and exploited by chemical engineers to engineer "Industrial Revolution II" processes.

In addition to this overview, the chapter contains the following:

1. Early History
2. Fundamentals and Basic Principles
3. Nanomaterials
4. Production Methods
5. Current Applications
6. Environmental Concerns
7. Future Prospects

20.2   Early History [3]

Nanoparticles arrived on scene immediately following the Big Bang some 13 billion years ago. However, it is not clear when humans first began to take advantage of nanosized materials. It is known that in the fourth century A.D. Roman glassmakers were fabricating glasses containing nanosized metals. Michael Faraday published a paper in 1857 in the *Philosophical Transactions of the Royal Society*, which attempted to explain how metal particles affect the color of church windows. Gustav Mie was the first to provide an explanation of the dependence of the color of the glasses on the kind of metal and its size. A century later, Richard Feynman presented a lecture titled, "There Is Plenty of Room at the Bottom," where he speculated on the possibility and potential of nanosized materials. He proposed manipulating individual atoms to make new small structures having different properties. Groups at Bell Laboratories and IBM fabricated the first two-dimensional quantum cells in the early 1970s. They were made by thin-film (epitaxial) growth techniques that build a semiconductor layer one atom at a time; the work was the beginning of the development of the zero-dimensional quantum dot, which is now one of the nanotechnologies in commercial applications. In 1996, a number of government agencies led by the National Science Foundation (NSF) commissioned a study to assess the current worldwide status of trends, research, and development in nanoscience and nanotechnology. This early NSF activity provided the necessary impetus for the future for this industry to expand and flourish.

20.3   Fundamentals and Basic Principles

Matter is anything that has mass and can be physically observed. All mater is composed of *atoms* and *molecules*; it consists of a finite number of elements, often represented as *building blocks*. *Atoms* are small particles that cannot be made smaller while *molecules* are groups of atoms bound together, but possessing properties different from an atom.

The aforementioned atom is composed of a small core defined as the *nucleus* that is surrounded by *electrons*. The nucleus is composed of two types of particles: *protons* and *neutrons*; however, their is significant space between the electrons and the nucleus. Protons and neutrons are themselves made up of even smaller particles, known as quarks. One generally depicts the changes in units of the electron charge, so that the charge of an electron is written as 1^- and that of a proton is written as 1^+.

The aforementioned matter has physical and chemical properties that are related to its size, i.e., the properties of most solids depend on the size range over which they are measured. The size range can be macroscopic, microscopic, or molecular; one may view these sizes as finite, differential and molecular, respectively. The object of this chapter is to discuss these characteristics at the molecular, or nanometer level. At the macro—or large—scale range ordinarily studied in traditional fields of physics such as mechanics, electricity, magnetism, and optics, the sizes of the objects under study range from millimeters to kilometers, i.e., finite sizes. The properties that one associates with these materials are averaged properties, such as density and thermal conductivity.

When familiar materials such as metals, metal oxides, ceramics and polymers, and novel forms of carbon such as the aforementioned carbon nanotubes and fullerenes (or buckyballs) are converted/produced into infinitesimally small particle sizes (and, in the case of carbon nanotubes and buckyballs, unique structural geometries, as well), the resulting particles have an order-of-magnitude *increase* in available surface area. It is this remarkable surface area of particles in the nanometer range that confers upon them some unique material properties, especially when compared to macroscopic particles of the same material [4].

One of the hallmarks of nanotechnology is the desire to produce and use nanometer-sized particles of various materials in order to explore the remarkable characteristics and performance attributes that many materials exhibit at these infinitesimally small (particle) sizes.

20.4   Nanomaterials

Nanomaterials have their own origin with what have come to be defined as prime materials. These prime materials essentially consist of (pure) elements and compounds. The range of these elements and compounds that have been successfully produced and deployed as nanometer-sized particles include:

1. Metals such as iron, copper, gold, aluminum, nickel, and silver
2. Oxides of metal such as iron, titanium, zirconium, aluminum, and zinc
3. Silica sols, and fumed and colloidal silica
4. Clays such as talc, mica, smectite, asbestos, vermiculite, and montmorillonite
5. Carbon compounds, such as fullerences, nanotubes, and carbon fibers

Information on each of these types of materials along with the manufacturing methods (see next section) used to render them into nanoscale particles, is available in the literature [5].

Six of the major metals that have served as prime materials include the following:

1. Iron
2. Aluminum
3. Nickel
4. Silver
5. Gold
6. Copper

Five of the major mixed oxides include the following:

1. Iron oxides (Fe_2O_3 and Fe_3O_4)
2. Silicon dioxide (silica; SiO_2)
3. Titanium dioxide (titania; TiO_2)
4. Aluminum oxide (alumina; Al_2O_3)
5. Zirconium dioxide (zirconia; ZrO_2 and zinc oxide; ZnO).

Additional details on both metals and metal oxides are available in the literature [6].

20.5   Production Methods

In general, there are 6 widely used methods for producing nanoscaled particles. These are listed below.

1. Plasma-arc and flame-hydrolysis methods (including flame ionization)
2. Chemical vapor deposition (CVD)
3. Electrodeposition techniques
4. Sol-gel synthesis
5. Mechanical crushing via ball milling
6. Use of natural occurring nanomaterials

Naturally occurring materials, such as zeolites, can be used as found or synthesized and modified by conventional chemistry. A zeolite is a caged molecular structure containing large voids that can admit molecules of certain size and deny access to other larger molecules. They find application as catalysts as well as adsorbents and other materials [2,4,7,8].

The ongoing challenge for industry is to continue to devise, perfect, and scale up viable production methodologies that can cost effectively and reliably produce the desired nanoparticles with the desired particle size, particle size distribution, purity and uniformity in terms of both composition and structure.

To summarize, there are 6 major methods to produce nanomaterials. These are plasma arcing, flame hydrolysis, chemical vapor deposition (CVD), electrodeposition, sol gel synthesis, ball milling, and the use of natural occurring nanoparticles.

20.6   Current Applications

Present day applications include chemical products, such as plastics, specialty metals, powders, etc., computer chips, and computer systems [1]. Specific examples of nanotechnology in actual commercial use today include [4]:

1. Semiconductor chips and other microelectronics applications
2. High surface-to-volume catalysts, which promote chemical reactions more efficiently and selectively
3. Ceramics, lighter-weight alloys, metal oxides and other metallic compounds
4. Coating, paints, plastic, fillers, and food-packaging applications
5. Polymer-composite materials, including tires with improved mechanical properties
6. Transparent composite materials such as sunscreens containing nanosized titanium dioxide and zinc oxide particles

7. Use in fuel cells, battery electrodes, communication applications, photographic film developing, and gas sensors
8. Nanobarcodes
9. Tips for scanning probe microscopes
10. Purification of pharmaceuticals and enzymes

A host of other applications are certain to emerge in the future.

20.7 Environmental Concerns [9,10]

Any technology can have various and imposing effects on the environment and society. Nanotechnology is no exception, and the results of these effects will be determined by the extent to which the technical community manages this technology. This is an area that has, unfortunately, been seized upon by a variety of environmental groups.

An environmental implication of nanotechnology has been dubbed by many in this diminutive field as "potentially negative". The reason for this label is as simple as obvious. The technical community is dealing with a significant number of unforeseen effects that could have disturbingly disastrous impacts on society. Fortunately, it appears the probability of such dire consequences actually occurring is near zero…but *not* zero. This finite, but differentially small, probability is one of the reasons this section was included in this chapter; and it is the key topic that is addressed in the material to follow.

Air, water, and land (solid waste) emissions from nano-technology operations in the future, as well as companion health and hazard issue are a concern. These issues arose earlier with the Industrial Revolution, the development/testing/use of the atomic bomb, the arrival of the internet, Y2K, etc.; and, all were successfully (relatively speaking) resolved by the engineers and scientists of their period.

To the authors' knowledge there are no documented nano human health hazards. Statements in the literature refer to *potential* health problems. The authors have also speculated on the need for future nanoregulations (see next paragraph and problem 2 in a later section). Suggestions and potential options are provided [11] while noting that the ratio of pollutant nanoparticles (from conventional sources such as power plants) to engineered nanoparticles being released into the environment may be as high as a trillion to one, [12] i.e., 10^{12}:1. If this be so, the environmental concerns for nanoparticles can almost certainly be dismissed.

Current environmental regulations, as they apply to nanomaterial, are provided by Stander and Theodore [13]. Many environmental concerns are addressed by existing health and safety legislation. Most countries including the U.S. require a health and safety assessment for any new chemical before it can be marketed. In addition, the European Union (EU) recently introducing the world's most stringent labeling system. Prior experience with materials such as PCBs and asbestos, and a variety of unintended effects of drugs such as thalidomide, mean that both companies and governments have incentives to keep a close watch on potential negative health and environmental effects [14].

It is very difficult to predict future nanoregulations. In the past, regulations have been both a moving target and confusing. What can be said for certain is that there will be regulations, and the probability is high that they will be contradictory and confusing. Past and current regulations provide a measure of what can be expected. And, it is for this reason that this section is included in this chapter. Detailed analysis of various U.S. and EU laws and regulations are available in the literature [15–19].

20.8   Future Prospects [6,8,20,21]

Ranking high among the challenges is the ongoing need to develop and perfect reliable techniques to produce (and mass produce) nanoscaled particles that have not only the desirable particle sizes and particle size distributions but also a minimal number of structural defects and produce acceptable purity levels since these latter attributes can drastically alter the anticipated behavior and properties of the nanoscaled particles. Experience to date indicates that scale-up issues associated with moving today's promising nanotechnology-related development from laboratory and pilot scale demonstrations to full scale commercialization can be considerable [6].

Most believe nanotechnology will have its major impact on war, crime, terrorism, and the massive companion industries, particularly security and law enforcement. The military has a significant interest in nanotechnology, including such areas as optical systems, nanorobotics, nanomachines, "smart" weapons, nanoelectronics, virtual reality, massive memory, specialty materials for armor, nanobased materials for stopping bullets, and bionanodevices to deter and destroy chemical and biological agents. Most of this activity is concerned with protection against attack and minimizing risk to military personnel, e.g., devices that may be able to repair defective airframes or the hulls of ships before major problems develop. But make no mistake, the rush is on (as with development of the atomic bomb) to conquer this technology; the individual or organization or country that successfully conquers this technology will almost certainly conquer the world. Society, as well as the technical community, has to understand that the misuse of this new technology can lead to and cause catastrophic damage; alternately, nanotechnology could be used to provide not only sophisticated sensor and surveillance systems to identify military threats but also weapons (or the equivalent) that will eliminate these threats [20].

Regarding crime, the techniques of nanoscience will have a lot to offer forensic investigations, both for biological analysis, and materials and chemical studies. Portable instruments with sophisticated nanosensors will be able to perform accurate high level analyses at crime sites. These instruments should greatly improve conviction rates and the ability to locate real clues. Nanotechnology will also stop money laundering by imprinting every computer digit [21].

Nanotechnology may open up new ways of making computer systems and message transfers secure using special hardware keys that are immune to any form of hacking. Very few current computer protection systems are able to keep out determined hackers. Nanoimprinting, which already exists, could be used to make "keys" or even special nano-based biosensors coded with a dynamic DNA sequence. Nano-imprinting is already used to make bank notes virtually impossible to forge by creating special holograms in the clear plastic; forgery would be possible only if the master stamps were actually stolen, but then a new hologram could be made [20].

In addition, nanoparticle-related developments are being actively pursued to improve fuel cells, batteries, solar devices, advanced data-storage devices such as computer chips and hard drives, magnetic audio and videotapes, and sensors and other analytical devices. Meanwhile, nanotechnology-related developments are also being hotly pursued in other medical applications, such as the development of more effective drug-delivery mechanisms and improved medical diagnostic devices, to name just a few.

A detailed and expanded treatment of nanotechnology is available in the following three references.

1. Adapted from, L. Theodore and R. Kunz, *Nanotechnology: Environmental Implications and Solutions*, John Wiley & Sons, Hoboken, NJ, 2005 [6].

2. Adapted from, L. Theodore, *Nanotechnology: Basic Calculations for Engineers and Scientists*, John Wiley & Sons, Hoboken, NJ, 2006 [8].

3. L. Theodore, *Chemical Engineering: The Essential Reference*, McGraw-Hill, New York City, NY, 2014 [21].

20.9   Illustrative Open-Ended Problems

This and the last section provide open-ended problems. However, solutions *are* provided for the three problems in this section in order for the reader to hopefully obtain a better understanding of these problems which differ from the traditional problems/illustrative examples. The first problem is relatively straightforward while the third (and last problem) is somewhat more difficult and/or complex. Note that solutions are not provided for the 26 open-ended problems in the next section.

Problem 1: Briefly define and discuss present-day atomic theory [8].

Solution: In ancient Greek philosophy, the word *atomos* was used to describe the smallest bit of matter that could be conceived. This "fundamental particle" was thought of as indestructible; in fact, atomos means "not divisible". Knowledge about the size and nature of the atom grew slowly throughout the centuries.

As discussed in the Overview, the atom consists of three subatomic particles: the *proton, neutron,* and *electron.* The charge of an electron is -1.602×10^{-19} C (coulombs), and that of a proton is $+1.602 \times 10^{-19}$ C. The quantity -1.602×10^{-19} is defined as the *electronic charge.* Note that the charges of these subatomic particles are expressed as multiples of this charge rather than in coulombs. As noted, the charge of any electron is 1−, and that of a proton is 1+. Neutrons carry no charge, i.e., they are electrically neutral. Since an atom has an equal number of electrons and protons, it has zero or no net electric charge.

Both the protons and neutrons reside in the nucleus of the atom, which is extremely small. Most of the atom's volume is the space in which the electrons reside. The external electrons are attracted to the protons in the nucleus because of their opposite electrical charge.

Problem 2: Provide a general overview of nanotechnology environmental regulations.

Solution: Completely new legislation and regulatory rulemaking may be necessary for the environmental control of nanotechnology. However, in the meantime, one may speculate on how the existing regulatory framework might be applied to the nanotechnology area as this emerging field develops over the next several years.

Commercial applications of nanotechnology are likely to be regulated under TSCA, which authorizes the U.S. EPA to review and establish limits on the manufacture, processing, distribution, use, and/or disposal of new materials that EPA determines to pose "an unreasonable risk of injury to human health or the environment." The term *chemical* is defined broadly by TSCA. Unless qualifying for an exemption under the law (a statutory exemption requiring no further approval by EPA), low-volume production, low environmental releases along with low-volume releases, or plans for limited test marketing, a prospective manufacturer is subject to the full-blown procedure. This requires submittal of said notice, along with toxicity and other data to EPA at least 90 days before commencing production of the chemical substance [13].

Approval then involves recordkeeping, reporting, and other requirements under the statute. Requirements will differ, depending on whether EPA determined that a particular application constitutes a "significant new use" or a "new chemical substance." The EPA can impose limits on production, including an outright ban when it deemed necessary for adequate protection against "an unreasonable risk of injury to health or the environment". The EPA may revisit a chemical's status under TSCA and change the degree or type of regulation when new health/ environmental data warrant [13].

Workplace exposure to a chemical substance and the potential for pulmonary toxicity is subject to regulation by OSHA, including the requirement that potential hazards be disclosed on a MSDS (Material Safety Data Sheet). (An interesting question arises as to whether carbon nanotubes, chemically carbon but with different properties because of their small size and structure, are to be considered the same as or different from carbon black for MSDS purposes.) Both governmental and private agencies can be expected to develop the requisite threshold limit values (TLVs) for workplace exposure. Also, the EPA may once again utilize TSCA to assert its own jurisdiction, appropriate or not, to minimize exposure in the workplace. This is almost definitely wishful thinking given the past performance of similar bureaucratic agencies. Adding to the dilemma is the breadth of the nano field and the lack of applicable toxicology and epidemiology data [22].

Another likely source of regulation would fall under the provisions of the Clean Air Act (CAA) for particulate matter less than 2.5 μm ($PM_{2.5}$). Additionally, an installation manufacturing nanomaterials may ultimately become subject as a "major source" to the CAA's Section 112 governing hazardous air pollutants (HAP) [13].

Wastes from a commercial-scale nanotechnology facility would be classified under RCRA, provided that it meets the criteria for RCRA waste. RCRA requirements could be triggered by a listed manufacturing process or the act's specified hazardous waste characteristics. The type and extent of regulation would depend on how much hazardous waste is generated and whether the wastes generated are treated, stored, or disposed of on site [6,13].

Problem 3: Your consulting firm has received a contract to develop, as part of research study, a mathematical model describing the concentration of a nanochemical in a medium-sized ventilated laboratory room. The following notation/data (SI) units are provided:

V = volume of room, m^3

v_0 = volumetric flow rate of ventilation air, m^3/min

c_0 = concentration of the nanochemical in ventilation air, $gmol/m^3$

c = concentration of the nanochemical leaving ventilated room, $gmol/m^3$

c_i = concentration of the nanochemical initially present in the ventilated room, $gmol/m^3$

r = rate of appearance of disappearance of the nanochemical in the room due to reaction and/or other effects, $gmol/m^3 \cdot min$.

As an authority in the field, you have been requested to:

1. Obtain the equation describing the concentration in the room as a function of time.
2. Calculate the minimum air ventilation flow rate containing $10 ng/m^3$ nano-particles flowing into the room to assure that the nanoagent concentration does not exceed $35.0 ng/m^3$. The nanoagents are appearing (internally) in the laboratory at a rate of 250 ng/min. Assume steady-state conditions.
3. Comment on the validity and limitations of both the model (1) and calculated results (2).

Solution: 1. Use the laboratory room as the control volume. Apply the concentration law for mass (see also chapter 3) to the nanochemical.

$$\begin{Bmatrix} rate\ of\ mass \\ in \end{Bmatrix} - \begin{Bmatrix} rate\ of\ mass \\ out \end{Bmatrix} + \begin{Bmatrix} rate\ of\ mass \\ generated \end{Bmatrix} = \begin{Bmatrix} rate\ of\ mass \\ accumulated \end{Bmatrix}$$

(20.1)

Employing the notation specified in the problem statement gives:

$$\{\text{rate of mass in}\} = v_0 c_0$$

$$\{\text{rate of mass out}\} = v_0 c$$

$$\{\text{rate of mass generated}\} = rV$$

$$\{\text{rate of mass accumulated}\} = \frac{dc}{dt} \tag{20.2}$$

Substituting above gives

$$v_0 c_0 - v_0 c + rV = \frac{dc}{dt} \tag{20.3}$$

If the laboratory room volume is constant, V may be taken out of the derivative term. This leads to

$$\frac{v_0}{V}\left(c_0 - c\right) + r = \frac{dc}{dt} \tag{20.4}$$

The term V/v_0 represents the average residence time the nanochemicals reside in the room and is usually designated as τ. The above equation may then be rewritten as

$$\frac{dc}{dt} = \frac{c_0 - c}{\tau} + r \tag{20.5}$$

2. The applicable model for this case is:

$$v_0\left(c_0 - c\right) + rV = V\frac{dc}{dt} \tag{20.6}$$

Under steady-state conditions, $dc/dt = 0$.

$$v_0\left(c_0 - c\right) + rV = 0 \tag{20.7}$$

Pertinent information incudes

$$rV = 250 \text{ ng} / \text{min}$$
$$c_0 = 10 \text{ ng} / \text{m}^3$$
$$c = 35 \text{ ng} / \text{m}^3$$

Substituting into equation 20.7 gives

$$
\begin{aligned}
v_0 &= \frac{-rV}{c_0 - c} \\
&= \frac{-rV}{c - c_0} \\
&= \frac{250}{35 - 10} \\
&= 10 \text{ m}^3 / \text{min} = 353 \text{ft}^3 / \text{min}
\end{aligned}
$$

3. There were two assumptions in the development. The assumption that the room volume is constant is reasonable. However, the development also assumes that room contents are perfectly mixed (similar to a continuous stirred tank reactor, i.e., CSTR) and this can significantly impact the analysis and results [2,3].

This open-ended problem will be revisited in Part III, Chapter 26, Environmental Health and Hazard Risk; Term Projects 26.4-26.5.

20.10   Open-Ended Problems

This last section of the chapter contains open-ended problems as they relate to nanotechnology. No detailed and/or specific solution is provided; that task is left to the reader, noting that each problem has either a unique solution or a number of solutions or (in some cases) no solution at all. These are characteristics of open-ended problems described earlier.

There are comments associated with some, but not all, of the problems. The comments are included to assist the reader while attempting to solve the problems. However, it is recommended that the solution to each problem should initially be attempted *without* the assistance of the comments.

There are 26 open-ended problems in this section. As stated above, if difficulty is encountered in solving any particular problem, the reader should next refer to the comment, if any is provided with the problem.

The reader should also note that the more difficult problems are generally located at or near the end of the section.

1. Describe the early history associated with nanotechnology.
2. Discuss the recent advances in nanotechnology.
3. Select a refereed, published article on nanotechnology from the literature and provide a review.
4. Provide some normal everyday domestic applications involving the general topic of nanotechnology.
5. Develop an original problem on nanotechnology that would be suitable as an illustrative example in a book.
6. Prepare a list of the various technical books which have been written on nanotechnology. Select the three best (hopefully, it will include a book written by one of the authors) and justify your answer. Also select the three weakest books and, once again, justify your answer.
7. Describe nanotechnology in layman terms.
8. Consider how to integrate pure nanomaterials into the periodic table.
9. Outline how to determine the number of potential nanomaterials that could possibly be derived from the 112 elements.
10. Describe, in technical detail, the different between true density, actual density, and bulk density.
11. Describe, in layman terms, the difference between true density, actual density, and bulk density.
12. Describe the impact of different shaped particles on true density, actual density, and bulk density.
13. Develop equations describing the area and volume of various shaped particles.
    Comment: Consider a sphere and cube initially and continue on from there. Also, refer to the literature [8].
14. Describe some of the procedures that are available to estimate the physical properties of nanomaterials.
15. Describe some of the procedures that are available to estimate the chemical properties of nanomaterials.
16. Attempt to develop an improved method to describe submicron particle behavior in air.
17. The bulk of material on the Cunningham Correction Factor (CCF) is based on particle behavior in air [7]. Quantitatively discuss how the CCF would be affected by other gases.

18. Develop equations describing the particle collection efficiency as a function of particle size for various air pollution control devices, e.g., baghouses, electrostatic precipitators, cyclones, etc.

    Comment: Refer to the literature [7] for additional details.

19. Discuss in layman terms the difference between collection efficiency and penetration.

    Comment: Refer to the literature [7] for additional details.

20. Refer to Problem 3 in the previous Section. Develop describing equations for the ventilation system if the nanomaterial is undergoing a reaction. Present the equations for reaction mechanisms of your choice.

    Comment: Refer to the literature [8] for additional details.

21. Refer to Problem 3 in the previous Section. Develop describing equations for the ventilation systems if the nanomaterial is being introduced into the system for different time-variable flowrates. Present the equations for reaction mechanisms of your choice.

22. Provide calculational procedures to estimate nanoparticle emissions from line and area sources [7].

23. As discussed in the Overview, there are six widely used methods for producing nanoscaled particles of various materials. These are listed below:
    - Plasma-arc and flame-hydrolysis methods (including flame ionization)
    - Chemical vapor deposition (CVD)
    - Electrodeposition techniques
    - Sol-gel synthesis
    - Mechanical crushing

    Promising technologies

    Describe these methods in layman terms.

24. Size distributions of nanoparticles are often characterized by a "mean" particle diameter. Although numerous "means" have been defined in the literature, the most common are the arithmetic mean and the geometric mean. The arithmetic mean diameter is simply the sum of the diameters of each of the particles divided by the number of particles measured. The geometric mean diameter is the nth root of the product of the n number of particles in the sample. In addition to the arithmetic and geometric means, a particle size distribution may also

be characterized by the "median" diameter. The median diameter is that diameter for which 50% of the particles are larger in size and 50% are smaller in size. Another important characteristic is the measure of *central tendency*. Is it sometimes referred to as the dispersion or variability. The most common term employed is the standard deviation. Present at least two other methods of describing particle size distributions [7,20]

25. Briefly describe the regulatory processes that opponents of nanotechnology could pursue in the future and comment whether (in your judgment) their efforts will be successful.

26. Determine the number of four, five, and six element chemical compounds that can theoretically be generated from a pool of 112 elements. Assume each element counts only once in the chemical formula. An example of a three-element compound is H_2SO_4 (sulfuric acid), or CH_3OH (methanol). An example of a four-element compound is $NaHCO_3$. Also perform the calculations if the element appears twice in the chemical formula. Comment on the results.

References

1. L. Theodore, *Nanotechnology I*, The Williston Times, East Williston, NY, April 16, 2004.
2. R. D'Aquino, personal communication, to L. Theodore, Bronx, NY, 2005.
3. Adapted from, C. Poole and F. Owens, *Introduction to Nanotechnology*, John Wiley & Sons, Hoboken, NJ, 2003.
4. S. A Shelley with G. Ondrey, *Nanotechnology- The Sky's the Limit*, Chemical Engineering, New York City, NY, December 2002.
5. A. Boxall, Q. Chaudhry, A. Jones, B. Jefferson, and C.D Watts, *Current and Future Predictions of Environmental Expose to Engineered Nanoparticles*, Central Science Laboratory, Sand Hutton, UK, 2008.
6. Adapted from, L. Theodore and R. Kunz, *Nanotechnology: Environmental Implications and Solutions*, John Wiley & Sons, Hoboken, NJ, 2005.
7. L. Theodore, *Air Pollution Control Equipment Calculations*, John Wiley & Sons, Hoboken, NJ, 2008.
8. Adapted from, L. Theodore, *Nanotechnology: Basic Calculations for Engineers and Scientists*, John Wiley & Sons, Hoboken, NJ, 2006.
9. M.K. Theodore and L. Theodore, *Major Environmental Issues Facing the 21st Century*, (originally published by Prentice-Hall), Theodore Tutorials, East Williston, NY, 1996.

10. G. Burke, B. Singh, and L. Theodore, *Handbook of Environmental Management and Technology*, 2nd edition, John Wiley & Sons, Hoboken, NJ, 2000.

11. L. Theodore, *Waste Management of Nanomaterials*, USEPA, Washington, DC, 2006.

12. Personal notes, L. Theodore, 2006.

13. L. Stander and L. Theodore, *Environmental Regulatory Calculations Handbook*, John Wiley & Sons, Hoboken, NJ, 2010.

14. National center for Environmental Research, *Nanotechnology and the Environment: Applications and Implications*, STAR Progress Review Workshop, Office of Research and Development, National Center for Environmental Research, Washington, DC, 2003.

15. D. Fiorino, *Voluntary Initiatives, Regulations, and Nanotechnology Oversight: Charting a Path*, Woodrow Wilson International Center for Scholars, Washington, DC, U.S.A, November 2010.

16. L. Breggin, R. Faulkner, N. Cheel, K. Prendergrass, and R. Porter, *Securing the Profits of Nanotechnologies: Towards Transatlantic Regulatory Cooperation*, Chatham House: London, UK, September, 2009.

17. Laws and Regulations; U.S EPA: Washington, DC, U.S.A, available online: http://www.epa.gov'lawsregs/laws/index.html (accessed on 23 November 2010).

18. L. Bergeson, *Nanotechnology Trend Draws Attention of Federal Regulators*, Manufacturing Today, location unknown, March/April 2004.

19. L. Bergeson, B. Auerbach, *The Environmental Regulation Implications of Nanotechnology*, BNA Daily Environment Reporter, pp. B-1-B-7, location unknown April 14, 2004.

20. L. Theodore, *Nanotechnology II*, The Williston Times, East Williston, NY, April 23, 2004.

21. L. Theodore, *Chemical Engineering: The Essential Reference*, McGraw-Hill, New York City, NY, 2014.

22. L. Theodore and R. Dupont, *Environmental Health and Hazard Risk Assessment: Principles and Calculations*, CRC Press/Taylor & Francis Group, Boca Raton, FL, 2012.

23. L. Theodore, *Chemical Reactor Analysis and Applications for the Practicing Engineer*, John Wiley & Sons, Hoboken, NJ, 2012.

21

Legal Considerations

This chapter is concerned with legal considerations. As with all the chapters in Part II, there are several sections: overview, several technical topics, illustrative open-ended problems, and open-ended problems. The purpose of the first section is to introduce the reader to the subject of legal considerations. As one might suppose, a comprehensive treatment is not provided although numerous references are included. The second section contains three open-ended problems; the authors' solution (there may be other solutions) is also provided. The third (and final) section contains 35 problems; *no* solutions are provided here.

21.1 Overview

This overview section is concerned with—as can be noted from the chapter title—legal considerations. As one might suppose, it was not possible to address all topics directly or indirectly related to legal considerations.

However, additional details may be obtained from either the references provided at the end of this Overview and/or at the end of the chapter.

Note: Those readers already familiar with the details associated with this subject may choose to bypass this Overview.

Technology affects almost every area of human activity in one way or another. Therefore, one can expect that legal relations between people will have to be taken into account. Even though unique developments in technology will continue to be in the realm of reality, those involved with any field of engineering and science plus those involved with law should consider how technology and law might interact.

It is incumbent upon those engaged in any area of technological development to acquire a basic understanding of patent law, and legal considerations in general, because the patent portfolio of a company, particularly one focused on research and development, may represent its most valuable asset(s). Certain activities, such as premature sale or public disclosure, can jeopardize one's right to obtain a patent; the reader should note that patents are creatures of the national law of the issuing country and are enforceable only in that country. Thus, a U.S. patent is enforceable only in the U.S. To protect one's invention in foreign countries, one must apply in the countries in which protection is sought. In addition, one can obtain a general idea of the development/progress of a new technological field by monitoring the number of patents issued in that field.

As with any invention, the qualities that make technology-related inventions patentable are *novelty*, *non-obvious*, and *utility*. While many unique aspects of technology are already known, more are being discovered as are new ways of exploiting known properties, all of which can lead to patentable inventions. The growth of patents in the chemical engineering industry is a clear indication of the recognition of its importance in this field.

Another area of concern is government regulations. Environmental regulations, in particular, have become a concern for new technological development. In fact, for many technologies it is the increasingly stringent clean air and water regulations that drive new technological development. This is particularly true, for example, in the automotive industry and in heavy industries where emissions are released into the atmosphere or water. What effects will new technology have on the environment? This is yet unknown because the developments are still, relatively speaking, in their infancy. But suppose, for example, new products are developed, products that can be ingested or inhaled, and that act upon the interior organs of the human body. Also, suppose that such products are released into the atmosphere. They might be carried by air within the state or across state lines. The state and federal environmental regulatory agencies would understandably be taking great interest in the environmental impact of such a release.

Although a host of topics are examined in this chapter, patents receive the bulk of the treatment. Other topics are also discussed as noted below

1. Intellectual Property Law
2. Contract Law
3. Patents
4. Infringement and Interferences
5. Copyrights
6. Trademarks
7. The Engineering Professional Licensing Process

21.2 Intellectual Property Law

An important area of law with which the chemical engineer must be concerned is *intellectual property law*. Technological development is all about ideas. Ideas have commercial value only if they can be protected by excluding others from exploiting those ideas. Typically, the way to protect ideas is through intellectual property rights such as *patents, trademarks, copyrights, trade secrets*, and *financial secrets*. Each of these concepts is briefly introduced in the next paragraph.

Patents can be used to protect useful inventions, ornamental designs, and even botanical plants. The patent allows the owner of the patent the right to prevent anyone else from making, using, or selling the "invention" covered by the claims of the patent. Trademarks are distinctive marks associated with a product or service (these are usually referred to as service marks), which the owners of the mark can use exclusively to identify themselves as the source of the product or service. Copyrights protect the expression of an idea, rather than the idea itself, and are typically used to protect literary works (such as this book) plus visual and performing arts, such as photographs, paintings and drawings, sculptures, movies, songs, etc. Trade secret laws protect technical or business information that a company uses to gain a competitive business advantage by virtue of the secret being unknown to others. Customer or client lists, secret formulations, or methods of manufacture are typical business secrets.

21.3   Contract Law

Another legal area that is relevant to chemical engineers is *contract law*. Any time two or more parties agree upon something, the principles of contract law come into play. The essential components of a contract are:

1. Parties competent to enter into a contractual agreement and subject matter, i.e., what the contract is about
2. Legal considerations (the inducement to contract such as the promise(s)) or payment exchanged, or some other benefit or loss or responsibility incurred by the parties.
3. Mutuality of agreement
4. Mutuality of contract

While verbal contracts can be legally binding, in the event of a dispute, it may be difficult to establish in court who said what. It is far better to memorialize the agreement in the form of a written contract. Who are the entities engaged in the contract? Typically, these are business entities. Also, one must also consider whether there may be some peripheral issues, or corporation or partnership law.

One of the basic principles of contract law is that the parties should have a meeting of minds, i.e., they should have a common understanding of what the terms of the contract mean. Sometimes it is not clear what particular terms mean or represent, or the meaning or its implications may change in time. What, for example, qualifies as "technology"? Not only is technology not well defined today, it may encompass that which are not even imagined today in the future.

Generally, contracts are employed with the sale and licensing of exclusive rights to a technology. Also there are agreements to fund technological research and development.

Finally, it should be noted that *most* chemical engineers leave *most* legal activities and decisions to the legal department. However, the chemical engineer *should* recognize that detailed contractual relation(s) should be maintained. Unfortunately, this is often not the case.

21.4   Tort Law

There is a branch of law that can retrospectively address certain situations in which property or people are harmed. That is *tort* law, a topic that continues to receive significant attention by politicians. A tort is a civil wrong, other than a break of contract, for which the law provides a remedy. One can recover damages under tort law if a legal duty has been breached that causes foreseeable harm. These duties are created by law other than duties created by criminal law, governmental regulations, or those agreed to under a contract. Tort law can be very encompassing.

Chemical engineers have to consider the possibilities of reasonably foreseeable harm arising from their developments and activities and take prudent precautions to avert such harm. In the event that a technology is inherently dangerous, chemical engineers and/or their company may be held to a standard of strict liability for any harm caused by the technology regardless of whether an accident or problem was foreseeable.

Most chemical engineers are not at all interested in deliberately causing harm. But some have been. Some are presently. And, some will be. Technology can include military applications, where governments may be interested in developing new weapons. Suppose that new weapons are developed that can invade the human body and do harm. Is a cloud of toxic gas considered a poison gas? Or is it a collection of antipersonnel objects like shrapnel? And suppose such a cloud drifts over, or is released over, a civilian population? How will new weapons be treated under the Geneva Convention? The devastating effect of land mines, which remain lethal long after hostilities are ended and that wreak havoc upon unsuspecting civilians wandering into minefields, has been amply documented. Will new weapons remain harmful years after their deployment? What responsibilities do government(s) have morally, and under international law?[2]

21.5   Patents

In order to encourage new discoveries for the benefit of society/mankind, the U.S. Constitution provides for *patents*. These are *limited monopolies* provided in exchange for the public disclosure of new products and inventions. Patents are an integral part of a free enterprise system. The process discourages secret behavior by rewarding the aforementioned "monopoly" for prompt and adequate disclosure. The U.S. patent system is responsible for much of the growth in the chemical engineering industry because it encourages research upon which the growth is based. Attaining a patent is a procedure requiring skilled and experiences guidance. The patent must be fully disclosed, and the essentials must be covered by the claims.

Of all the intellectual property rights, the most pertinent for products and inventions developed by chemical engineers are patents. A patent can protect, for example, a composition of matter, an article of manufacture, or a method of doing something. Patent rights are private property rights. *Infringement* (see next section) of a patent is a civil offense, not criminal. The patent owner must come to his or her own defense through litigation, if necessary. And, this is often a very expensive undertaking. Lawsuits costing more than a million dollars are not unusual. But at stake can be exclusive rights to a technology worth several orders of magnitude more.

Patents pertaining to any new product, process, equipment, use, or application should be reviewed by a patent attorney. The product or invention of concern may be involved in other patents. Patents available for purchase and lease as well as participation in patent pools should be reviewed by a legal expert.

As with any invention, the qualities noted earlier that make technology-related inventions patentable are *novelty*, *non-obvious*, and *utility*. While many "processes" of technology are already known, more are being discovered as are new ways of exploiting these, all of which can lead to patentable intentions. As noted above, the growth of patents is a clear indication of the industry's recognition of the potential in this field.

It is fair to say that a patent is essentially a contract between an inventor and the public. By full disclosure of the invention to the public, the patentee is given exclusive rights to control the use and practice of his/her product or invention. A patent gives the holder the power to prevent others from using or practicing the invention for a period of years from the date of granting such patent. In contrast, *trade secrets* (see later section) receive protection as long as the information is not public knowledge.

As noted earlier, a patent may be obtained on any new and useful process, method of manufacture, composition of matter, etc., provided it has not been known or used by others before the patentee made the invention or discovery. The invention must not have been described in a printed publication or been in public use or on sale for more than one year prior to the patent application. A patentable item must result from the use of *creative* ability above and beyond that which would be expected of a person working the particular field. A patentable item cannot be something requiring merely mechanical skill. Furthermore, a patent will not be granted for a change in a previously known item or process unless the change involves something entirely *new*.

A patent application consists of the following:

1. A petition, directed to the Commissioner of Patents requesting the grant of a patent
2. An oath, sworn to before a notary public or other designated officer

3. Specifications and claims, in which the claims to be patented are indicated along with detailed specifications including drawings (if applicable) and other pertinent information
4. The application filing fee

When the application is then examined and after a period of time, official action on the claim is taken [3].

21.6 Infringement and Interferences

Two topics directly related to patents are *infringement* and *interferences*. Infringement is of greater concern to the chemical engineer. However, both subject areas are briefly discovered below.

21.6.1 Infringement

The infringement of a patent may consist of making, using, or selling the invention covered by the patent without permission of the patentee. A *contributory infringement* involves the assistance or cooperation with another in the unauthorized making, using, or selling of a patented invention. If the infringement is deliberate, the court may award the patentee as much as three times the actual damages caused *plus* three times the earned profits. The award to the plaintiff is no more than the actual loss. The infringement process is conducted by a search in the Patent Office.

21.6.2 Interferences

A situation can arise in which two or more independent patent applications, covering essentially the same invention, is on file in the U.S. Patent Office. Although this rarely occurs, a procedure called *interference* is instituted to determine who is entitled to the patent. For example, an interference may also be instituted between a pending application and granted patent. Generally, interferences are decided on the basis of priority. The patent is granted to the applicant who was first to conceive the idea for the invention.

Because of the role of *priority* in any type of interference process, it is very important for an inventor to maintain complete records. A written description and sketches should be prepared by an inventor as soon as possible after the conception of an idea that might eventually be patented. This material should preferably be disclosed to one or more witnesses who should indicate in writing that they understand the purpose, method, and structure of the invention. The disclosure should be signed and dated by the inventor and the witnesses. Additional details are available in the Patent Office [4].

21.7   Copyrights

Copyright is a form of protection provided by U.S. law to the author of "original works of authorship" fixed in any tangible medium of expression. The manner and medium of fixation are virtually unlimited. Creative expression may be captured in words, numbers, notes, pictures, or any other graphic or symbolic media. The subject matter of copyright is extremely broad. Although protection is available to both published and unpublished works, the authors are of the opinion that this rarely arises in the publication of science and engineering works since the law of gravity is the law of gravity, the heat transfer equation is the heat transfer equation, and the multiplication and/or log tables are just that; as one of the authors once said: "if you've read one thermodynamic textbook, you've read them all" [5].

Under the 1976 Copyright Act, the copyright owner has the exclusive right to reproduce, adapt, distribute, publicly inform, and publicly display the work. These exclusive rights are transferable and may be licensed, sold, donated to charity, or bequeathed to one's heirs. It is illegal for anyone to violate any of the exclusive rights of the copyright owner. If the copyright owner prevails in an infringement claim, the available remedies include preliminary and permanent injunctions (court orders to stop current or prevent future infringements), impounding, and destroying the infringing articles.

The exclusive rights of the copyright owner, however, are limited in a number of important ways. Under the *fair use* doctrine, which has long been a part of U.S. copyright law and was expressly incorporated in the 1976 Copyright Act, a judge may excuse unauthorized uses that may otherwise be infringing. Section 107 of the Copyright Act lists criticism, comments, news reporting, teaching, scholarship, and research as examples of uses that may be eligible for the fair use defense. In other instances, the limitation takes the form of a *compulsory license* under which certain limited uses of copyrighted works are permitted upon payment of specified royalties and compliance with statutory conditions. The Copyright Act also contains a number of statutory limitations covering specific uses for education purposes.

Chemical engineers should also be aware that there is no such thing as an "intentional copyright" that will automatically protect an author's works in countries around the world. Instead, copyright protection is "territorial" in nature, which means that copyright protection depends on the national laws where protection is sought. However, most countries are members of the Berne Convention on the Protection of Literary and Artistic Works and/or Universal Copyright Convention, the two leading international copyright agreements, which provide important protections for foreign authors. The Patent Office provides additional details [6].

21.8   Trademarks

A *trademark* is a brand name. A trademark includes any word, name, symbol, device, or any combination, used, or intended to be used in commerce to identify and distinguish the goods of one manufacturer or seller from goods manufactured or sold by others, and to indicate the source of the goods. A *service mark* is any word, name, symbols, device, or any combination, used, or intended to be used in commerce to identify and distinguish the services of one provider from services provided by others, and to indicate the source of the services. Not all trademarks need to be registered. But, federal registration has several advantages, including a notice to the public of the registrant's claim of ownership of the mark, a legal presumption of ownership nationwide, and the exclusive right to use the mark on or in connection with the goods or services set forth in the registration.

The trademark process initially consists of the following 7 steps:

1. Determines whether protection is required
2. Determines whether to hire a trademark attorney
3. Identifies trademark format
4. Clearly identifies the precise goods or services to which the trademark will apply
5. Determines whether anyone is already claiming trademark rights in a particular mark through a federal registration
6. Identifies the proper "basis" for filing a trademark application
7. Files the application online through the Trademark Electronic Application System

Filing is relatively simple process. Additional details are provided by the Patent Office [7].

21.9   The Engineering Professional Licensing Process

Becoming a licensed professional engineer (PE) was really not that important in the "old days", particularly for non-civil engineers. In fact, both authors of this text are not a PE. Interestingly both authors have claimed in all honesty, that it has not affected their professional development. However, this situation has changed. The chemical engineer that is not licensed will eventually experience significant constraints on his or her professional development. In effect, licensing has become a necessity for the chemical engineer.

The four requirements that must be satisfied for one to become a licensed professional engineer are listed below: [8]

1. Obtaining a degree from a four-year engineering program accredited by the Accreditation Board for Engineering and Technology, Inc. (ABET)
2. Passing the Fundamentals of Engineering (FE) examination
3. Completion of four years of acceptable engineering experience
4. Passing the Principles and Practice of Engineering (PE) examination.

The reader should note that it appears that the engineering profession is moving toward a goal of requiring a license for all individuals who so desire to practice engineering. This statement particularly applies to chemical engineers in recent years. In addition, one of the authors [9] has prepared a series of tutorials addressing the various licensing exams.

21.10   Illustrative Open-Ended Problems

This and the last section provide open-ended problems. However, solutions *are* provided for the three problems in this section in order for the reader to hopefully obtain a better understanding of these problems which differ from the traditional problems/illustrative examples. The first problem is relatively straightforward while the third (and last problem) is somewhat more difficult and/or complex. Note that solutions are not provided for the 35 open-ended problems in the next section.

Problem 1: No matter how small the organization or how much resistance to change there may be, every person in a supervisory capacity can take steps to minimize exposure to personal liability suits. These steps have the added effect of decreasing the likelihood of incidents that could lead to accidents. Consider the following scenario:

"If I'm every hurt in this place, I'm going to sue the living daylights out of the teacher [or supervisor]!" So said a student in an undergraduate science laboratory [or an employee in an industrial quality control laboratory]. Later, the student [or employee] is injured in an accident and brings a personal liability suit against the teacher [or supervisor] alleging negligence. While the outcome of the jury trial will probably depend most heavily upon the events that led immediately to the accident, a defense against negligence could include testimony to the level of care and supervision regularly provided by the teacher [or supervisor]. Discuss the retroactively verifiable actions of an individual that could be relevant to such a defense.

Solution: A coherent, understandable safety manual that deals specifically with the hazardous circumstances and safety procedures unique to the specific location involved is evidence of forethought and concern. A safety pledge signed by the employee (or student) acknowledging receipt of such a manual can be kept with personal records. Evidence of adequate training in the operation of laboratory or work place equipment is an additional verification of the concern for safety exhibited by a supervisor/ organization.

A written plan for safety instruction that should increase in depth and scope in proportion to the hazards encountered by the employee (or student) can be complemented by a dated record of when such instruction was given. The record should also include a summary of the results of all routine training and safety drills. A syllabus documenting the contents of all safety training courses should also be a part of the permanent employee file.

Testimony to compliance with local laws, e.g., that safety glasses must be worn in the laboratory, etc., is very important. A jury is not likely to let a supervisor off who neglected to enforce safety practices that are legal requirements.

An up-to-date record of accident investigations of accidents and "near misses" with recommendations for changes *and* evidence of implementation of those recommendations are also very helpful. Posted signs and warnings, memoranda on safety issues, and records of meetings of an active safety committee are all physical evidence of a climate that is consistent with noncontributory negligence on the part of supervisory personnel.

Problem 2: Criminal prosecutions for violations of business-related laws are conducted by the Department of Justice. List as many factors you think the Department of Justice may consider when deciding whether or not to conduct a *criminal* prosecution against a violator.

Comment: Think of factors that show how irresponsible or negligent the violator was rather than what laws or regulations were violated.

Solution: The following are some of the criteria used by the Department of Justice to assess the potential for criminal prosecution. You may have though of some that are not listed here.

1. Was the violation discovered by an internal audit or by a government investigation? Did the company commit adequate resources to its internal self-audit program? Does the self-audit program appear likely to uncover or prevent future violations? Does the self-audit program contain safeguards to ensure its integrity?

2. Assuming the violation was discovered by the company, was it disclosed
   - voluntarily
   - promptly upon discovery
   - with sufficient data of good quality and
   - before the regulatory agency discovered it?

3. What was the extent and quality of the cooperation with any government investigation of the violation?

4. Was the violation an isolated incident or one of many violations? How serious was the violation? How long did it go on? How often did it happen?

5. How many employees were involved in the violation? What were the levels of the employees involved in the violation? Was environmental compliance a criterion by which employees were judged? Were employees ever disciplined for environmental violations?

6. What efforts were made to remedy the results of the violation? What steps were taken to prevent a repetition of the violation?

7. What was the extent of good faith efforts to reach compliance agreements with government agencies? Were any such agreements fully carried out?

Problem 3: A chemical company's internal environmental audit uncovered a significant pollution problem involving repeated releases of a toxic chemical into the sewer system. The company took immediate action and completely corrected the problem but did not inform any regulatory agency. Discuss both the legality and the ethics (see next chapter) of the company's actions.

Solution: Failure to report a spill could be a violation of a law, rule or regulation. It could result in fines or even criminal prosecution. The focus of this question, however, is strictly on the ethical aspects of the company's actions.

The following are some of the ethical considerations of this action. You may have thought of others.

While the company's prompt action in correcting the problem is praiseworthy, its failure to report it is *unethical*. Environmental regulatory programs are largely self-policing. They rely heaving on voluntary reporting. Failure to report such a problem lessens the integrity of the entire system and gives a false picture of the need to protect the environment.

Failure to report the release also places an unfair burden on other companies that use the same chemical. This may arise because the sewer authority, having observes high concentrations of the chemical, places additional monitoring and treatment requirements on all the companies. Had the company in question reported the releases, the other companies would not have had to implement these additional, potentially expensive, monitoring and treatment requirements.

Sewer workers and members of the public may have been harmed by direct exposure to the toxic chemicals. Exposure could continue for some time even after the releases have ended. If the chemical is odorless and tasteless, the releases may have gone undetected. Failure to report the releases may deprive these people of the ability to protect themselves from further exposure and from receiving proper medical help.

Even though the releases have stopped, much damage may have occurred to sewage treatment plants, streams, rivers, fish and other aquatic life. People may have been harmed by eating contaminated fish, clams, lobsters, and the like. It is unethical not to inform the public of such dangers. If a cleanup is required to remediate any damage caused to the environment, it would be unethical to remain silent while the taxpayers foot the bill.

See also Chapter 22 for additional information dealing with ethics.

21.11   Open-Ended Problems

This last section of the chapter contains open-ended problems as they relate to legal considerations. No detailed and/or specific solution is provided; that task is left to the reader, noting that each problem has either a unique solution or a number of solutions or (in some cases) no solution at all. These are characteristics of open-ended problems described earlier.

There are comments associated with some, but not all, of the problems. The comments are included to assist the reader while attempting to solve the problems. However, it is recommended that the solution to each problem should initially be attempted *without* the assistance of the comments.

There are 35 open-ended problems in this section. As stated above, if difficulty is encountered in solving any particular problem, the reader should next refer to the comment, if any is provided with the problem. The reader should also note that the more difficult problems are generally located at or near the end of the section.

1. Describe the early history associated with laws.
2. Select a refereed, published article on legal activities/ actions from the literature and provide a review.
3. Provide some normal everyday domestic applications involving the general topic of legal consideration.
4. Develop an original problem concerned with legal matters that would be suitable as an illustrative example in a book.
5. Prepare a list of the various technical books which have been written on law. Select the three best and justify your answer. Also select the three weakest books and, once again, justify your answer.
6. Refer to Problem 1 in the previous Section. Make a list of retroactively verifiable actions of an employee that could be relevant to such a defense.
7. The federal government considers its employees to be *personally* responsible for violations of environmental laws. The government reasons that since it is not possible for an employee to be required in his/her job description to violate any law, rule or regulation, then any such violations must have been committed while the employee was acting "outside the scope of his or her employment". If the employee acted outside the scope of his or her employment, then he/ she is not entitled to legal help from the government and must personally pay any fines levied. If the employee is sent to jail, he/she would almost certainly be fired. Managers who directed or allowed employees to violate environmental laws, rules, or regulations would be held even more

responsible than the employees that he or she supervised. Give at least one benefit of this policy.

8. Refer to the previous problem. Give at least one drawback to this policy.

9. Refer to the previous two problems. Suggest some ways that the incentive program could be improved.

10. Refer to the previous three problems. Comment on the legality of the program.

11. There was an anti-technology movement in the 1960s in which engineers were blamed for the ills of our society. Engineers were blamed for nuclear bombs, pesticides, crashes, etc. This is sometimes described as the "Existential Pleasure of Engineering." Explain this description, and discuss how it relates to legal conduct.

12. Discuss the differences between a law and a regulation. Confine the discussion to federal laws and regulations.

13. Most chemical engineers will someday submit and obtain a patent. Discuss the advantages and disadvantages of applying for a U.S. patent and a foreign patent.

14. Most chemical engineers will someday submit and obtain a patent. Discuss the difference between a U.S. patent and a foreign patent.

15. Most chemical engineers will someday submit and obtain a copyright. Discuss the advantages and disadvantages of applying to a U.S. copyright and a foreign copyright.

16. Most chemical engineers will someday submit and obtain a copyright. Discuss the differences between a U.S. copyright and a foreign copyright.

17. Most chemical engineers will someday submit and obtain a trademark. Discuss the advantages and disadvantages of applying for to a U.S. trademark and a foreign trademark.

18. Most chemical engineers will someday submit and obtain a trademark. Discuss the differences between a U.S. trademark and a foreign trademark.

19. Provide your interpretation of intellectual property law.

20. Provide your interpretation of tort law.

21. Provide your interpretation of contract law.

22. Provide a layman's definition of a patent.

23. Provide a layman's definition of a trademark.

24. How does infringement come into play in Patent Office activities?

25. Provide detailed information on how the Patent Office operates.
26. Based on your experience, outline a potential patentable item/thought/concept.
27. Based on your experience, outline a potential trademark that represents you and/or your activities.
28. The question that often arises in trademark litigation is "how similar is it". Provide your interpretation of this comment.
29. The term "confusingly similar" is normally judged a violation in trademark litigation. Provide your interpretation of this comment.
30. The question that often arises in patent litigation is concerned with "obviousness". Provide your interpretation of this term.
31. Describe in layman terms "joint and several liability".
32. Describe in layman terms "cradle-to-grave".
33. One of the key legal standards is based on "what would a reasonable person do if…". What is your interpretation of this quote?
34. You recently conceived of how to increase the sales of a specialty product of a pharmaceutical company. The plan would involve both the company and those marketing (selling the product). Contracts are about to be signed when the pharmaceutical decides to go at it alone with your idea and essential bypass your "middleman" involvement. What legal options are available to you? Also indicate how someone in this position could better protect themselves in any future similar activity.
35. Theodore Engineers, in conjunction with an equipment manufacture, designed a boiler for a power plant to operate at or above an overall thermal efficiency of 34%. Once the boiler was installed and running, the unit operated with an efficiency of 32.5%. What legal options are available to the power plant?

References

1. Adapted from: A. Calderone , L. Theodore and R. Kunz, *Nanotechnology: Environmental Implications and Solutions*, John Wiley & Sons, Hoboken, NJ, 2005.

2. L. Theodore, *Chemical Engineering: The Essential Reference*, McGraw-Hill, New York City, NY, 2014.
3. M. Peters, *Plant Design and Economics for Chemical Engineers*, McGraw-Hill, New York City, NY, 1958.
4. http://www.uspto.gov/ip/boards/bpai/index.jsp
5. Personal notes: L. Theodore, East Williston, NY, 1974.
6. http://www.uspto.gov/web/offices/dcom/olia/copyright/copyright
7. http://www.uspto.gov/trademarks/process/index.jsp
8. Personal notes: L. Theodore, East Williston, NY, 1994.
9. Various Theodore Tutorials, Theodore Tutorials, East Williston, NY.

22

Ethics

This chapter is concerned with ethics. As with all the chapters in Part II, there are several sections: overview, several technical topics, illustrative open-ended problems, and open-ended problems. The purpose of the first section is to introduce the reader to the subject of ethics. As one might suppose, a comprehensive treatment is not provided although several technical topics are also included. The next section contains three open-ended problems; the authors' solution (there may be other solutions) are also provided. The final section contains 34 problems; *no* solutions are provided here.

22.1 Overview

This overview section is concerned with—as can be noted from the chapter title—ethics. As one might suppose, it was not possible to address all topics directly or indirectly related to ethics. However, additional details may be obtained from either the references provided at the end of this Overview and/or at the end of the chapter.

Note: Those readers already familiar with the details associated with this subject may choose to bypass this Overview.

The following was adapted from Theodore [1] and obviously represents this author's opinion on what has become a controversial issue.

"Unfortunately, ethics has come to mean different things to different people. Ethics is a philosophical discipline that draws on human reason and analysis to assess moral choices; Webster talks of *conforming to moral standards* where moral relates to decisions that have an impact on the lives of other people.

There is currently a renaissance or grass roots movement in academia to make students aware of the FOLD (Fabricating data, Omitting information, Lying and acting in a Deceitful manner) principle. It is happening because many colleges and/or universities are now including ethics training in their curriculum. The Accreditation Board of Engineering and Technology (ABET), which accredits engineering schools, now requires ethics training to be incorporated in the curricula. In fact, all engineering programs need to address ethical issues in order to receive accreditation.

One needs to examine what is happening out in the real world. The federal government is nearly totally corrupt. Elected officials are primarily concerned with getting reelected and not representing the electorate; they regularly apply the aforementioned FOLD principle [11].

The chapter contains sections concerned with

1. The Present State
2. Moral Issues
3. Engineering Ethics
4. Environmental Justice

A significant portion of the above sections have been drawn from the work of Wilcox and Theodore [2].

22.2   The Present State

Sine this chapter is on ethics, there is a need for information on definitions. Here are four that might help the reader.

1. *Ethics* is defined as that branch of philosophy dealing with the rules of right conduct.

2. *Environmental ethics* deal with the moral issues of conduct with respect to the environment.

3. *Ethical theory* is an attempt to answer certain questions about standards of conduct or ethics. It also attempts to provide a framework for decisions regarding what moral principles are correct and how one should treat one another, the environment, other species, etc.

4. *Morality* is concerned with reasons for the desirability of certain kinds of actions and the undesirability of others. To say that an act is right is not to express a mere feeling or bias, but instead to asset the best moral reasons supporting. Moral reason is a reason that requires individuals to respect other people. In addition, moral reasons are such that they set limits to the legitimate pursuit of self-interest. They can be used to evaluate, praise, and criticize laws.

With respect to the above, MBAs and CPAs of industry have spawned financial abuses and excesses. These individuals have turned what used to be moral and ethics questions into legal technicalities. Industry executives are more likely to ask what one can get away with legally rather than to be concerned with what is right, honest and/or fair.

The above abuses and excesses have spread like a cancer to society, particularly with lawyers and individuals in government. Lawyers have become adept at creating controversy, smokescreens, and/or obstruction of the truth, and using the complexity of the laws for their own aggrandizement. Elected officials *at all levels of government* on the other hand, have used their entrusted position of power to further and maintain their career... rather than to represent their constituency. And, as most know, nearly all elected officials are lawyers [1].

Consider if a chemical engineer, acting as an expert witness in a court of law, knowingly conjures up an idea that creates reasonable doubt even though it is highly unlikely. This type of conduct is clearly defined as fraud in any engineering Code of Ethics. Given the above, the reader should consider a lawyer's present-day conduct in and out of court [1].

Well, what's the answer? Is there an answer? The reality may be that the above-referenced corruption is just too widespread and amorphous—like humidity in South Florida in the summer. It seems that the only thing that can ultimately turn things around on this issue is for each individual to take a stand for basic values and virtues, and return to the likes of integrity, responsibility and selflessness. If a grass-roots movement is to succeed, it will ultimately rest on each individual's ability and willingness to discard the practices of so many lawyers and elected officials. As Albert Schweitzer put it, "Man must cease attributing his problems to his environment, and learn again to exercise his will and his personal responsibility in the realm of faith and morals" [2].

22.3   Moral Issues

It is generally accepted that any historical ethic can be found to focus on one of four different underlying moral concepts [2–4]

1. *Utilitarianism* focuses on good consequences for all
2. *Duties Ethics* focus on one's duties
3. *Rights Ethics* focus on human rights
4. *Virtue Ethics* focus on virtuous behavior

(Note that Duties and Rights Ethics are often combined and referred to as *Deontological Ethics*)

Utilitarians hold that the most basic reason why actions are morally right is that they lead to the greater good for the greatest number. "Good and bad consequences are the only relevant consideration, and, hence all moral principles reduce to one: 'we ought to maximize utility'" [5].

Duties Ethicists concentrate on an action itself rather than the consequences of that action. To these ethicists there are certain principles of duty such as "Do not deceive" and "Protect innocent life" that should be fulfilled even if the most good does not result. The list and hierarchy of duties differs from culture to culture and religion to religion. For Judeo-Christians, the Ten Commandments provide an ordered list of duties imposed by their religion [5].

Often considered to be linked with Duties Ethics, Right Ethics also assesses the act itself rather than its consequences. Rights Ethicists emphasize the rights of the people affected by an act rather than the duty of the person(s) performing the act. For example, because a person has a right to life, murder is morally wrong. Rights Ethicists propose that duties actually stem from a corresponding right. Since each person has a "right" to life, it is everyone's "duty" not to kill. It is because of this link and their common emphasis on the actions themselves that Rights Ethics and Duty Ethics are often grouped under the common heading: Deontological Ethics [6].

The display of virtuous behavior is the central principle governing *Virtue Ethics*. An action would be wrong if it expressed or developed vices, e.g., bad character traits. Virtue Ethicists, therefore, focus upon becoming a morally good person.

To display the different ways that these moral theories view the same situation one can explore their approach to the following scenario that Martin and Schinzinger [5] present:

"On a midnight shift, a botched solution of sodium cyanide, a reactant in organic synthesis, is temporarily stored in drums for reprocessing. Two weeks later, the day shift foreperson cannot find the drums. The plant manager finds out that the batch has been illegally dumped into the sanitary sewer. He severely disciplines the night shift foreperson. Upon making discreet inquiries he finds out that no apparent harm has resulted from the dumping. Should the plant manager inform government authorities, as is required by law in this kind of situation?"

If a representative of each of the four different theories on ethics just mentioned were presented with this dilemma, their decision-making process would focus on different principles based on the definition above.

Numerous case studies of this nature are provided by Wilcox and Theodore [2].

22.4 Engineering Ethics [3,7]

The ethical behavior of chemical engineers as well as other professionals is more important today than at any time in the history of the profession. The chemical engineers' ability to direct and control the technologies they master has never been stronger. In the wrong hands, the scientific advances and technologies of today's engineer could become the worst form of corruption, manipulation, and exploitation. Chemical engineers, however, are bound by a code of ethics that carry certain obligations associated with the profession. A baker's dozen of these obligations follow.

1. Support one's professional society
2. Guard privileged information and data
3. Accept responsibility for one's actions
4. Employ proper use of authority
5. Maintain one's expertise in a state-of-the-art world
6. Build and maintain public confidence
7. Avoid improper gifts and/or gift exchange(s)
8. Avoid conflict of interest
9. Apply equal opportunity employment
10. Maintain honesty in dealing with employers, clients and government
11. Practice conservation of resources, pollution prevention, and sustainability

12. Practice energy conservation
13. Practice health, safety, and accident prevention and management

There are many codes of ethics that have appeared in the literature. The preamble for one of these codes is provided below:

"Engineers in general, in the pursuit of their profession, affect the quality of life, for all people in our society. Therefore, an Engineer, in humility and with the need for divine guidance, shall participate in none but honest enterprises. When needed, skill and knowledge shall be given for the public good without reservation. In the performance of duty and in fidelity to the profession, Engineers shall give utmost" [5].

Regarding environmental ethics, Taback [8] defined ethics as, "the difference between what you have the right to do and the right thing to do." More recently, he has added that the engineer/scientist should recognize situations encountered in professional practice with conflicting interest that test one's ability to take the "right" action. Then, take each situation to a trusted colleague to determine the best course of action consistent with the above precepts and which would have the least adverse impact on all stakeholders.

Because of the Federal Sentencing Guidelines, the Defense Industry Initiative, as well as a move from compliance to a value-based approach in the marketplace, corporations have inaugurated companywide ethics programs, hotlines, and senior line positions responsible for ethics training and development. The sentencing Guidelines allow for mitigation of penalties if a company has taken the initiative in developing ethics training programs and codes of conduct.

Regarding education, the ABET 2000 accreditation guidelines require academic programs to clearly demonstrate that students are exposed to an ethics education; and they also have to do outcome assessments. In spite of indicators that reveal the value of an ethics education, few large universities require an ethics course. Ideally, a student would take an ethics course and would also be exposed to ethics in several other courses each year.

22.5   Environmental Justice

The environmental policy of the EPA has historically had two main points of focus: defining an acceptable level of pollution and creating the legal rules to reduce pollution to a specified level. Understandingly, it seems that the program has also been concerned with economic costs and efficiency [9]. Thus, it appears to some to lack consideration of equity, both distributional and economic. While EPA's two main points of focus are important considerations, relying on such criteria in the formation of environmental protection policy can potentially neglect to account for potential inequalities of capitalism and its effects throughout the policy process.

The history of environmental policymaking illustrates to some the incompatibility of equity and efficiency. Economic pressures of environmental regulations have motivated corporations to seek new ways to reduce costs. Industries have attempted to maximize profits by "externalizing" the environmental costs [10]. It has been suggested that this redistribution of costs is more regressive in its effects than the general sales tax [11]. To date, big corporate polluters sometimes have more to gain financially by continuing pollution practices than in obeying regulations. In some instances, the result of increased environmental costs has paradoxically caused negative impacts on environmental degradation. In effect,they have little incentive other than altruism to end these debilitating practices.

22.5.1   The Case For and Against Environmental Justice

Environmental protection policy has attempted to reduce environmental risks overall; however, in the process of protecting the environment, risks have been redistributed and concentrated in particular segments of society. Although federal regulations to protect the environment are not explicitly discriminatory, some argue that environmental protection policies have not been sensitive to distributional inequalities. Other insists that they have not adequately addressed specific minority environmental concerns. Like many programs of reform and activism, environmental justice was principally started with good intentions. However, ground rules need to be set before any meaningful discussions regarding environmental justice can be presented. One of the problems is that environmental justice has come to mean different things to different people, at different times, at different locations, and for different situations. There appears to be no clear-cut decision regarding this term but the EPA defines it as "the fair treatment of people of all races, cultures, and incomes with respect to development, implementation, and enforcement of environmental laws and policies, and their meaningful involvement in the decision-making process of government [11]. Based on the EPA's definition, there appears to be three major components of environmental justice:

1. Environmental racism
2. Environmental equity
3. Environmental health

Details of each are available in the literature [12–14].

A detailed and expanded treatment of ethics is available in the following two references

1. J. Wilcox and L. Theodore, *Environmental and Engineering Ethics: A Case Study Approach*, John Wiley & Sons, Hoboken, NJ, 1998 [2].
2. L. Theodore, *Chemical Engineering: The Essential Reference*, McGraw-Hill, New York City, NY, 2014 [14].

22.6   Illustrative Open-Ended Problems

This and the last section provide open-ended problems. However, solutions *are* provided for the three problems in this Section in order for the reader to hopefully obtain a better understanding of these problems, which differ from the traditional problems/illustrative examples. The first problem is relatively straightforward while the third (and last problem) is somewhat more difficult and/or complex. Note that solutions are not provided for the 34 open-ended problems in the next section.

Problem 1: Vincenzo has a Ph.D. in chemical engineering, a Professional Engineer (P.E.) License, and a job as an assistant professor at a local university. To learn more about ISO 14000 and to earn some extra money, Vincenzo became a certified ISO 14000 auditor. During an audit of a facility, Vincenzo observed a significant violation of an EPA regulation. (Note that ISO 14000 is a voluntary standard of environmental management systems, i.e., it is not a regulation and ISO 14000 audits are confidential.)

What does Vincenzo do? Does he forget about the significant violation since he is not an EPA inspector and it does not violate the ISO 14000 standard he is there to audit? Does he insist it be part of the audit report? Does he call it to the attention of the facility management? Does he report it to the EPA?

Solution: The question of professional ethics presented here can be generalized as follows: does a professional have an ethical duty to report any violation of which he/she becomes aware? Most licensing organization have codes of ethics to which their professionals must subscribe. These codes generally call upon the professional to serve their client/employer with loyalty and to protect the welfare of the public. The codes, however, are sufficiently vague so as to not provide a clear-cut answer to Vincenzo's situation. In the absence of such a clear-cut code, there is no one right answer and one's answer may vary from the one presented below.

Vincenzo must first investigate further to make certain of his facts. If the violation is confirmed and has never been reported, he must insist that it be part of the findings presented to management at the exit meeting and in the report to the ISO Registrar who makes the final decisions on awarding certification. Although ISO 14000 does not require compliance, knowingly failing to comply could affect the evaluation of the facility's commitment to good environmental management system practices that the standard does require.

Having done all that, does Vincenzo go the last step and report it to the U.S. EPA? Such a report would violate the confidentiality of the audit and might even precipitate a lawsuit. Vincenzo is also concerned that he may never be hired again as an auditor. He senses a conflict between two sets of ethics: the ethics of a professional bound to protect the public, and the ethics of a professional bound to confidentiality. Keeping silent would be easier but ethics is not easy.

It would be easy to answer "report it!" but the real question is what would YOU (the reader) do?

Problem 2: The passage of a number of environmental laws in recent times, coupled with educational campaigns on the part of environmental groups and a supportive citizenry, have resulted in a substantial reduction in the quantity of waste produced. It is not likely, however, that the production of waste can be completely eliminated. As long as waste is generated, it will be necessary to dispose of that waste as safely as possible.

Incineration is an effective method of waste disposal and it has become safer than ever. Yet, local communities (often with the help of vocal environmental groups) have successfully opposed construction of waste incinerators in their areas. Discuss the ethics of opposing the construction of a waste incinerator in your local community.

Solution: This is obviously an open-ended question with no single correct answer. Your answer may have addressed points that are not made here, and vice versa.

On the issue of opposition to the construction of the hazardous waste incinerator, it is not unethical to protect one's self, family, and community from a risk. This is especially true concerning the risk from, for example, a hazardous waste incinerator, over which one may have little control. Even with the best design and pollution controls, no one can absolutely guarantee that the incinerator will have zero emissions. Nor can anyone absolutely guarantee that the low level of emissions will not pose a hazard to anyone.

Although the risk may be estimated to be as low as one excess cancer or other disease in each one million exposed people annually, this is no consolation if that one additional affected party is you or a member of your family.

It would, however, be unethical to avoid your risk by shifting it to someone else. The manufacture and use of products, such as furniture, rugs, clothing, cars, and electronics, have become an integral part of one's way of life. Wastes are produced by the manufacture of all these products. As long as the products continue to be used, there is a moral obligation to share the risk that results from the disposal of the waste generated in making these products.

Sometimes, widespread opposition to hazardous waste disposal results in industry finding creative ways to eliminate the production of a waste that they cannot dispose of. Those who are successful in advocating the Not-In-My-Backyard (NIMBY) policy can then claim the highest of ethical behavior since all society had benefitted from the improved production processes. All too often, unfortunately, industry simply moves its operations to a state or country where environmental restrictions are less severe. In addition to possibly shifting the risk to someone else, the NIMBY policy also causes the harm of loss of jobs. Since one cannot ascertain whether opposition to the hazardous waster incinerator will lead to the elimination of the production of the waste or the shifting of the risk to someone else, this behavior cannot be classified as ethical.

On the issue of the problem raised by such opposition, a possible (but not perfect) solution may be a national hazardous waste incinerator siting program. Each region would be required to accept its share of incinerators. A government agency, or some other neutral body, could evaluate all potential sites in each region based on factors such as availability of facilities, accessibility, number of people affected, etc. A lottery system could be used to make the final choice among the top, and roughly equal sites. To mitigate the risk imposed on the selected site, benefits, such as free health care or reduced taxes, could be provided to the people in the selected area. In this way, the risks and alternatives would have to be squarely faced by virtually everyone. It would be possible to simply evade the problem by shifting it to someone else.

Problem 3 [2]: Bill is a senior chemical engineer in the biomedical division of a major corporation and the head of a research department that specializes in the construction of artificial organs. About a year ago, an artificial heart was created and tested in a human patient. (One of the authors, who suffers from cardiomyopathy may someday be the recipient of one of these hearts.) Unfortunately, the patient survived only nine months. Although it was a huge step in the fight against heart disease, it brought world-wide recognition to his company and himself.

One day, Mary, his top research assistant, reports to him that a problem has been detected in the tricuspid valve of the artificial heart model. With further testing, it is discovered that the rate at which this valve allows blood to pass tends to slow down after eight months of continuous usage. The coroner's report states that the patient's death was due to the body's rejection of the artificial heart. However, it is very likely that the patient's death was brought on by this flaw in the artificial heart.

Bill becomes extremely worried and is confronted with a dilemma. If he tells his superiors about this piece of information there is a great possibility that the project will be terminated. And, if this knowledge becomes public, not only will it bring humiliation to the company and probably cause his dismissal, but also the company will be highly susceptible to a million-dollar lawsuit by the patient's family. If Bill decides to withhold this information, a new model could be created with the flaw corrected, without anyone knowing about the dilemma.

Bill decides to ask for advice from two older colleagues. He first asks Bob, a fellow chemical engineer and someone who understands the technical aspects of the project.

"You have no choice Bill," replies Bob. "You made a mistake and now have to suffer the consequences. Also, if you withhold this information and it is discovered later, the situation will worsen."

Bill then phones his sister, Sheila (a cardiologist), and explains the situation to her. "It's a tough call Bill," replies Sheila. "Ordinarily, I would say let the truth be known, but it is your decision."

Bill also contacted the authors of this book for critical advice. They suggested he attempt to address the following four questions:

1. What are the facts in this case?
2. What ethical dilemma has been encountered?
3. Are you being selfish if you do not report the flaw in the artificial heart?
4. Should your loyalty to the company play a major role?

22.7   Open-Ended Problems

This last section of the chapter contains open-ended problems as they relate to ethics. No detailed and/or specific solution is provided; that task is left to the reader, noting that each problem has either a unique solution or a number of solutions or (in some cases) no solution at all. These are characteristics of open-ended problems described earlier.

There are comments associated with some, but not all, of the problems. The comments are included to assist the reader while attempting to solve the problems. However, it is recommended that the solution to each problem should initially be attempted *without* the assistance of the comments.

There are 34 open-ended problems in this section. As stated above, if difficulty is encountered in solving any particular problem, the reader should next refer to the comment, if any is provided with the problem. The reader should also note that the more difficult problems are generally located at or near the end of the section.

1. Describe the early history associated with the ethics movement.
2. Discuss the recent advances in implementing ethical programs at the industrial level.
3. Select a refereed, published article on ethics from the literature and provide a review.
4. Provide some normal everyday domestic applications involving the general topic of ethics.
5. Develop an original problem on ethics that would be suitable as an illustrative example in a book.
6. Prepare a list of the various technical books which have been written on ethics. Select the three best (hopefully, it will include a book written by one of the authors [2]) and justify your answer. Also select the three weakest books and, once again, justify your answer.
7. The following writeup on plagiarism applied to the preparation of a laboratory report at the academic level.

   "Plagiarism occurs when you represent someone else's work as being your own, whether the source is another student's laboratory report or a copyrighted publication. College policy requires that plagiarism be reported by the instructor to a judiciary committee. If the committee finds the student guilty of plagiarism, the penalty is an automatic "F" in the course and a notation in the student's academic files. Plagiarism includes borrowing a computer disk from a friend "to look at the experimental trends," then submitting the same report, graphs, spreadsheet, and/

or MathCAD calculations (perhaps after some minor revisions such as substituting your own data). If this disk swapping is caught, both the borrower and owner of the disk will be severely penalized."

Prepare a similar statement that a company could prepare for its chemical engineering employees.

8. Discuss the difference between moral obligation and personal responsibility.
   Comment: Refer to the literature [2] for more information.
9. Discuss the problems associated with recognizing ethical issues.
10. Discuss the relationship (if any) between honesty and integrity.
    Comment: Refer to the literature [2] for additional details.
11. Much has been written about "the guardians of the system". Provide your interpretation of this phrase
    Comment: Refer to the literature [2] for additional details.
12. How can ethics best be taught at the academic level?
    Comment: Refer to the literature [2] for additional details.
13. How can ethics best be taught at the domestic level?
    Comment: Refer to the literature [2] for additional details.
14. How can ethics best be taught at the industrial level?
    Comment: Refer to the literature [2] for additional details.
15. How can ethics best be taught at the business level?
    Comment: Refer to the literature [2] for additional details.
16. Most regulations depend on voluntary compliance. Hence, ethics becomes a concern. In dealing with ethics there are several terms that one may encounter: ethical theory, consequentialist theory, deontological theory, utilitarianism, ethical egoism, nationalism, and retributivism. Briefly explain these terms.
    Comment: Refer to the literature [2] for additional details.
17. As noted in the Overview, there was an anti-technology movement in the 1960s in which engineers were blamed for the ills of society. Engineers were blamed for nuclear bombs, pesticides, crashes, etc. This is sometimes described as the "Existential Pleasure of Engineering." Explain this description, and discuss how it relates to ethical conduct.
18. How are ethical values determined by you?
19. It has been generally accepted despite cultural variations that any historical ethic can be found to focus on one of

four different underlying moral concepts. Identify and define each of these in your own words.

Comment: Refer to the Overview.

20. Aldo Leopold made the following observation on personal ethics in his 1949 *A Sand County Almanac,...*"The scope of one's ethics is determined by the inclusiveness of the community with which identifies oneself." In terms of the concept of environmental ethics, how does the concept of "community" expand beyond that captured in the cited concept of personal ethics?

21. If a consortium of investors from California and Germany was proposing to build a ski resort on U.S. Forest Service land near your property and, as a rule, you like to ski, would you be likely to use the facilities? If so, do you think this is "evidence" that you prefer development of public lands over forest resource preservation? Do citizens always "vote with their pocketbooks?" Why?

22. The U.S. Forest Service has a "multiple-use" directive, i.e., the lands and resources it administers provide a variety of goods and services to the public. For example, this means that the land and resources provide not only recreational opportunities but revenues from timber harvests and sales as well. Should government agencies such as the U.S. Forest Service strive to make a profit? Why or why not? How do you think your ethics affect your opinion on this issue?

23. As a chemical engineering student, how do you incorporate environmental awareness into your everyday ethics?

24. ISO 14000 is a voluntary standard for environmental management systems. A company can declare itself in conformity with the standard, but third-party certification of conformity is available and is generally regarded more highly by purchasers. The company seeking third-party certification contacts a "registrar" who sends an audit team of three trained professionals to review the company's management system. The audit usually takes three days.

On the third day of an audit of a facility, one member of the team reveals to the lead auditor that he has worked for the facility recently as a consultant. Is this an ethical problem? What should the lead auditor do?

Comment: See also the earlier chapter on Environmental Management.

25. You are a young chemical engineer and have just completed an interview for a new environmental management position with a relatively small company that manufactures metal specialty products. Until now, this position has been non-existent within the company. During your interview, the company's president openly praised the company's "responsible attitude toward the environment," citing an example of employees organizing the recycling of aluminum and other metals from certain lathing operations, as well as the company's "clean record" with the state regulatory agency.

    While touring the facility's metal degreasing process area, you notice two employees through an open doorway in the rear of the building pouring liquids from two 5-gallon waste solvent containers directly onto the ground. The ground is extremely discolored in the area where the liquids are being poured and void of vegetation. The vegetation nearby is visibly stressed.

    You have been offered the position at an attractive salary rate and are asked to make a decision within 2 weeks. What would be your decision regarding employment with this company?

26. Refer to the previous problem. If you were to take the job, what would be your approach to the environmental management attitude you witnessed during the interview? Would you report the improper discharge of solvent to the local regulatory authorities? Support your decision from either a professional or personal ethical standpoint, or both, and be explicit.

27. Refer to Problem 2 in the previous Section. Propose a solution to the problems raised by such opposition.

28. Consider the following hypothetical scenario where three adjoining states have conflicting needs for disposal of wastes.

    The citizens of New Jersey (population of 12 million, an urban state with little available land) are reluctant to dispose of their waste in their own state and are forbidden under federal law to dump it at sea. They have the political will and the financial ability to dispose their waste out of state. To do this, they must transport their waste through the state of Pennsylvania, and store it there temporarily until completion of a treatment, storage and disposal facility (TSDF) planned for construction in the state of West Virginia. About half the waste is currently accumulating under conditions that are not considered safe for storage beyond a period of 10 years. Land in New Jersey is either too scarce, too expensive, or too close to densely populated areas for a TSDF, and waste incineration was voted down as an option two years ago.

The state of Pennsylvania (population of 16 million) has decided to dispose of its wastes in several engineered landfills that are permitted to contain properly containerized hazardous wastes. The landfills were sized to allow for a 50-year operating life before closure based on projected waste generation rates within the state. New Jersey's wastes would be transported along a corridor that would expose about one million additional people to risks of transportation accidents, and then temporarily stored north of a large city in the western part of the state. Pennsylvania does not want New Jersey's waste temporarily stored on its soil and has threatened to pass legislation banning the temporary facility. It has also threatened to sue New Jersey in federal court if the latter tries to contract for out-of-state disposal of its wastes that involves interim storage in Pennsylvania.

The state of West Virginia (population of 3 million) is relatively poor and rural. Its economy has been devastated by layoffs from steel mills and coal mines, and its citizens and elected officials are very interested in attracting new industries to the state. A TSDF capable of accepting wastes from several states in the region, including New Jersey's, would produce several hundred jobs during the construction and operation phases and would be a potential source of tax revenue to the impoverished state. The largely blue-collar labor pool is used to industrial hazards and is receptive to the opportunity to be retained to treat, store, and dispose of containerized wastes. There is widespread support for the construction of a regional facility that would dispose of both West Virginia's and New Jersey's wastes. There is a small vocal opposition that is concerned about the environmental risks and does not want West Virginia to be perceived as a dumping ground for other states. The governor of West Virginia has contacted his influential U.S.

senator, who is holding hostage a federal water quality bill that funds projects for water quality improvements that are administered jointly with Pennsylvania until Pennsylvania becomes more compliant on the waste transportation and interim storage issue.

The federal government, in the persons of the Congress, various regulatory agencies, and the courts may be called upon to participate in the resolution of the dispute among these states.

There are several political rationales that could be used to resolve this dispute:

- A "greatest good for the greatest number" rationale could be applied, i.e., reducing the risk to the 12 million citizens of New Jersey outweighs the increased risk to the one million potentially exposed citizens of Pennsylvania.
- Alternatively, a "minority rights" rationale could be applied, where large populations are not allowed to worsen the quality of life of smaller populations. In this case, the minority is numerical, not ethnic.
- An "each takes care of their own" rationale could be applied, where each state must locate and develop, no matter what the cost, a waste TSDF within its own boundaries. This rationale deprives West Virginia of economic benefits, reduces risks for Pennsylvania, and increases disposal costs for the taxpayers of New Jersey.

List the major risks associated with allowing wastes to remain in New Jersey under partially unsafe conditions in a scenario where New Jersey "loses" its bid to transport its waste out of state.

29. Refer to the previous problem. List the risks of transporting New Jersey's waste via Pennsylvania (with interim storage in Pennsylvania) under a scenario where New Jersey "wins" its dispute. In formulating the answer, consider the increase in volume of waste transportation and storage that would occur in Pennsylvania.

30. Refer to the previous two problems. List the risks of disposing of New Jersey's wastes at a RCRA-permitted facility in West Virginia.

31. Refer to the previous three problems. List how New Jersey can reduce the risks associated with transporting its waste through Pennsylvania.

32. Consider the rationales described in the previous four problems. Each rationale above is cast in "win-lose" terms. Are there alternative rationales that are "win-win" or "lose-lose"? For example, under the first rationale, could New Jersey compensate Pennsylvania for the increased risk New Jersey exposes Pennsylvania to? The payments could be used for road, rail or river improvements and increased surveillance of hazardous waste shipments. What would be involved in determining a fair price for this compensation?

33. Consider the previous problem. Find and summarize some case studies in the library that involve actual disputes among state governments. In reading the literature on a particular dispute, try to identify the nature of the disposal problem (type of waste and disposal method), and the motivating interests of the parties in conflict, i.e., who stands to gain from having the proposed facility and why, and who stands to be hurt by the facility and why.

34. Refer to the previous problems. List the pluses and minuses, or "winners" and "losers" for each of the rationales described in the earlier problem statements.

References

1. L. Theodore, *On Ethics I*, Williston Times, Williston Park, NY, Nov. 28, 2003.

2. Adapted from: J. Wilcox and L. Theodore, *Environmental and Engineering Ethics: A Case Study Approach*, John Wiley & Sons, Hoboken, NJ, 1998.

3. L. Theodore, *On Ethics II*, Williston Park, NY, March 26, 2004.

4. M.K. Theodore and L. Theodore, *50 Major Issues Facing the 21st Century*, Theodore Tutorials (originally published by Prentice Hall), East Williston, NY, 1995.

5. H. Rolston, *Philosophy Gone Wild*, Prometheus Books, Buffalo, NY, 1986.

6. A. Leopold, *A Sand County Almanac*, Oxford University Press, New York City, NY, 1949.

7. I. Barbour, *Ethics in an Age of Technology*, Harper, San Francisco, CA, 1993.

8. H. Taback, *AWMA's Code of Ethics*, EM, Pittsburgh, pp. 42-43, September 2007.

9. www.awma.org/about/index/html.

10. R. Lazarus, *Pursuing Environmental Justice: The Distributional Effects of Environmental Protection*, Northwestern University Law Review, Chicago, IL, Spring 1987.

11. L. Cole, *Empowerment as the Key to Environmental Protection: The Need for Environmental Poverty Law*, Ecology Law Quarterly, 1992.

12. B. Marquez, Lecture, University of Wisconsin-Madison, Madison, WI, May 5, 1994.

13. M.K. Theodore and L. Theodore, *Introduction to Environmental Management*, CRC Press/ Taylor & Francis Group, Boca Raton, FL, 2010.

14. L. Theodore, *Chemical Engineering: The Essential Reference*, McGraw-Hill, New York City, NY, 2014.

Part III
TERM PROJECTS

This last Part (III) is concerned with open-ended term projects where the phrase *term project* is defined as a study that requires an order of magnitude more effort than that associated with the normal problem (as presented in the last Part).

Each project in this portion of the book is presented in a stand-alone manner. No solution of any form is provided for any of the projects, although several potential solutions are available (for those who adopt the book for classroom and/or training purposes) in the authors' files. A reference section complements each term project area.

There are 12 term project subject areas. They are provided below with the accompanying number of term projects and chapter number.

Chapter 23: Applied Mathematics (2)

Chapter 24: Stoichiometry (2)

Chapter 25: Thermodynamics (2)

Chapter 26: Fluid Flow (6)

Chapter 27: Heat Transfer (4)

Chapter 28: Mass Transfer Operations (5)

Chapter 29: Chemical Reactors (2)

Chapter 30: Plant Design (4)

Chapter 31: Environmental Management (4)

Chapter 32: Health and Hazard Risk Assessment (4)

Chapter 33: Unit Operations Design Projects (3)

Chapter 34: Miscellaneous Topics (4)

23

Term Projects (2): Applied Mathematics

23.1 Simplified Procedure for Solving Differential Equations
23.2 The Weighted Sum Method of Analysis

Term Project 23.1

Simplified Procedure for Solving Differential Equations

There are essentially two classes of differential equations that the chemical engineer occasionally encounters: ordinary and partial. Generally speaking, an equation containing one or more derivatives is called a *differential equation*. If these are ordinary derivative, the equation is said to be an *ordinary differential equation* (ODE); i.e., it contains only one independent variable, and as a consequence, is referred to as a total or ordinary derivative. Partial differential equations contain several independent variables, and as a consequence, are referred to as partial derivative, and the equation is termed a *partial differential equation* (PDE).

An ordinary differential equation, which is of the first degree in the dependent variable and all its derivatives, is termed a *linear equation*. Correspondingly, an equation in which the dependent variable or any of its derivatives is of the second or higher degree, is termed a *nonlinear equation*. However, the order of an equation is the order of the highest derivative contained therein.

For many chemical engineering applications, the dependent variable is defined in terms of more than one independent variable. Variation with respect to the independent parameters is possible and the derivatives describing these independent variations are defined as the aforementioned partial derivatives and the differential equations involving partial derivatives are known as partial differential equations. The three most often encountered PDEs are the *elliptic, parabolic, and hyperbolic* equations.

Providing specific details on the various methods of solving both ODEs and PDEs are beyond the scope of both this term project and this book. However, it may be possible to develop simple procedures to solve either or both of these two classes of differential equations. With this possibility in mind, you have been requested to prepare a sample; e.g., cookbook; method of solving:

1. ordinary differential equations, and
2. partial differential equations.

Comment: One very simple approach to solving differential equations is to guess a solution [1]. See also Illustrative Problem 2 in Part II Chapter 2. See also material available in the literature [2,3].

Term Project 23.2

The Weighted Sum Method of Analysis

The Weighted Sum Method (WSM) is a semi-quantitative method for screening and ranking process and design options. This method can and has been applied previously in quantifying the important criteria that affect equipment selection [4], energy resources [5], waste management at a particular facility [6], and pollution prevention [7], etc. This method involves three steps.

1. Determine what the important criteria are in terms of the program goals and constraints, and the other corporate goals and constraints. Example criteria as applied to pollution prevention are listed below.
   a. reduction in waste quantity
   b. reduction in waste hazard (e.g., toxicity, flammability, reactivity)
   c. reduction in waste treatment/disposal costs
   d. reduction in raw material costs
   e. reduction in liability and insurance costs
   f. previous successful use within the company
   g. previous successful use in industry
   h. not detrimental to product quality
   i. low capital cost
   j. low operating and maintenance costs
   k. short implementation period with minimal disruption of plant operations
   l. improved public relations
   m. reduced workman's compensation
   n. improved employee morale
   o. reductions or elimination of liability
   p. reduction or elimination of regulatory concerns.
   A weight factor for each criteria is assigned. This is defined as the *weight of the criteria*. The weights (on a scale of 0 to 10, for example) are determined for each of the criteria in relation to their importance. For example, if reduction in waste treatment and disposal costs are very important, while previous successful use within the company is of minor importance, then the reduction in waste costs is given

a high weight factor of 9 or 10, and the previous use within the company is given a low weight factor of either 1 or 2. Criteria that are not important are either not included, or are given a weight factor of 0.

2. Each criteria is then rated for its effect on the various options to be investigated. These ratings are defined as *effectiveness factors*. Again, a scale of 0 to 10 can be used (0 for low and 10 for high).

3. Finally, the *effectiveness factor* (step 2) for a particular criterion is multiplied by the *weight factor of the criterion* (step 1). An option's *overall rating* is the sum of the *effectiveness factor* and the *weight factor criterion*.

The option(s) with the best overall rating(s) is (are) then selected for further technical and/or economic feasibility analyses.

Apply the Weighted Sum Method to a process/design/project of your choice. Comment on the results.

References

1. Personal Notes, L. Theodore, East Williston, NY, 1969.
2. R. Ketter and S. Prawler, *Modern Methods of Engineering Computation*, McGraw-Hill, New York City, NY 1969.
3. L. Theodore, *Transport Phenomena for Engineers*, Theodore Tutorials, East Williston, NY, originally published by International Textbook CO., Scranton, PA, 1971.
4. A.J. Buonicore, J. Reynolds, and L. Theodore, *Control Technology for Fine Particulate Emissions*, Argonne National Laboratory, Chicago, IL, 1979.
5. K. Skipka and L. Theodore, *U.S. Energy Resources: Past, Present, and Future Management*, CRC Press/Taylor & Francis Group, Boca Raton, FL, 2014.
6. J. Santoleri, J. Reynolds, and L. Theodore, *Introductions to Hazardous Waste Incineration*, 2nd edition, John Wiley & Sons, Hoboken, NJ, 2004.
7. R. Dupont, L. Theodore, and K. Ganesan, *Pollution Prevention: The Waste Management Option for the 21st Century*, CRC Press/ Taylor & Francis Group, Boca Raton, FL, 2000.

24

Term Projects (2): Stoichiometry

24.1 A Vapor Pressure Equation
24.2 Chemical Plant Solid Waste

Term Project 24.1

A Vapor Pressure Equation

Consider the molecules of a pure liquid substance such as water standing in a container. The attractive forces acting among them tend to keep them together. The kinetic energy associated with their thermal motions tends to separate them from one another. As long as the average energy of attraction exceeds the average kinetic energy of translation, the molecules remain mainly in the condensed, liquid phase. Although the *average* velocity and *average* kinetic energy of translation per molecule is fixed for any temperature, the velocities of individual molecules differ over wide limits owing to increments of energy momentarily lost or acquired during the chaotic inter-particle collisions.

A molecule in the interior of the liquid is acted upon from all sides by the attractive forces exerted by its neighbors, but conditions are quite different at the surface. There is nothing above the liquid's surface except vapor and, perhaps, air. Based on the fact that at room temperature a gmol of water occupies roughly 18 ml in the liquid state and over 24,000 ml in the vapor state, one realizes just how sparsely populated the space above the surface or interface is. As a consequence, there is a net inward pull exerted on any molecule occupying the surface, which accounts for the phenomenon of surface tension. It follows that in order for a molecule to be at the surface at all it must acquire energy in excess of the average. This extra energy is the so called *free surface energy* and it may be visualized as the work required to move the molecule from the interior to the surface against the net inward pull.

If, in the course of the thermally induced intermolecular collisions, several molecules converge by chance of one being near the surface and hit from below simultaneously, it may impart sufficient kinetic energy to propel the other through the interface and out into the relatively empty space above. It follows that whatever molecules escape into the vapor phase are highly energized compared with those in the liquid phase. Thus, one can visualize the latent heat of vaporization from the fact that even under conditions where liquid represents the normal physical state of aggregation, some of the molecules nonetheless attain the vapor state.

If the space above the liquid is sealed off and the system is held at constant temperature, equilibrium is soon established between the liquid and its vapor, with molecules leaving and returning from each phase at the *same* rate from each unit of surface area. The space above the liquid now holds the greatest concentration of vapor molecules possible at the temperature in question. In other words, it is saturated with vapor and there is, exerted on the walls on the container a characteristic and constant pressure called *vapor pressure* or *equilibrium vapor pressure or saturation pressure*. This pressure is independent of the size of the area exposed, the amount of liquid present just so long as *some* is present, or the shape of the container. Only upon changing the temperature does this characteristic equilibrium vapor pressure change; it increases with rise in temperature and decreases with fall in atmosphere pressure, the result is a temperature. When the vapor pressure of a liquid in an open vessel reaches atmospheric pressure, the result is a turbulent escape of molecules from the liquid, called *ebullition* or *boiling*. The temperature at which the liquid attains a vapor pressure of 760 mm Hg is defined as the *normal boiling point* [1].

The vapor pressure of a liquid depends on the magnitude of the forces of attraction acting among its molecules. These forces in turn depend upon molecular size, shape, and composition. At any given temperature, every liquid possesses a unique latent heat of vaporization as well as a unique vapor pressure; and although, as one would predict from molecular theory—the heat of vaporization decreases and the vapor pressure increases with rise in temperature—the function relating these variable is complex.

Vapor pressure data are available in the literature [2,3]. However, there are two equations that can be used in lieu of actual vapor pressure information—the Clapeyron equation and the Antoine equation. The two-coefficient Clapeyron equation is given by

$$\ln p' = A - (B/T) \tag{24.1}$$

where p' and T are the vapor pressure and temperature, respectively. The three-coefficient Antoine equation is given by

$$\ln p' = A - B/(T + C) \tag{24.2}$$

Note that for both equations, the units of p' and T *must* be specified for given values of A and B and/or C. Values of the Clapeyron equation coefficients—A and B—are provided in Table 24.1 for some compounds [4]. Some Antoine equation coefficients—A, B, and C—are listed in Table 24.2. Additional values for these coefficients for both equations are available in the literature [2,3].

There is debate as to which of the two equations is "better" from a statistical [4,5] and calculation perspective. These questions suggest that a better model/equation(s) can be developed to describe vapor pressure variation with temperature.

Table 24.1 Approximate Clapeyron Equation Coefficients*

	A	B
Acetaldehyde	18.0	3.32×10^3
Acetic anhydride	20.0	5.47×10^3
Ammonium Chloride	23.0	10.0×10^3
Ammonium cyanide	22.9	11.5×10^3
Benzyl alcohol	21.9	7.14×10^3
Hydrogen peroxide	20.4	5.82×10^3
Nitrobenzene	18.8	5.87×10^3
Nitromethane	18.5	4.43×10^3
Phenol	19.8	5.96×10^3
Tetrachloroethane	17.5	4.38×10^3

*T in K, p′ in mmHg.

Table 24.2 Antoine Equation Coefficients*

	A	B	C
Acetone	14.3916	2795.82	230.00
Benzene	13.8594	2773.78	220.07
Ethanol	16.6758	3674.49	226.45
n-Heptane	13.8587	2911.32	216.64
Methanol	16.5938	3644.30	239.76
Toluene	14.0098	3103.01	219.79
Water	16.2620	3799.89	226.35

*T in °C, p′ in kPa.

Based on the above, you are requested to develop another (and hopefully better) vapor pressure equation. Once the equation(s) has been generated, perform a comparative analysis (for numerous species) between your equation and that of Clapeyron and Antoine. Also, comment on the results.

Term Project 24.2

Chemical Plant Solid Waste

There are four general categories of solid waste sources in chemical plants [6].

1. *Inorganic by-products or unused portions of raw materials.* These wastes are most common in plants that process inorganic materials. For instance, the production of phosphoric acid leaves behind large quantities of minerals from which the phosphates have been leached by sulfuric acid.
2. *Solid organic waste materials.* These include such things as off-spec polymers and impure or off-spec crystalline organic products.
3. *Heavy residues and sludges.* Though not strictly solids, these are heavy materials which accumulate in tank bottoms and still bottoms, or materials that are created in treating waste water.
4. *Auxiliary solid waste materials.* These are materials not directly involved in the chemical processes as products or reactants. They include such things as spent catalysts, packaging materials, old pipe and process equipment, etc.

In the wet-acid process, phosphoric acid and by-product hydrofluoric acid are made by treating phosphate rock with sulfuric acid. The overall chemical reaction is:

$$CaF_2 \cdot [3Ca_3(PO_4)_2] + 10H_2SO_4 + 20H_2O \rightarrow$$
$$10CaSO_4 \cdot 2H_2O + 2HF + 6H_3PO_4 \qquad (24.3)$$

The rock feed contains 90% by weight $CaF_2 \cdot [3Ca_3(PO_4)_2]$, the remainder is insoluble rock of varying composition.

You have been hired to address the separation problem associated with the three products produced in the reaction. In addition, you are requested to develop a plan/procedure to convert and/or use the waste generated in the process.

Reference

1. Personal communication to L. Theodore, source unknown, approximately 1965-1970.
2. L. Theodore, *Chemical Engineering: The Essential Reference*, McGraw-Hill, New York City, NY, 2014.

3. L. Theodore, F. Ricci, and T. VanVliet, *Thermodynamics for the Practicing Engineer*, John Wiley & Sons, Hoboken, NJ, 2009.
4. J. Kibuthu: Thermodynamics class assignment submitted to L. Theodore, Manhattan College, Bronx, NY, 2004.
5. S. Shaefer and L. Theodore, *Probability and Statistics Applications in Environmental Science,* CRC Press/ Taylor & Francis Group, Boca Raton, FL, 2007.
6. D. Kauffman, *Process Synthesis and Design*, A Theodore Tutorial, Theodore Tutorials, East Williston, NY, originally published by the USEPA/APTI, RTP, NC, 1992.

25

Term Projects (2): Thermodynamics

25.1 Estimating Combustion Temperatures
25.2 Generating Entropy Data

Term Project 25.1

Estimating Combustion Temperatures

As described in Part II, Chapter 3, Illustrative Open-ended Problem 2, some reasonable assumptions can be made to simplify the rigorous approach to calculating combustion temperatures. These were detailed at that time. When compared to the rigorous approach, a simpler (and in many instances, a more informative) set of equations resulted, which that are valid for purposes of engineering calculation [1,2]. An approximate enthalpy balance ultimately produced Equation 25.1.

$$T = 60 + \frac{\text{NHV}}{(0.3)\left[1 + (1 + \text{EA})(7.5 \times 10^{-4})(\text{NHV})\right]} \tag{25.1}$$

Based on the above assumptions and development, you are requested to extend the work of Theodore [1–4]. to include other combustible mixtures. In effect, you are being asked to remove the constraint that the combustible mixture consists of only hydrocarbons.

Term Project 25.2

Generating Entropy Data

With unites constant will those provided in Chapter 2. Based on extensive experimental data, Walther Hermann Nernst (1864 – 1941), a German physicist, postulated a "heat theorem" that the entropy change for any chemical reaction at absolute zero temperature; i.e., 0°R or 0K, was zero. Thus,

$$\Delta S = 0 \text{ at } T = 0 \tag{25.2}$$

Planck, another German physicist, further proposed that at absolute zero temperature

$$S = 0 \tag{25.3}$$

Experimental data has verified these two conclusions. The third law of thermodynamics is based on Equation (25.4); i.e., third law of thermodynamics is concerned with the absolute values of entropy. Thus, by definition, the entropy of all pure crystalline materials at absolute zero temperature is exactly zero and in line with Equation (25.4)

$$S_{-273°C} = S_{0K} = 0 \tag{25.4}$$

Note, however, that engineers are usually concerned with *changes* in thermodynamic properties. The third law allows one to calculate the entropy at another temperature via the following equation:

$$\Delta S = \int_0^{T_f} \frac{(C_P)_s}{T} dT + \frac{\Delta H_f}{T_f} + \int_{T_f}^{T_v} \frac{(C_P)_l}{T} dT + \frac{\Delta H_v}{T_v} + \int_{T_v}^{T} \frac{(C_P)_g}{T} dT \tag{25.5}$$

where the first, third, and fifth terms on the right hand side (RHS) represent the entropy change associated with temperature changes for the solid (*s*), liquid (*l*), and gas (*g*) phases, respectively, and the second and fourth RHS terms represent the entropy change associated with fusion and vaporization, respectively. The terms T, C_p and ΔH represent the absolute temperature, and heat capacity at constant pressure, and latent enthalpy change, respectively, in consistent units. Since the entropy is zero at absolute zero:

$$\Delta S = S_T - S_0 = S_T \tag{25.6}$$

The above equation provides the entropy at the temperature in question, T [2,5].

Based on the above (see also Chapter 3), you are requested to obtain *entropy* values for five chemicals (including water) over the temperature range 0K to the boiling point of the material at 1.0 atmospheric pressure plus saturated vapor values above the boiling point. The chemicals can include oxygen (O_2), chlorine (Cl_2), calcium (Ca), sodium chloride (NaCl), etc. Employ both English and SI units.

References

1. L. Theodore and J. Reynolds, *Thermodynamics*, A Theodore Tutorial, East Williston, NY, originally published by the USEPA/APTI, RTP, NC, 1994.
2. L. Theodore, F. Ricci, and T. VanVliet, *Thermodynamics for the Practicing Engineer*, John Wiley & Sons, Hoboken, NJ, 2009.
3. L. Theodore, *Chemical Engineering: The Essential Reference*, McGraw-Hill, New York City, NY, 2014.

4. Personal notes: L. Theodore, East Williston, NY 1976.
5. L. Theodore and J. Reynolds, *Thermodynamics*, A Theodore Tutorial, East Williston, NY, originally published by the USEPA/APTI, RTP, NC, 1991.

26

Term Projects (6):
Fluid Flow

26.1 Pressure Drop – Velocity – Mesh Size Correlation
26.2 Fanning's Friction Factor: Equation Form
26.3 An Improved Pressure Drop and Flooding Correlation
26.4 Ventillation Model I
25.5 Ventilation Model II
25.6 Two-Phase Flow

Term Project 26.1

Pressure Drop – Velocity – Mesh Size
Correlation

Figure 26.1 below provides information on pressure drop variation with velocity for different mesh-sized particles [1–4]. Using any suitable statistical technique [5], convert Figure 26.1 into equation form.

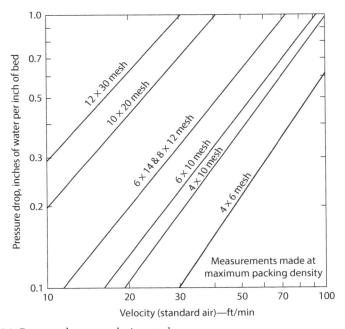

Figure 26.1 Pressure drop – mesh size graph

Term Project 26.2

Fanning's Friction Factor: Equation Form

The effect of the Reynolds number on the Fanning friction factor is provided in Figure 26.2. In the turbulent regime, the "roughness" of the pipe becomes a consideration. In his original work on the friction factor, Moody [6] defined the term k as the roughness and the ration, k/D, as the relative roughness. Thus, for rough pipes/tubes in turbulent flow

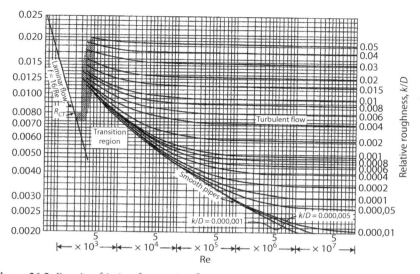

Figure 26.2 Fanning friction factor; pipe flow

$$f = f(\text{Re}, k/D) \qquad (26.1)$$

This equation reads that the friction factor is a function of *both* the Re and k/D. However, as noted in chapter 5, the dependency on the Reynolds number is a weak one. Moody [6] provided one of the original friction factor charts. His data and results as applied to the Fanning friction factor, and as presented in Figure 26.2 also contains friction factor data for various relative roughness values.

The reader should note the following:

1. Moody's original work included a plot of the Darcy (or Moody) friction factor, not the Fanning friction factor. His chart has been adjusted to provide the Fanning friction factor i.e., the plot in Figure 26.2 is for the Fanning friction factor. Those choosing to work with the Darcy friction factor need only multiply the Fanning friction factor f by 4, sin

$$f_D = 4f \qquad (26.2)$$

2. The intermediate regime of Re between 2100 and 4000 is indicated by the shaded area in Figure 26.2.
3. The average "roughness" of commercial pipes is given in Table 26.1 [7,8].

4. Note in Figure 26.2 that the relative roughness lines are nearly horizontal in the fully turbulent regime to the right of the dashed lines.

5. Roughness is a function of a variety of effects—some of which are difficult, if not impossible, to quantify. In effect, the roughness of a pipe resembling a smooth sine wave exhibits different frictional effects than one resembling a sharp saw-tooth or step function.

In summary, for Reynolds numbers below 2100, the flow will always be laminar and the value of f should be taken from the line at the left in Figure 26.2. For Reynolds number above 4000, the flow will practically always be turbulent and the values of f should be read from the lines at the right. Between Re = 2100 and Re = 4000, no accurate calculations can be made because it is generally impossible to predict flow type in this range. If an estimate of friction loss must be made in this range, it is recommended that the figures for turbulent flow should be used, as that provides an estimate on the high side [9].

Theodore [10] provides additional information.

Abulencia and Theodore [8] provide a number of models that have attempted to convert Figure 26.2 into equation form. This term project is

Table 26.1 Average Roughness of Commercial Pipes

	Roughness, k	
Material (new)	ft	mm
Riveted steel	0.003 – 0.03	0.9 – 9.0
Concrete	0.001 – 0.01	0.3 – 3.0
Wood stove	0.0006 – 0.003	0.18 – 0.9
Cast iron	0.00085	0.26
Galvanized iron	0.0005	0.15
Asphalted cast iron	0.0004	0.12
Commercial steel (wrought iron)	0.00015	0.046
Drawn tubing	0.000005	0.0015
Glass	"smooth"	"smooth"

concerned with developing an improved model to describe Fanning's friction factor in terms of both the Reynolds number and the roughness factor.

Term Project 26.3

An Improved Pressure Drop and Flooding Correlation

Consider a packed column operating at a given liquid rate and the gas rate is then gradually increased. After a certain point, the gas rate is so high that the drag on the liquid is sufficient to keep the liquid from flowing freely down the column. Liquid begins to accumulate and tends to block the entire cross section for flow (a process referred to as *loading*). A further increase in the gas flow rate increases both the pressure drop and prevents the packing from mixing the gas and liquid effectively, and ultimately some liquid is even carried back up the column. This undesirable condition, known as *flooding*, occurs fairly abruptly and the superficial gas velocity at which it occurs is called the *flooding velocity*. The calculation of column diameter is usually based on flooding considerations, with the usual operating range being taken as 50 – 75% of the flooding rate.

One of the more commonly used flooding correlations is U.S. Stoneware's [11] generalized pressure drop correlation, as presented in Figure 26.3. The procedure to determine the column diameter is as follows: [12-14]

1. Calculate the abscissa, $(L/G)(\rho_G/\rho_L)^{0.5}$; mass basis for all terms.
2. Proceed to the flooding line and read the ordinate (design parameter).
3. Solve the ordinate equation for G_f at flooding.
4. Calculate the column cross-sectional area, S, for the fraction of flooding velocity chosen for operation, f, by the equation:

$$S = \frac{W}{fG_f} \qquad (26.3)$$

where $W(\dot{m})$ is the mass flow rate of the gas in lb/s and S is the area in ft^2.

5. The diameter of the column is then determined by

$$D = (\frac{4}{\pi}S)^{0.5} = 1.13S^{0.5}; \text{ft} \qquad (26.4)$$

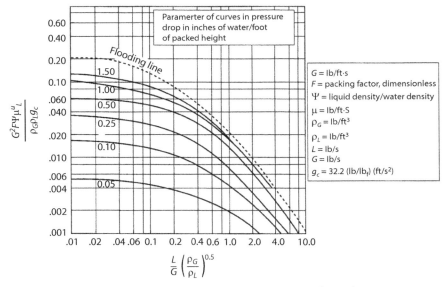

Figure 26.3 Generalized pressure drop correlation to estimate column diameter

Note that the proper units, as designated in the correlation, must be used as the plot is not dimensionless. The flooding rate is usually evaluated using total flows of the phases at the bottom of the column where they are at their highest value.

6. The pressure drop may be evaluated directly from Figure 26.3 using a revised ordinate that contains the actual, not flooding, value of G [14].

Chen [15] later developed the following equation from which the tower diameter can easily be obtained:

$$D = 16.28 \left(\frac{W}{\phi L} \right)^{0.5} \left(\frac{\rho_L}{\rho_G} \right)^{0.25} \tag{26.5}$$

where

$$\log_{10} \phi = 32.5496 - 4.1288 \log_{10} \left(\frac{L^2 A_v \mu_L^{0.2}}{\rho L^2 \varepsilon^3} \right) \tag{26.6}$$

and (employing Chen's notation) A_v is the specific surface area of dry packing (ft²/ft³ packed column), L is the liquid flux (gal/min·ft² of superficial tower cross section), W is the mass rate of flow of gas (lb/h), ε is the void fraction, μ_L is the liquid viscosity (cP), and the density terms are in lb/ft³.

You have received the following assignment. Convert the U.S. Stoneware correlation for the flooding conditions and pressure drop into equation form. Use any suitable statistical technique [16,17] in solving the project.

Term Project 26.4

Ventilation Model I

Refer to Part II, Chapter 21, Illustrative Open-ended Problem 3.

Your consulting firm has received a contract from Theodore Industries (TI) to develop mathematical models describing the concentration of a chemical in a medium-sized ventilated laboratory room. The following information/data (SI units) all provided by TI:

V = volume of room, m³
q_0 = volumetric flow rate of ventilation air, m³/min
c_0 = concentration of the chemical in ventilation air, gmol/m³
c = concentration of the chemical leaving ventilated room, gmol/m³
c_1 = concentration of the chemical initially present in ventilated room, gmol/m³
r = rate of disappearance of the chemical in the room due to reaction and/or other effects, gmol/m³·min.

Using the laboratory room as the control volume, one may apply the conservation law (see also chapter 3) for mass to the chemical [18-20], produced following the equation.

$$\frac{q_0}{V}\left(c_0 - c\right) + r = \frac{dc}{dt} \qquad (26.7)$$

As noted earlier, the term V/q_0 represents the average residence time that the chemicals reside in the room and is usually designated as τ. The above equation may then be written as

$$\frac{dc}{dt} = \frac{c_0 - c}{\tau} - r \qquad (26.8)$$

As an authority in the field, you have been requested to:

1. Obtain the equation describing the concentration in the room as a function of time if there are no "reaction" effects, i.e. $r = 0$.

2. Obtain the equation describing the concentration in the room as a function of time if $r = -k$. Note once again that the minus sign is carried since the chemical is disappearing.
3. Obtain the equation describing the concentration in the room as a function of time if $r = -kc$. Note once again that the minus sign is carried since the chemical is disappearing.
4. For Part 2, discuss the effect on the resultant equation if k is extremely small, i.e., $k \to 0$.
5. Provide your comments on the validity of the above results.

Term Project 26.5

Ventilation Model II

Refer to the previous term project [21,22].

1. Qualitatively discuss the effect on the results if the volumetric flow rate, v_0, varies sinusoidally.
2. Qualitatively discuss the effect on the final equation if the inlet concentration, c_0, varies sinusoidally.
3. Discuss the effect on the results if both v_0 and c_0 vary sinusoidally.
4. Arbitrarily select a mode of variation for both v_0 and c_0, and solve the problem
5. Present the results if both v_0 and c_0 are allowed to vary in an arbitrary manner.

Term Project 26.6

Two – Phase Flow

The simultaneous flow of two phases in pipes (as well as other conduits) is complicated by the fact that the action of gravity tends to cause settling and "slip" of the heavier phase with the result that the lighter phase flows at a different velocity in the pipe than does the heavier phase. The results of this phenomena are different depending on the classification of the two phases, the flow regime, and the inclination of the pipe (conduit).

As one might suppose, the major industrial application in this area is gas (G) – liquid (L) flow in pipes. The extension of much of this material to flow in various conduits can be accomplished by employing the equivalent diameter [23,24] of the conduit in question.

The general subject of flashing and boiling liquids is also a consideration. However, when a saturated liquid flows in a pipeline from a given point at a given pressure to another point at a lower pressure, several processes can take place. As the pressure decreases, the saturation or boiling temperature decreases, leading to the evaporation of a portion of the liquid. The net results is that a one-phase flowing mixture is transformed into a two-phase mixture with a corresponding increase in frictional resistance in the pipe. Boiling liquids arise when liquids are vaporized in pipelines at approximately constant pressure. Alternatively, the flow of condensing vapors in pipes is complicated due to the properties of the mixture constantly changing with changes in pressure, temperature, and fraction condensed. Further, the condensate, which forms on the walls, requires energy in order to be transformed into spray, and this energy must be obtained from the main vapor steam, resulting in an additional pressure drop. Additional information is available in the literature [23].

The suggested method of calculating the pressure drop of gas – liquid mixtures flowing in pipes is essentially that originally proposed by Lockhart and Martinelli [25] nearly 60 years ago. The basis of their correlation is that the two-phase pressure drop is equal to the single-phase pressure drop for either phase (G or L) multiplies by a factor that is a function of the single-phase pressure drops of the two phases. That development was presented in Part II, Chapter 5, Illustrative Open-ended Problem 3. You have been requested to improve on the outdated method developed by Lockhart and Martinelli.

References

1. L. Theodore and A. Buonicore, *Control of Gaseous Emissions*, USEPA/APTI, RTP, NC, 1982.
2. L. Theodore and J. Barden, *Mass Transfer Operations*, A Theodore Tutorial, Theodore Tutorials, East Williston, NY, originally published by the USEPA/APTI, RTP, NC, 1992.
3. S. Ergun, *Chem. Eng. Progr.* 48, 89, 1952; *Ind. Eng. Chem.,* New York City, NY, 41, 1179, 1949.
4. Union Carbide Corp., Linde division, Molecular Sieve Department, New York City, NY, Bulletin F-34.
5. S. Shaefer and L. Theodore, *Probability and Statistics Applications in Environmental Science,* CRC Press/ Taylor & Francis Group, Boca Raton, FL, 2007.

6. L. Moody, *Friction Factors for Dye Flow*, Trans. Am. Soc. Mech. Engrs., 66, 67 1-84, New York City, NY 1944.

7. I. Farag, *Fluid Flow*, A Theodore Tutorial, Theodore Tutorials, East Williston, NY, originally published by the USEPA/APTI, RTP, NC, 1996.

8. P. Abulencia and L. Theodore, *Fluid Flow for the Practicing Chemical Engineer*, John Wiley & Sons, Hoboken, NJ, 2009.

9. W. Badger and J. Banchero, *Introduction to Chemical Engineering*, McGraw-Hill, New York City, NY, 1955.

10. L. Theodore, *Chemical Engineering: The Essential Reference*, McGraw-Hill, New York City, NY, 2014.

11. Generalized Pressure Drop Correlation, Chart No. GR-109, Rev. 4, U.S. Stoneware Co., Akron, OH, 1963.

12. Personal notes: L. Theodore, East Williston, NY, 1979.

13. L. Theodore and J. Barden, *Mass Transfer Operations*, A Theodore Tutorial, Theodore Tutorials, East Williston, NY, originally published by the USEPA/ APTI, RTP, NC, 1995.

14. Adapted from: L. Theodore and F. Ricci, *Mass Transfer Operations for the Practicing Engineer*, John Wiley & Sons, Hoboken, NJ, 2010.

15. N. Chen, *New Equation Gives Tower Diameter, Chem. Eng.*, New York City, NY, May 2, 1962.

16. S. Shaefer and L. Theodore, *Probability and Statistics Applications in Environmental Science*, CRC Press/ Taylor & Francis Group, Boca Raton, FL, 2007.

17. L. Theodore, *Chemical Engineering: The Essential Reference*, McGraw-Hill, New York City, NY, 2014.

18. L. Theodore, *Chemical Reaction Kinetics*, A Theodore Tutorial, Theodore Tutorials, East Williston, NY, originally published by the USEPA/APTI, RTP, NC, 1992.

19. J. Reynolds, J. Jeris, and L. Theodore, *Handbook of Chemical and Environmental Engineering Calculations*, John Wiley & Sons, Hoboken, NJ, 2004.

20. P. Abulencia and L. Theodore, *Fluid Flow for the Practicing Chemical Engineer*, John Wiley & Sons, Hoboken, NJ, 2008.

21. L. Theodore, *Chemical Reaction Kinetics*, A Theodore Tutorial, Theodore Tutorials, East Williston, NY, originally published by the USEPA/APTI, RTP, NC, 1992.

22. J. Reynolds, J. Jeris, and L. Theodore, *Handbook of Chemical and Environmental Engineering Calculations*, John Wiley & Sons, Hoboken, NJ, 2004.

23. P. Abulencia and L. Theodore, *Fluid Flow for the Practicing Chemical Engineer*, John Wiley & Sons, Hoboken, NJ, 2008.

24. L. Theodore, *Chemical Engineering: The Essential Reference*, McGraw-Hill, New York City, NY, 2014.

25. R. Lockhart and R. Martinelli, *Generalized Correlation of Two-Phase, Two-Component Flow Data*, CEP, New York City, NY, 45, 39-48, 1949.

27

Term Projects (4): Heat Transfer

27.1 Wilson's Method
27.2 Heat Exchanger Network I
27.3 Heat Exchanger Network II
27.4 Heat Exchanger Network III

Term Project 27.1

Wilson's Method

There is a laboratory procedure for evaluating the *outside* film coefficient, h_o, for a double pipe unit. Wilson's method [1] is a graphical technique for evaluating this coefficient. The inside coefficient, h_i, is a function of the Reynolds and Prandtl numbers via the Dittus – Boelter equation [2,3]; i.e.,

$$h_i = f(\mathrm{Re}^{0.8}\mathrm{Pr}^{0.3})$$

(27.1)

A series of experiments can be carried out on a double pipe exchanger where all conditions are held relatively constant except for the velocity (V) of the cooling (in this case) inner stream. Therefore, for the proposed experiment:

$$h_i = f\left(\mathrm{Re}^{0.8}\right) = f\left(V^{0.8}\right) = f\left(\dot{m}\right); \ \mathrm{Re} = DV\rho/\mu$$

(27.2)

or, in equation form,

$$h_i = aV^{0.8}$$

(27.3)

where a is a constant. Equation (27.3) can be substituted into the overall coefficient equation

$$\frac{1}{h_o A_o} = R_o + R_w + R_i$$

(27.4)

where R_o = outside resistance
R_w = wall resistance
R_i = inside resistance
so that [2]

$$\frac{1}{U_o A_o} = \frac{1}{h_o A_o} + \frac{\Delta x}{kA_{\mathrm{lm}}} + \frac{1}{aV^{0.8} A_i}$$

(27.5)

Data can be taken at varying velocities. By plotting $1/U_o A_o$ versus $1/V^{0.8}$, a straight line should be obtained since the first two terms of Equation (27.5) are constants. The intercept of this line corresponds to an infinite velocity where the inside resistance is zero. Thus,

$$\left(\frac{1}{U_o A_o}\right)_{intercept} = \frac{1}{h_o A_o} + \frac{\Delta x}{k A_{lm}}$$

(27.6)

The second term on the right-hand side is known and/or can be calculated and h_o can then be evaluated from the intercept; details on this calculational scheme are provided in an illustrative example provided by Theodore [3], fouling coefficient, h_o, can be predicted or is negligible. Note that the fouling resistance is normally included in the intercept value.

Based on the above analysis, you have been requested to develop a similar-type procedure to evaluate the inside heat transfer coefficient in a shell-and-tube (tube-and-bundle) heat exchanger. Also describe the experimental equipment and method to be employed.

Term Project 27.2

Heat Exchanger Network I

In many chemical and petrochemical plants there are cold streams that must be heated and hot steams that must be cooled. Rather than use steam to do all the heating and cooling water to do all the cooling, it is often advantageous to have some of the hot streams heat the cold ones.

Highly interconnected networks of exchangers can save a great deal of energy in a chemical plant. The more interconnected they are, however, the harder the plant is to control, start-up and shut-down. Often auxiliary heat sources and cooling sources must be included in the plant design in order to ensure that the plant can operate smoothly.

A plant has three streams to be heated and three streams to be cooling. Cooling water (90°F supply, 155°F return) and steam (saturated at 250 psia) are available. Devise a network of heat exchangers that will make full use of heating and cooling streams against each other, using utilities only if necessary.

The three streams to be heated are provided in Table 27.1
The three streams to be cooled are provided in Table 27.2
The reader is referred to Illustrative Open-ended Problem 2 in Chapter 6 for calculations related to the above streams.

There are many possible combinations of interconnected heat exchangers that will work. Detailed cost analyses would be needed to determine which one was best. For each exchanger in the network, one needs to make sure that the duties on each side are equal and that there are no *temperature crossovers*. These occur when the stream being cooled is colder than the lowest temperature of the stream being heated at some point in the exchanger. This is (of course) forbidden by the second law of thermodynamics [6,7].

Table 27.1 Heated Stream

Stream	Flowrate, lb/h	c_p, Btu/lb·F	T_{in}, °F	T_{out}, °F
1	50,000	0.65	70	300
2	60,000	0.58	120	310
3	80,000	0.78	90	250

Table 27.2 Cooled Stream

Stream	Flowrate, lb/h	c_p, Btu/lb·F	T_{in}, °F	T_{out}, °F
4	60,000	0.70	420	120
5	40,000	0.52	300	100
6	35,000	0.60	240	90

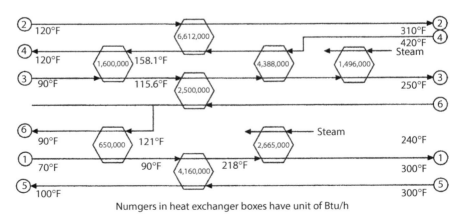

Numgers in heat exchanger boxes have unit of Btu/h

Figure 27.1 Kauffman's [4] heat exchanger network.

Comment: Since this and the next two term projects are concerned with the general subject of heat exchanger networks, Kauffman's solution [4] is provided in Figure 27.1. There are, of course, other solutions. Your assignment is to provide another solution.

Term Project 27.3

Heat Exchanger Network II

Any discussion of energy conservation above leads to an important second-law consideration—energy has "quality" as well as quantity. Because work is 100% convertible to heat whereas the reverse situation is not true, work is a more valuable form of energy than heat. Although it is not as obvious, it can also be shown through second law arguments that heat also has "quality" in terms of the temperature at which it is discharged from a system. The higher the temperature, the greater the possible energy transformation into work. Thus, thermal energy stored at high temperatures is generally more useful to society than that available at lower temperatures. While there is an immense quantity of energy stored in rivers and oceans, for example, its present availability to society for performing useful tasks is quite low. This implies, as noted above, that thermal energy loses some of its "quality" or is degraded when it is transferred by means of heat transfer from one temperature to a lower one. Other forms of energy degradation include energy transformation due to frictional effects and electrical resistance. Such effects are highly undesirable if the use of energy for practical purposes is to be maximized [8–10].

The second law provides a means of measuring this energy degradation through a thermodynamic term referred to as *entropy* and it is the second law (of thermodynamics) that serves to define this important property. It is normally designated as S with units of energy per absolute temperature (e.g., Btu/°R or cal/K). Furthermore, entropy calculations can provide quantitative information on the "quality" of energy and energy degradation [9,10].

There are a number of other phenomena that cannot be explained by the first law of conservation of energy. It is the second law of thermodynamics that provides an understanding and analysis of these diverse effects. However, among these considerations, it is the second law that can allow the measuring of the aforementioned "quality" of energy, including its effect on the design and performance of heat exchangers.

One of the areas where the aforementioned meaningful energy conservation measures can be realized is in the design and specification of process (operating) conditions for heat exchangers. This can be best accomplished by the inclusion of second law principles in the analysis. The quantity of heat recovered in an exchanger is not alone in influencing size and cost. As the energy temperature difference driving force (LMTD) in the exchanger approaches zero, the "quality" heat recovered increases [11,12].

Most heat exchangers are designed with the requirements/specification that the temperature difference between the hot and cold fluid be at all times positive and be at least 20°F. This temperature difference or driving force is referred to by some as the *approach temperature*. However, the corresponding entropy change is also related to the driving force, with large temperature difference driving forces resulting in large irreversibilities and the associated large entropy changes [11,12].

The individual designing a heat exchanger is faced with two choices. He/she may decide to design with a large LMTD that results in both a more compact (smaller area) design and large entropy increase that is accompanied by the loss of "quality" energy. Alternately, a design with a small driving force results in both a larger heat exchanger and a smaller entropy change/larger recovery of "quality" energy [11,12].

Refer to the previous term project (27.2) on heat exchanger networks. Solve the problem if the network of exchangers must be such that the *entropy* increase in the proposed network is minimized

The reader should consider reviewing second law principles presented in Part II, Chapters 4 and 6.

Term Project 27.4

Heat Exchanger Network III

Refer to the two previous term projects concerned with heat exchanger networks [13,14].

1. *Outline* how to solve the first problem if the heat exchanger cost is a function of the heat exchanger area. The network should be such that the total cost of the resulting network is *minimized*. In effect, the economics involved are to be included with the outline to the solution.

2. *Outline* how to solve the two previous problems so that the network's entropy increase *and* the network's cost are *minimized*. Is there a unique solution? Justify your answer to this question. If there is not a unique solution, *outline* how a practicing chemical engineer could arrive at a *reasonable* solution to the problem.

References

1. E. Wilson, *Trans, ASME*, 34, 47, New York City, NY, 1915.
2. R. Perry, and D. Green, (editors), *Perry's Chemical Engineers' Handbook*, 7th edition, McGraw-Hill, New York City, NY, 1997.
3. L. Theodore, *Heat Transfer for the Practicing Engineer*, John Wiley & Sons, Hoboken, NJ, 2011.
4. M. Theodore and L. Theodore, *Introduction to Environmental Management*, CRC Press/Taylor & Francis Group, Boca Raton, FL, 2009.
5. J. Smith, H. Van Ness, and M. Abbott, *Chemical Engineering Thermodynamics*, 6th edition, McGraw-Hill, New York City, NY, 2001.
6. L. Theodore and J. Reynolds, *Thermodynamics*, A Theodore Tutorial, East Williston, NY, originally published by the USEPA/APTI, RTP, NC, 1994.
7. L. Theodore, F. Ricci, and T. VanVliet, *Thermodynamics for the Practicing Engineer*, John Wiley & Sons, Hoboken, NJ, 2009.
8. L. Theodore, *Heat Transfer for the Practicing Engineer*, John Wiley & Sons, Hoboken, NJ, 2011.
9. D. Kauffman, *Process Synthesis and Design*, A Theodore Tutorial, Theodore Tutorials, East Williston, NY, originally published by the USEPA/APTI, RTP, NC, 1992.
10. L. Theodore, *Heat Transfer for the Practicing Engineer*, John Wiley & Sons, Hoboken, NJ, 2011.
11. L. Theodore, F. Ricci, and T. VanVliet, *Thermodynamics for the Practicing Engineer*, John Wiley & Sons, Hoboken, NJ, 2009.
12. L. Theodore, *Chemical Engineering: The Essential Reference*, McGraw-Hill, New York City, NY, 2014.
13. L. Theodore, F. Ricci, and T. VanVliet, *Thermodynamics for the Practicing Engineer*, John Wiley & Sons, Hoboken, NJ, 2009.
14. L. Theodore, *Heat Transfer for the Practicing Engineer*, John Wiley & Sons, Hoboken, NJ, 2011.

28

Term Projects (5): Mass Transfer Operations

28.1 An Improved Absorber Design Procedure
28.2 An Improved Adsorber Design Procedure
28.3 Multipcomponent Distillation Calculations
28.4 A New Liquid-Liquid Extraction Process
28.5 Designing and Predicting the Performance of Cooling Towers

Term Project 28.1

An Improved Absorber Design Procedure

One of the authors [1–3] has developed a procedure that enables one to quantitatively outline how to size (diameter, height) a packed tower to achieve a given degree separation without any information on the physical and chemical properties of a gas to be absorbed. To calculate the height, one needs both the height of a gas transfer unit H_{OG} and the number of gas transfer units N_{OG} (see also Chapter 7). Since equilibrium data are not available, assume that m (slope of equilibrium curve) approaches zero. This is not an unreasonable assumption for most solvents that preferentially absorb (or react with) the solute. For this condition:

$$N_{OG} = \ln\left(\frac{y_1}{y_2}\right) \tag{28.1}$$

where y_1 and y_2 represent inlet and outlet concentrations, respectively. Since it also reasonable to assume the scrubbing medium to be water or a solvent that effectively has the physical and chemical approaching that properties of water, H_{OG} can be assigned values usually encountered for water systems. These are given in Table 28.1. For plastic packing, the liquid and gas flow fluxes are both typically in the range of 1500 – 2000 lb/(h·ft² of cross-sectional area). For ceramic packing, the range of flow rates is 500 – 1000 lb/h·ft². For difficult-to-absorb gases, the gas flow rate is usually lower and liquid flow rate higher. Superficial gas velocities (velocity of the gas if the column is empty) are in the 3 – 6 ft/s range. The height Z may then be calculated from

Table 28.1 Packing Diameter versus H_{OG}

Packing diameter, inches	Plastic packing H_{OG}, feet	Ceramic packing H_{OG}, feet
1.0	1.0	2.0
1.5	1.25	2.5
2.0	1.5	3.0
3.0	2.25	4.5
3.5	2.75	5.5

$$Z = (H_{OG})(N_{OG})(\text{SF}) \tag{28.2}$$

where SF is a safety factor, the value of which can range from 1.25 – 1.5. Pressure drops can vary from 0.15 – 0.40 inch H_2O/ft packing. Packing size increases with increasing tower diameter. Packing diameters of 1 inch are recommended for tower diameter in the 3 ft range. One should use large packing for larger-diameter units; for smaller towers, smaller packing is usually employed [4,5].

You have been requested to improve the procedure originally proposed by Theodore [2]. In effect, you are being asked to convert the semi-qualitative procedure; i.e., the calculation for N_{OG}, H_{OG}, ΔP, and D need to be improved.

Term Project 28.2

An Improved Adsorber Design Procedure

There are several factors to be considered in the design of an adsorber [6].

1. The adsorbent particle size
2. The physical adsorbent bed depth
3. The gas velocity
4. The temperature of the inlet gas stream and the adsorbent
5. The solute concentration to be adsorbed
6. Any solute concentration(s) not to be adsorbed, including moisture
7. The required or desired removal efficiency
8. Possible polymerization on the adsorbent
9. The frequency of operation
10. Regeneration conditions
11. The system pressure.

Refer to the Overview in Part II, Chapter 7. Employing the above considerations/factors, Theodore [7,8] developed a rather simplified overall design procedure for a carbon system adsorbing an organic that consists of two horizontal units (one on/one off) that are regenerated with steam [9,10].

Dissatisfied with the semi-qualitative nature of Theodore's [7] design procedure, your boss has requested that the procedure not only be improved both technically and quantitatively but also be applicable for vertical columns, more than two units, and all methods of regeneration.

Term Project 28.3

Multicomponent Distillation Calculations

The Fenske-Underwood-Gilliland (FUG) procedure for sizing multicomponent distillation columns is well documented in the literature. Theodore and Ricci [11] provide the following.

The Fenske equation is used to calculate the minimum number of theoretical stages when the column is being operated under total reflux. While many forms of the Fenske equation have been presented in the literature, Equation (28.3) is preferred because it is highly useful from a design perspective.

$$N_{min} = \frac{\ln\left[\left(\frac{r_{LK}}{1-r_{LK}}\right)\left(\frac{r_{HK}}{1-r_{HK}}\right)\right]}{\ln \overline{a}_{LK}} \tag{28.3}$$

where N_{min} = minimum number of theoretical stages (including partial reboiler)

$\overline{a}_{LK} = \sqrt{a_{D,LK} a_{B,LK}}$, a geometric mean of light key relative volatilities

r_{LK}, r_{HK} = fractional recoveries

While the Fenske equation calculates the minimum number of equilibrium stages required for separation at total reflux, Underwood developed equations that estimate the minimum reflux ratio. The Underwood equations are a set of two mathematical expressions that are generally solved sequentially (unless there are one or more components in between the light and heavy keys, in which case they should be solved simultaneously in order to determine the correct root). These equations are listed below:

$$N_{min} = \sum_{i=1}^{N} \frac{\overline{a}_i x_{F,i}}{\overline{a}_i - \Theta} = 1 - q; \ (1 < \Theta < \overline{a}_{LK}) \tag{28.4}$$

$$R_{min} = \sum_{i=1}^{N} \frac{\overline{a}_i x_{D,i}}{\overline{a}_i - \Theta} - 1 \tag{28.5}$$

where q = the q-factor, dependent upon the thermal condition of the feed
Θ = root of the first Underwood equation (28.4)

$\overline{\alpha}_i = \sqrt{\alpha_{D,i}\alpha_{D,i}}$, mean relative volatility of the ith component

Now that both Fenske and Underwood equations have been utilized to determine N_{min} and R_{min}, respectively, the last step in the FUG procedure is to employ the Gilliland correlation in order to determine the number of theoretical trays. The Gilliland correlation is shown in Figure 28.1 [12].

As is evidenced by Figure 28.1, one need only know the minimum reflux ratio and the operating reflux ratio in order to compute the abscissa of the correlation. The corresponding ordinate is then read from the plot. Since the minimum number of theoretical stages is known, the actual number of theoretical stages, N_t, may be calculated. However, it should be noted that the Gilliland correlation was derived for systems with nearly constant relative volatilities throughout the column. Therefore, the Gilliland correlation may not be a suitable short-cut method for non-ideal systems, which relative volatilities may vary drastically.

While the Gilliland correlation itself is immensely useful for use as a shortcut method in column quick-sizing, it is inconvenient in that the graph must be read manually. In order to make the Gilliland correlation more applicable to computer programming, several analytical expressions

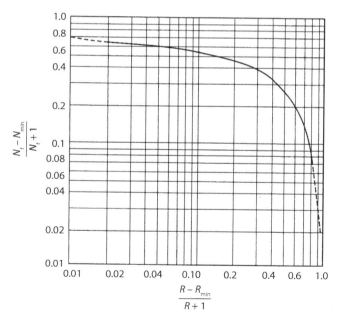

Figure 28.1 Gilliland correlation.

have been developed. One such outstanding correlation, as suggested by Chang, [13] is provided in Equation 28.6 below:

$$Y = 1 - \exp\left[1.490 + 0.315X - \frac{1.805}{X^{0.1}}\right]$$

(28.6)

where X is the abscissa (x-axis) of this correlation, given by

$$X = \frac{R - R_{min}}{R + 1}$$

(28.7)

and Y is the ordinate (y-axis) of the correlation,

$$Y = \frac{N_t - N_{min}}{N_t + 1}$$

(28.8)

which may be rearranged to:

$$N_t = \frac{N_{min} + Y}{1 - Y}$$

(28.9)

where N_t = the number of theoretical stages.
Thus, Gilliland's graphical correlation may be replaced by using Chang's convenient mathematical expression.

The calculation of the number of theoretical trays is the last step in the FUG procedure. However, the sizing of distillation columns does not end there. To the contrary, there are several other considerations which must be taken into account in order to completely quick-size a distillation column, including: theoretical location of the feed tray (Kirkbride equation), calculation of the actual number of trays, calculation of column diameter, calculation of column height, etc.

In the addition to the FUG, the Kirkbride equation may be employed to determine the location of the *theoretical feed tray* and the O'Connell correlation [14] may be used to obtain a column's overall efficiency.

You have been assigned the task of developing an improved multicomponent distillation calculational procedure. Your revised equations need not necessarily be based on either the Fenske, or the Underwood, or the Gilliland correlations.

Term Project 28.4

A New Liquid-Liquid Extraction Process

Extraction is a term that is used for an operation in which a constituent of a liquid or a solid is transferred to another liquid (the *solvent*). The term *liquid – liquid extraction* describes the processes in which both phases in the mass transfer process are liquids. The term *liquid – solid extraction* is restricted to those situations in which a solid phase is present and includes those operations frequently referred to as *leaching, lixiviation,* and *washing* [15–18].

Extraction involved the following two steps: contact of the *solvent* with the liquid or solid to be treated so as to transfer the *solute* (soluble component) to the solvent, and separation or washing of the resulting solution. The complete process may also include a separate recovery procedure involving the solute and solvent; this is normally accomplished by another operation such as evaporation, distillation, or stripping. Thus, the streams leaving the extraction system usually undergo a series of further operations before the finished "product" is obtained; either one or both solutions may contain the desired material. In addition to the recovery of the desired product or products, recovery of the solvent for recycling is also an important consideration [15–18].

Liquid – liquid extraction is used for the removal and recovery of primarily organic solutes from aqueous and nonaqueous streams. Concentrations of solute in these streams range from either a few hundred parts per million to several mole/mass percent. Most organic solutes may be removed by this process. Extraction has been specifically used in removal and recovery of phenols, oils, and acetic acid from aqueous streams, and in removing and recovering freons and chlorinated hydrocarbons from organic streams [15–18].

Treybal [15] has described the liquid – liquid extraction process in the following manner. If an aqueous solution of acetic acid is agitated with a liquid such as ethyl acetate, some of the acid but relatively little water will enter the ester phase. Since the densities of the aqueous and ester layers are different at equilibrium, they will settle on cessation of agitation and may be decanted from each other. Since the ratio of acid to water in the ester layer is now different from that in the original solution and also different from that in the residual water solution, a certain degree of separation has occurred. This is an example of stage-wise contact and it may be carried out either in a batch or continuous fashion. The residual water may be repeatedly extracted with more ester to additionally reduce the acid content. One may arrange a countercurrent cascade of stages to accomplish

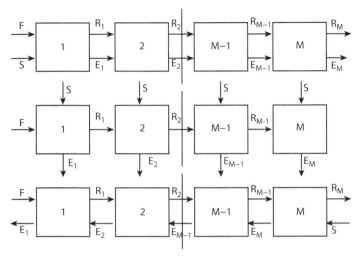

Figure 28.2 Multistage extractors.

the separation. Another possibility is to use some sort of countercurrent or crosscurrent continuous-contact device where discrete stages are not involved. These are pictured in Figure 28.2.

The solution whose components are to be separated is the *feed* (F) to the process. The feed is composed of a *diluent* and *solute*. The liquid contacting the feed for purposes of extraction is referred to as the *solvents* [19]. The solvent-lean, residual feed solution, with one or more constituents removed by extraction, is referred to as the *raffinate* (R). The solvent-rich solution containing the extracted solute(s) is the *extract* (E).

There are two major categories of equipment for liquid extraction. The first is *single-stage* units, which provide one stage of contact in a single device or combination of devices. In such equipment, the liquids are mixed, extraction occurs, and the insoluble liquids are allowed to separate as a result of their density differences. Several separate stages may be used in an application. Second, there are *multistage* devices, where many stages may be incorporated into a single unit. This type is normally employed in practice. (See also Figure 28.2).

In addition to *concurrent* flow (rarely employed) provided in Figure 28.2 for an *n*-stage system, *crosscurrent* extraction is a series of stages in which the raffinate from one extraction stage is contacted with additional fresh solvent in a subsequent stage. *Crosscurrent* extraction is usually not economically appealing for large commercial processes because of the high solvent usage and low solute concentration in the extract. Figure 28.2 also illustrates countercurrent extraction in which the extraction solvent enters

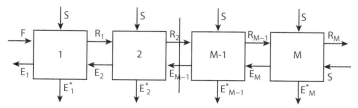

Figure 28.3 A hybrid extraction process

a stage at the opposite end from where the feed enters and the two phases pass each other countercurrently.

You have been assigned to a project that is concerned with developing a new liquid-liquid extraction process that integrates/combines both features of the countercurrent and crosscurrent processes. A diagram of one proposed process is provided in Figure 28.3 [19].

Term Project 28.5

Designing and Predicting the Performance of Cooling Towers

Cooling towers find application in industry. The same operation that is used to humidify air may also be used to cool water. There are many cases in practice in which warm water is discharged from condensers or other equipment and where the value of this water is such that it is more economical to cool it and reuse it than to discard it. Water shortages and thermal pollution have made the cooling tower a vital part of many plants in the chemical process industry. Cooling towers are normally employed for this purpose and they may be destined to have an increasingly important role in almost all phases in industry. Modern (newer) power-generating stations remain under construction or in the planning stage, and both water shortages and thermal pollution are serious problems that must be addressed [20,21].

The cooling of water in a cooling tower is accomplished by bringing the water into direct contact with unsaturated air under such conditions that the air is humidified and the water is brought approximately to the wet-bulb temperature. This method is applicable only in those cases where the wet-bulb temperature of the air is below the desired temperature of the exit water [22].

Qualitatively speaking, water is cooled in cooling towers by the exchange of sensible heat, latent heat, and water vapor with a stream of relatively cool dry air. The basic relationships developed for dehumidifiers also apply to cooling towers although the transfer is in the opposite direction since the unit acts as a humidifier rather than as a dehumidifier of air [23].

Brown and Associates [24] have provided empirical correlations from the literature [25,26] for estimating (roughly) sizes and capacities of conventional towers.

One of the authors [27] has repeatedly claimed that no simple, easy-to-use procedure is available for either predicting the performance or the design of cooling towers. Employing the citations above and sound mass transfer theory, develop a cooling tower model/equation(s) that the practicing chemical engineer can apply.

References

1. L. Theodore and A. Buonicore, *Control of Gaseous Emissions*, USEPA/APTI, RTP, NC, 1982.
2. Personal notes: L. Theodore, East Williston, NY 1979.
3. L. Theodore and J. Barden, *Mass Transfer Operations*, A Theodore Tutorial, Theodore Tutorials, East Williston, NY, originally published by the USEPA/APTI, RTC, NC, 1992.
4. Adapted from: L. Theodore, *Air Pollution Control Equipment Calculations*, John Wiley & Sons, Hoboken, NJ, 2008.
5. Adapted from: L. Theodore, *Mass Transfer Operations for the Practicing Engineer*, John Wiley & Sons, Hoboken, NJ, 2010.
6. L. Theodore and A. Buonicore, *Control of Gaseous Emissions*, USEPA/APTI, RTP, NC, 1982.
7. Personal notes: L. Theodore, 1979.
8. L. Theodore and J. Barden, *Mass Transfer Operations*, A Theodore Tutorial, Theodore Tutorials, East Williston, NY, originally published by the USEPA/APTI, RTC, NC, 1992.
9. L. Theodore, *Engineering Calculations: Adsorber Sizing Made Easy*, CEP, New York City, NY, March 2005.
10. Adapted from: L. Theodore, *Chemical Engineering: The Essential Reference*, McGraw-Hill, New York City, NY, 2014.
11. L. Theodore and F. Ricci, *Mass Transfer Operations for the Practicing Engineer*, John Wiley & Sons, Hoboken, NJ, 2010.
12. E.R. Gilliland, *Ind. Eng. Chem.*, New York City, NY, 32, 1220,1940.
13. H.Y. Change *Hydrocarbon Processing*, New York City, NY, 64(3), 48, 1985.
14. H.E. O'Connell, *Trans. AIChE*, New York City, NY, 42, 741, 1946.
15. L. Theodore and J. Barden, *Mass Transfer Operations*, A Theodore Tutorial, Theodore Tutorials, East Williston, NY, originally published by the USEPA/APTI, RTP, NC, 1995.
16. R. Treybal, *Mass Transfer Operations*, McGraw-Hill, New York City, NY, 1955.

17. R. Treybal, *Liquid Extractions,* McGraw-Hill, New York City, NY, 1951.
18. L. Theodore and F. Ricci, *Mass Transfer Operations for the Practicing Engineer,* John Wiley & Sons, Hoboken, NJ, 2010.
19. Personal notes: L. Theodore, East Williston, NY, 2012.
20. L. Theodore and J. Barden, *Mass Transfer Operations,* A Theodore Tutorial, Theodore Tutorials, East Williston, NY, originally published by the USEPA/APTI, RTP, NC, 1995.
21. L. Theodore and F. Ricci, *Mass Transfer Operations for the Practicing Engineer,* John Wiley & Sons, Hoboken, NJ, 2010.
22. Author unknown, New York University, Chemical Engineering Unit Operations Laboratory Report, Bronx, NY, 1960.
23. Author unknown, Manhattan College, Bronx, NY, 1961.
24. G. Brown and Associates, *Unit Operations,* John Wiley & Sons, Chapmann & Hall Limited, Hoboken, London, 1950.
25. Fluor Corp Ltd., *Bulletin T* 337, location unknown,1939.
26. R.C. Kelly, paper published by location unknown, the Fluor Corp., presented before the California Natural gasoline Association, Dec. 3, 1942.
27. Personal notes: L. Theodore, Theodore Tutorials, East Williston, NY, 1973.

29

Term Projects (2): Chemical Reactors

29.1 Minimizing Volume Requirements for CSTRs in Series I
29.2 Minimizing Volume Requirements for CSTRs in Series II

Term Project 29.1

Minimizing Volume Requirements for CSTRs in Series I

CSTRs in series are usually designed so that the volumes of the individual reactors are equal. For almost all reactions, the total volume requirement for achieving a given conversion decreases as the number of reactors in series increases. This can significantly impact the economics, particularly the capital cost. However, the total volume requirement to achieve a particular conversion can be further reduced, particularly for non-elementary reactions, if the constraint of equal reactor volumes is removed, i.e., the volumes of each reactor need not be the same. Although these systems can be designed to lower volume requirements, the impact on the overall economics can be negative [1,2].

Consider the elementary irreversible reactions between

$$A + B \rightarrow \text{products} \tag{29.1}$$

If one employs a feed containing equimolar concentrations of reactants, the reaction rate expression can be written as

$$-r_A = k_A C_A C_B = k C_A^2; \quad k = k_A \tag{29.2}$$

One can calculate the reactor size requirements for either one CSTR or for a cascade composed of n identical CSTRs. Assume isothermal operation at 25°C where the reaction rate constant is equal to 9.92 m³/(kgmol·min). Reactant concentrations in the feed are each equal to 0.08 kgmol/m³, and the liquid feed rate is equal to 0.278 m³/min. The desired degree of conversion is 87.5% [3].

Outline a procedure to calculate the volume requirement of a cascade of n CSTRs, that *differ* in size for the *minimum* total volume and the manner in which the total volume should be distributed between the n reactors.

Also use the equation(s) developed for n reactors to calculate the minimum volume requirement for three different n values of your choice, e.g., 2, 5, and 7. Comment on the results.

Term Project 29.2

Minimizing Volume Requirements for CSTRs in Series II

Refer to the previous term project concerned with CSTRs in series. *Outline* how to solve the problem if the rate of reaction is once again given as

$$-r_A = k_A C_A C_B = kC_A^2; \quad k = k_A \tag{29.3}$$

but with the variation of k_A with temperature dictated by the Arrhenius equation, i.e.,

$$k = Ae^{-E/RT} \tag{29.4}$$

Assign an enthalpy of reaction (assume it to be exothermic) with the Arrhenius equations coefficients given by A and E for the reaction under consideration and redesign the n-stage CSTR system taking enthalpy of reaction effects into account. Assume the enthalpy of reaction is constant (and independent of temperature) and the operation is adiabatic. This effectively means that the temperature in the n reactors will *not* be 25°C. Once again, the n CSTRs will differ in size in order to satisfy the *minimum* total volume requirement [4,6].

After the solution has been outlined, assign at least three sets of values to A and E; i.e.,

$$\begin{array}{c} A_1, E_1 \\ A_2, E_2 \\ A_3, E_3 \\ \cdot \quad \cdot \\ \cdot \quad \cdot \end{array} \tag{29.5}$$

and solve the problem. The A and E values should be such that one can provide a meaningful analysis of the effect of A and E on the results.

References

1. L. Theodore, *Chemical Reaction Kinetics*, A Theodore Tutorial, Theodore Tutorials, East Williston, NY, originally published by the USEPA/APTI, RTP, NC, 1992.

2. L. Theodore, *Chemical Reactor Analysis and Applications for the Practicing Engineer*, John Wiley & Sons, Hoboken, NJ, 2012.

3. Adapted from: J. Reynolds, J. Jeris, and L. Theodore, *Handbook of Chemical and Environmental Engineering Calculations*, John Wiley & Sons, Hoboken, NJ, 2004.

4. L. Theodore, *Chemical Reaction Kinetics*, A Theodore Tutorial, Theodore Tutorials, East Williston, NY, originally published by the USEPA/APTI, RTP, NC, 1992.

5. L. Theodore, *Chemical Reactor Analysis and Applications for the Practicing Engineer*, John Wiley & Sons, Hoboken, NJ, 2012.

6. Adapted from: J. Reynolds, J. Jeris, and L. Theodore, *Handbook of Chemical and Environmental Engineering Calculations*, John Wiley & Sons, Hoboken, NJ, 2004.

30

Term Projects (4): Plant Design

30.1 Chemical Plant Shipping Facilities
30.2 Plant Tank Farms
30.3 Chemical Plant Storage Requirements
30.4 Inside Battery Limits (ISBL) and Process Flow Approach

Term Project 30.1

Chemical Plant Shipping Facilities

A large chemical plant complex is likely to have facilities for shipping by pipeline, sea, inland waterway, rail, and road. Smaller plants may have facilities for only one or two of these modes. In any case, they must be designed for safe loading and unloading. In addition, there is often a major economic incentive to load and unload ships, barges, rail cars and trucks as quickly as possible.

The cost of shipping materials, both raw materials and products, is often a very significant factor in the chemical industry. The energy cost associated with shipping can range from 0.001 gallons of fuel per ton·mile of cargo moved for supertankers to 0.2 for cargo jet aircraft.

A new process plant on the Ohio River expects to receive 500,000 gallons per day of a single raw material and ship out 100,000 gallons per day of each of two liquid products and 500,000 lbs each of four solid products. Estimate the total shipping cost per day for this plant's operations if all shipments are by barge and if all shipments are by rail. Data is provided

Table 30.1 Shipping Cost Data

Material		Distance shipped
Feed		800 miles
Liquid product A		250 miles
Liquid product B		200 miles
Solid product C		40 miles
Solid product D		300 miles
Solid product E		45 miles
Solid product F		120 miles
Shipping costs:		
Barge:	20 to 80 miles	$0.03 / ton·mile
	80 to 300 miles	$0.02 / ton·mile
	300 to 1000 miles	$0.015 / ton·mile
Rail:	10 to 100 miles	$0.08 / ton·mile
	100 to 1000 miles	$0.04 / ton·mile

in Table 30.1. Assume the liquids all have a specific gravity of 1.0. Also, attempt to devise an optimum shipping schedule [1].

Comment: Transport by ship and/or barge is usually much cheaper than transport by rail or truck. It often requires longer travel times, however, and thus more inventory is tied up in shipping.

Term Project 30.2

Plant Tank Farms

Liquid feeds, products, intermediates, and fuels at chemical and petrochemical plants are stored in tanks, which are usually located in a *tank farm* adjacent to the process plant area. The tank capacities are most often expressed in gallons or barrels (one barrel = 42 gallons). The individual tanks may range in size from a thousand gallons or less to several million gallons.

Whenever possible, liquids are stored at ambient temperature and pressure. Volatile liquids may have to be stored under pressure. In addition, they often require vapor recovery systems or other devices to prevent releases to the atmosphere as the tanks are filled and emptied. Liquids which are very viscous at ambient temperature or which would solidify at ambient temperature are kept in heated tanks.

Standard practice calls for each tank to be surrounded by a dike or berm sufficiently high to contain all the liquids stored in a tank in case it should rupture. Fire fighting equipment is permanently located near tanks containing flammable liquids.

Consider the same process plant described in the previous term project (30.1). The new process plant expects to receive 500,000 gallons per day of a single raw material and ship out 100,000 gallons per day of each of two liquid products and 500,000 lbs each of four solid products. If shipments are all made by barge, the plant will require tankage for a 15 day supply of feed and a 10 day supply of each product. If shipments are by rail, which can be schedules more reliably, the plant will require a 7 day supply of feed and a 5 day supply of each product. For ease of maintenance, there should be at least three tanks for each liquid, with one tank being "off-line" at any given time [2].

1. Determine at least one set of storage tank requirements for the liquid feed and products. Available standard tank sizes and capacities are provided in Table 30.2.

TABLE 30.2 Available Standard Tanks

Capacity Gallons	Height Feet	Diameter Feet
216,000	30	35
429,000	36	45
1,040,000	36	70
2,110,000	36	100
4,060,000	48	120

2. Resolve the problem for various inventory tie-up costs. In effect, assign a time factor to each mode of travel and a cost associated with the lost time.
3. Also consider the following: Storage tanks can present major risks both in terms of fire and environmental pollution. In both cases this is primarily due to the large inventories of materials that could be involved in any accident. There has been a major effort in industry in the last several decades to reduce the amount of storage at process plants, especially for flammable and/or toxic materials. Suggest how the inventory of both feed and products could be reduced.

   Comment: Transport by ship and/or barge is usually much cheaper than transport by rail or truck. It often requires longer travel times, however, and thus more inventory is tied up in shipping.

Term Project 30.3

Chemical Plant Storage Requirements

A chemical plant uses two liquid feeds of different densities;

Table 30.3 Feed Data, Project 30.3

Feed 1:	110,000 lb/day	$\rho = 49 \text{ lb/ft}^3$
Feed 2:	50,000 lb/day	$\rho = 68 \text{ lb/ft}^3$

produces four different liquid chemical products of varying density. Refer to Table 30.3. and 30.4

Table 30.4 Product Data, Term project 30.3

Product A:	40,000 lb/day	$\rho = 52$ lb/ft^3
Product B:	25,000 lb/day	$\rho = 62$ lb/ft^3
Product C:	10,000 lb/day	$\rho = 52$ lb/ft^3
Product D:	95,000 lb/day	$\rho = 47$ lb/ft^3

The plants storage requirements call for maintaining 4 to 5 weeks supply of each feed, 4 to 6 weeks supply of products A, B, and C; and 1 to 2 weeks supply of product D. The plant operates year round, but each tank must be emptied once a year for a week of maintenance. Tanks are normally "dedicated" to one feed or product, but one or two could be used as "swing" tanks, with one day of cleaning required between use with different liquids.

Specify several efficient set of tanks from the "standard" sizes given below in Table 30.5 to meet this plant's needs. Storage requirements are provided in Table 30.6. Select which set you consider to be most efficient and explain why. The number of tanks required will be quite large. If market forces, such as fluctuating demand, require this much storage, they may all be necessary. More modern commercial operations, such as "just-in-time" manufacturing, call for reducing in-plant inventory to the absolute minimum possible [3].

Comment: There is no single, simple method for determining the optimum mix of storage tanks for a chemical plant. Most often, estimates are made of the minimum and maximum amounts of feeds, intermediates, and products that must be kept on hand. Then some additional allowance is made to permit periodic cleaning and maintenance of the tanks. The minimum number of tanks may not always be optimum if the tanks are extremely large. Several smaller tanks may cost somewhat more initially but they offer more flexibility in use [4].

Table 30.5 Tank Sizes; Term Project 30.3

Standard Tank Sizes, Gallons		
2,800	16,800	281,000
5,600	28,100	561,000
8,400	56,100	1,123,000
11,200	140,000	

Table 30.6 Storage Requirements; Term Project 30.3

Material	lb/day	Gal/day	Days	Gals storage required
Feed 1	110,000	16,800	28 – 35	470,200 – 587,800
Feed 2	50,000	5500	28 – 35	154,000 – 192,500
Product A	40,000	5750	28 – 42	161,100 – 241,700
Product B	25,000	3020	28 – 42	84,500 – 126,700
Product C	10,000	1440	28 – 42	40,300 – 60,400
Product D	95,000	15,120	7 – 14	105,800 – 211,700

Term Project 30.4

Inside Battery Limits (ISBL) and Process Flow Approach

The physical layout of chemical and petrochemical plants may appear to be complex at first glance, but it generally follows a very logical pattern. Process units within a major chemical plant complex each deal with one or two major products. The individual pieces of equipment involved in their manufacture—reactors, columns, heat exchangers, compressors, etc.—are grouped together and operated from a control room located either within the unit or nearby. These individual process units are often referred to as *inside battery limits*, or ISBL, plants. Separate support units are referred to as *outside battery units*, or OSBL. Steam plants, cooling towers, electrical substations, storage facilities, shipping facilities, and waste treatment plants are considered OSBL. They are usually operated from different control rooms than the ISBL plants.

Within the ISBL units themselves, there are two basic approaches to plant layout: the *process flow* approach and the *common equipment* approach. In the process flow approach, equipment is arranged in the same order as it would be on a well-planned flowsheet, i.e., in the order of the main material flows in the plant. This is similar to the common "assembly line" approach in many factory installations. In the common equipment method, all pieces of similar equipment are grouped together. Pumps are all adjacent to one another, heat exchangers are grouped together, etc.

The process flow approach has two main advantages. First, piping lengths from one piece of equipment to the next are minimized. This saves both in capital cost for piping and in energy cost to move materials from one piece of equipment to the next. Second, it is an easier plant to "learn". Engineers, operators and maintenance staff can learn the function of each piece of equipment faster.

The common equipment approach also has advantages. Piping and wiring from utility services is minimized. Initial construction may be simpler because similar pieces of equipment can be installed at the same time by the same crews. Maintenance may be simplified, particularly for major plant overhauls. Pump maintenance crews, for instance, can do all their work in one place rather than have to move from pump to pump.

Most real process plants are laid out in a combination of the two basic approaches. The process flow approach seems to predominate, but features of each can be found in almost any large plant.

Regardless of the layout approach used, each plant layout design represents a compromise involving cost, convenience, and safety. The least expensive plant, both to build and to operate, would be one with the absolute minimum of space between pieces of equipment and between various process units. The safest plant would be one in which equipment and process units are very widely separated, so that there is no chance of a fire or explosion at one piece of equipment spreading to another. Convenience, both in terms of construction and operation, usually calls for something in between in terms of equipment spacing, but it does not necessarily lead to a safer or less expensive plant layout [5].

Based on the above descriptions, prepare an ISBL process plant layout for the plant sketches below in Figure 30.1 using the process flow approach [5].

Sizes are as follows (See also Table 30.7):

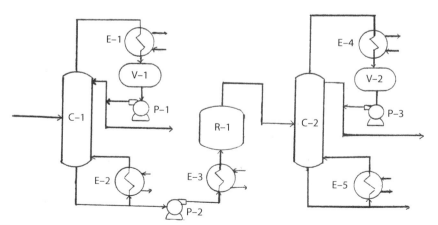

Figure 30.1 ISBL process plant layout

Table 30.7 Equipment Description; Term Project 30.4

Column 1	7 ft diameter x 83 ft tall, vertical cylinder
Exchanger 1	3 ft diameter x 9 ft long, horizontal cylinder
Exchanger 2	4 ft diameter x 8 ft long, horizontal cylinder
Vessel 1	4 ft diameter x 10 ft long, horizontal vessel
Pump 1	Approximately 2 ft x 2 ft x 4 ft
Pump 2	Approximately 2 ft x 2 ft x 4 ft
Exchanger 3	2 ft diameter x 8 ft long, horizontal cylinder
Reactor 1	8 ft diameter x 18 ft tall, vertical cylinder
Column 2	5 ft diameter x 70 ft tall, vertical cylinder
Exchanger 4	3 ft diameter x 7 ft long, horizontal cylinder
Exchanger 5	2 ft diameter x 6 ft long, horizontal cylinder
Vessel 2	3 ft diameter x 8 ft long, horizontal vessel
Pump 3	Approximately 1.5 ft x 1.5 ft x 3 ft

Vessels 1 and 2 are located underneath exchangers 2 and 5, respectively, to permit gravity flow into the vessels. Otherwise all equipment is located at ground level, or just high enough above ground level for maintenance access. Common rules-of-thumb for spacing between equipment, based on both safety and maintenance convenience considerations, call for the following minimum clearances around equipment (see Table 30.8). Note

Table 30.8 Equipment Spacing; Term Project 30.4

Columns	6 ft or one column diameter, whichever is less
Exchangers	5 ft
Pumps	4 ft
Vessels	3 ft
Reactors	12 ft

that these are typical spacing; they are not valid for all types of process plants. Much closer spacing and more vertical stacking of equipment is common in places where available land is limited. Such is the case in many large European chemical manufacturing locations.

References

1. D. Kauffman, *Process Synthesis and Design*, A Theodore Tutorial, Theodore Tutorials, East Williston, NY, originally published by the USEPA/APTI, RTP, NC, 1992.
2. D. Kauffman, *Process Synthesis and Design*, A Theodore Tutorial, Theodore Tutorials, East Williston, NY, originally published by the USEPA/APTI, RTP, NC, 1992.
3. D. Kauffman, *Process Synthesis and Design*, A Theodore Tutorial, Theodore Tutorials, East Williston, NY, originally published by the USEPA/APTI, RTP, NC, 1992.
4. J. Reynolds, J. Jeris, and L. Theodore, *Handbook of Chemical and Environmental Engineering Calculations,* John Wiley & Sons, Hoboken, NJ, 2004.
5. D. Kauffman, *Process Synthesis and Design*, A Theodore Tutorial, Theodore Tutorials, East Williston, NY, originally published by the USEPA/APTI, RTP, NC, 1992.

31

Term Projects (4): Environmental Management

31.1 Dissolve the USEPA
31.2 Solving Your Town's Sludge Problem
31.3 Benzene Underground Storage Leak
31.4 An Improved MSDS Sheet

Term Project 31.1

Dissolve The USEPA

One of the authors of this book has offered the following comments regarding the US Environmental Protection Agency (USEPA) [1].

The problems associated with the regulatory framework of the federal environmental management program have always been questioned. As with any government-controlled operation, many steps must often be taken before anything meaningful can be accomplished (this appears to apply to many activities, with the exception of war, where the president can exclusively command the armed forces for immediate action).

To implement an environmental regulation, the problem must be first identified (often in an EPA report), then data must be collected and analyzed (usually in another EPA report), and a goal has to be set, ultimately by congressional legislation. Once the law is in effect, it must be enforced by the EPA. The law has often been amended because of unreasonable goals and lax enforcement.

The present problem that exists with the EPA is an intricate one, consisting of primarily four main concerns:

1. Economically efficient measures are seldom, if ever, adopted, causing little progress in achieving environmental goals.
2. Data collection often has limitations, and when insufficient data is used for legislation, an ongoing string of amendments is attached.
3. The legal issues involving environmental problems have rocketed, brought on mainly by the complex legislation.
4. The EPA is presently primarily a legal organization that is serving the best interests of the law profession rather than the environment.

Consumer and political interest movements led earlier by Ralph Nader and growing groups of engineers, scientists, and other so called environmental experts, including some lawyers, influenced many of the new initiatives on the environmental legislation agenda. Events of the later 1960s, such as the oil burning on the Cuyahoga River in the center of Cleveland and the washing up of dead birds on the oil-slicked shores of Santa Barbara, reflected a sense of crisis and dissatisfaction within society.

The EPA was formed by bringing together 15 components from 5 executive departments and independent agencies. Air pollution control, solid waste management, radiation control, and the drinking water program were transferred from the Department of Health, Education, and Welfare (now the Department of Health and Human Services). The federal water pollution control program was taken from the Department of the Interior, as was part of a pesticide research program. EPA acquired authority to register pesticides and to regulate their use from the Department of Agriculture, and inherited the responsibility to set tolerance levels for pesticides in food from the Food and Drug Administration. EPA was assigned some responsibility for setting environmental radiation protection standards from the Atomic Energy Commission, and absorbed the duties of the Federal Radiation Council. Unfortunately, these groups were, and today essentially remain, compartmentalized [1]. The EPA was set up where each office dealt with a specific problem, and new offices were often created sequentially as individual environmental problems were identified and responded to by legislation.

The EPA's first administrator, William Ruckelshaus, initially sought to convey the impression that his agency would aggressively enforce the new policies, and adopted a systems approach by forming two primary program offices to handle the variety of issue areas and legislative mandates under its jurisdiction. Several function-oriented divisions were designed to be more responsive to White House concerns, as well as fulfill certain agency wide objectives, such as enforcement and research. The new agency, however, was quickly overwhelmed by its rapidly expanding regulatory responsibilities, the conflicting signals from the Nixon, and later Ford Administrations on how aggressively it should pursue such regulations, and effective industry maneuvering, which used scientific uncertainty in the regulation process to delay or counter the establishment and enforcement of standards [1].

A major criticism of the present regulatory approach to solving environmental problems (and pollution) is its economic inefficiency. The EPA's Annual Performance Plan and Congressional Justification request budget for 2013 is approximately $10 billion in discretionary budget authority and nearly 29,000 Full Time Employees (FTE) [2].

The problems of the environment need to be examined from an engineering perspective,. If an environmental concern arises, passing regulations before a good scientific basis and peer review are achieved can result in enormous expenditures in legalities, something that this country is presently burdened with. When environmental legislation is passed, it is often so ambiguous that an array of lawyers is needed to translate them. The main reason for this problem is that amendments are made based on premature or simply ill-defined findings. As mentioned previously, scientific data is not always featured predominantly when politics and emotion flare.

Complicated legislation passed based on insufficient data is by no means a solution to the environmental problem. Costly control measures are taken, and in some cases, the public's risk is increased. Constant amendments are needed, often doing little to alleviate problems. As noted, regulations can only help if they are based on sound scientific data. When the legislation is unclear, lawyers are often brought in to "clarify" it. Instead, they usually complicate the problems further.

The predictable bureaucratic tendency which feeds on the professional ambitions of "dedicated" staff and inevitably generates calls for larger budgets, is reinforced by the high costs of litigation and the long delays associated with the process. This centralizing effect feeds the political machinery to Congress. EPA is the whipping boy, never meeting the impossible deadlines and not doing enough to satisfy the politicians. Industry is the villain, and the flaming emotions of innocent people are fanned by the rhetoric that ensues. Heating hearings, more proposed laws, larger budgets, more lawyers, and limited progress is the result. Political demand continues to outstrip political supply [3].

When the EPA was formed in 1970, it was—in a very real sense—a technical organization. The Agency was manned primarily with engineers and scientists. Most of these individuals were dedicated to a common cause: correcting the environmental problems facing the nation and improving the environment. The problems these individuals tackled were technical, and there were little or no legal complications or constraints. The EPA was indeed a technical organization, run and operated by technical people, attempting to solve technical problems. Much was accomplished during these early years ... but something happened on the way to the forum [4].

Over 40 years later, the EPA is no longer a technical organization—it is now a legal organization. The EPA is no longer run by engineers and scientists. It is run and operated by lawyers. And, the EPA is no longer attempting to solve technical problem; it is now stalled in a legal malaise [5].

How in the world did this occur? It happened because it served the best interest of the career bureaucrats, in and out of Congress, most of whom are lawyers, and it happened because the technical community did nothing to stop it. The result is that this nation is now paying the price for an environmental organization with 20,000 employees and a monstrous annual budget that is not serving the best interests of either the nation or the environment [5].

Interestingly, all of the administrators to the EPA have been lawyers. Though lawyers are required in every industry for helping to settle disputes over legalities, protecting the environment is generally beyond their scope. In the EPA today, for every three engineers there is one lawyer; it is indeed (as described above) a legal organization, serving the legal profession and not the environment. Actual proposals for regulations and control, based on good scientific data, should be designed by scientists and engineers, or those who have come to be defined as problem solvers. They can analytically break down a problem, initially assess the damages, then fix them [4].

Creating problems and not solving them has become the mode of operation for the EPA. One need only look at Superfund (see earlier discussion in Part II Chapter 14 and 15) for an example of what the professional bureaucrats have accomplished. When one talks about wasting tax dollars, Superfund is at the top of the list, with nearly $10 billion down the drain.

Something has gone afoul. In this society, engineers are the problem solvers, but rarely the decision makers. Although the world known today has been called a product of engineering, engineers play a minor role in important decision making.

The environmental problem is one that developed over many years of civilization by many different sources. To think that the EPA, with its present mode of operation, can solve this problem is ludicrous. However, there is a solution. *Dissolve the EPA now*! No reorganization will work, since the lawyers and career bureaucrats have a stronghold in the Agency with their ties to Congress and the White House. What is needed is to make the present EPA disappear and start anew. The nation needs an environmental administration that will solve, not create problems [4]. The nation needs technically competent people who can lead an organization in making cost-effective decisions based on the public well-being, not on politicians whose goal is to get reelected or lawyers who cost the nation billions of dollars annually proposing and enforcing ill-defined legislation.

Based on the above comments, draft and propose a bill to Congress that would accomplish the following:

1. Dissolve the present EPA.
2. Form another environmental organization with another name (of your choice) that will be directed to serve the best interests of the environment, society, and the nation.

Term Project 31.2

Solving Your Town's Sludge Problem

Most waste water treatment plants use primary sedimentation to remove readily settleable solids from raw wastewater. In a typical plant, the dry weight of *primary sludge* solids (those removed by filtration, settling or other physical means) is roughly 50% of that for the total sludge solids. Primary sludge is usually easier to manage than *biological and chemical sludges*—which are produced in the advanced or secondary stages of treatment—for several reasons. First, primary sludge is readily thickened by gravity, either within a primary sedimentation tank or within a separate gravity thickener. In comparison with many biological and chemical sludges, primary sludge with low requirements can rapidly be mechanically dewatered. Furthermore, the dewatering device will produce a drier cake and give better solids capture than it would for most biological and chemical sludges.

Primary sludge always contains some grit, even when the wastewater has been processed through degritting. Typically, it also contains different anaerobic and facultative species of bacteria, such as sulfate-reducing and oxidizing bacteria. Primary sludge production is typically within the range of 800 – 2500 lbs per million gallons (100 – 300 mg/L) of wastewater. A basic approach to estimating primary sludge production for a particular plant is to compute the quantity of total suspended solids (TSS) entering the primary sedimentation tanks.

Biological sludges are produced by secondary treatment processes such as activated sludge, trickling filters, and rotating biological contactors. Quantities and characteristics of biological sludges vary with the metabolic and growth rates of the various microorganisms present in the sludge. Biological sludge that contains debris such as grit, plastics, paper, and fibers is produced at plants lacking primary treatment. Plants with primary sedimentation normally produce a fairly pure biological sludge. Biological sludges are generally more difficult to thicken and dewater than are primary sludge and most chemical sludges.

Ensuring the safe disposal of municipal sludge and other residues, such as grits, and skimmings, is considered an integral part of good planning, design, and management of municipal wastewater treatment facilities. Acceptable sludge disposal practices include conversion processes such as: incineration; wet oxidation; pyrolysis and composting, and land disposal by *land application* and *landfilling*.

Landfilling is probably the most popular disposal method and is generally used on wastes in the form of sludges. There are two types of landfilling: *area fill* and *trenching*. Area fill is essentially accomplished above ground, whereas trenching involves burying the waste. Trenching is the better-established and more popular form of the two. Yet, since trenching requires excavation, area fill has the advantage that is requires less manpower and machinery. Area fill is also less likely to contaminate groundwater since the filling is above ground. Trenching, however, may be used for both stabilized and unstabilized sludges and makes more efficient use of the land. Both techniques require the use of lime and other chemicals to control odors, and cold and wet weather can cause problems with either. Both methods also produce gas, which can cause explosions or harm vegetation, and leachate, which can contaminate ground and surface water [6].

Most wastes must be subjected to one or more pretreatments such as solidification, degradation, volume reduction, and detoxification before being landfilled. This practice stabilizes the waste and helps decrease the amount of gas and leachate produced from the landfill. Landfilling is similar to *landfarming* in that both ultimate disposal methods combine wastes and soil. Landfarming, as described above, involves the biochemical reaction between solid nutrients and wastes to degrade and stabilize the waste; as a result, only specific types of wastes can be landfarmed. A larger variety of waste may be handled by landfilling [6].

Determine the sludge disposal practices of your home town, village, or city. Write a report describing your findings. The report should include a discussion of the following items [7]:

1. The quantity of sludge produced as well as seasonal variations, if any.
2. On-site temporary storage. If the entity does provide for temporary storage describe the capacity, whether it is covered or not, and any management techniques that are utilized to control drainage from the sludge storage areas, used to treat the drainage, and used to control odor problems. Also discuss whether the capacity varies with seasons.
3. Discuss sludge disposal options. Where is the sludge disposed of and how? What quantity of sludge is disposed of? What regulations control its disposal? What are the costs of disposal?

Term Project 31.3

Benzene Underground Storage Tank Leak

One topic not reviewed in any detail earlier is underground storage tanks (USTs). Environmental contamination from leaking USTs poses a significant threat to human health and the environment. These leaking USTs contaminate the nation's groundwater, which a major source of drinking water. Nationally, there are over 500,000 USTs. Originally placed underground as a fire prevention measure, these tanks have substantially reduced the damage from stored flammable liquids. However, underground tanks are thought to be leaking now, and many more will begin to leak in the near future. Products released from these leaking tanks can threaten groundwater supplies, damage sewer lines and buried cables, poison crops, and lead to fires and explosion [9].

The primary reason for regulating underground storage tanks is to protect water, especially groundwater that is used for drinking water. This is one of the nation's greatest natural resources and one which is extremely difficult to remediate once it is contaminated. Approximately fifty percent of the U.S. population depends on groundwater for drinking water. Rural areas would be seriously affected if their groundwater were contaminated since it provides 95% of their total water supplies. Groundwater drawn for large-scale agricultural and industrial uses also can be adversely affected by contamination from leaking underground tanks [10].

Owners and operators of petroleum and hazardous substance UST systems must respond to a leak or spill within 24 hours of release or within another reasonable period of time as determined by the implementing agency. The responses to releases from USTs depend on several different factors, most of which is site-specific. Owners and operators can comply with the financial responsibility requirements in a number of ways that can include: self-insurance (which requires a financial test), guarantees, insurance and risk-retention group coverage, surety bonds, letter of credit, use of state-required mechanisms, state funds, or other state assurances, trust funds, and standby trust funds. Owners and operators can use a single means or a combination of methods to satisfy the required coverage of financial requirements [9].

A total of 400 L of pure benzene leaks from an underground storage tank before the leak is discovered. The water table lies a few feet below the tank. Discuss the following items related to this release [11]:

1. What is the possibility of recovering some of the pure product benzene, and how might this product recovery be accomplished.
2. What is the maximum benzene concentration expected in the groundwater?
3. What is the dissolved benzene retardation factor assuming that the soil organic carbon fraction (f_{oc}) = 0.5%?
4. What is the rate of biodegradation expected for this benzene?
5. Propose your solution to this problem.

Comment: The retardation factor R is given by

$$R = 1 + f_{oc} (K_{oc}) (\rho_B)/n$$

where f_{oc} = the fraction of organic carbon in the soil
   K_{oc} = the organic carbon normalized soil/water partition coefficient
   ρ_B = the bulk density of the aquifer solids
   n = the aquifer solid total porosity.
For benzene, K_{oc} is reasonably approximated by its octanol/water partition coefficient, $K_{ow} \approx 100$ (mL water/g octanol). Typical values of ρ_B and n for aquifer solids are 2 g/mL and 0.3, respectively.

Term Project 31.4

An Improved MSDS Sheet

As noted in the Overview in Part II, Chapter 15, the following information is generally provided on a typical MSDS sheet [12–14].

1. Product or chemical identity used on the label
2. Manufacturer's name and address
3. Chemical and common names of each hazardous ingredient
4. Name, address, and phone number for hazard and emergency information
5. The hazardous chemical's physical and chemical characteristics, such as vapor pressure and flashpoint
6. Physical hazards, including potential for fire, explosion, and reactivity
7. Known health hazard
8. Exposure limits
9. Emergency and first-aid procedure
10. Toxicological information
11. Precautions for safe handling and use
12. Control measures such as engineering controls, work practices, hygienic practices or personal protective equipment required
13. Procedures for spills, leaks, and clean-up

Develop an outline for a new and improved MSDS sheet. Provide specific details and information. In effect, improve on the above MSDS writeup.

References

1. R. Gottlieb, *Forcing the Spring*, Island Press, Washington, DC, 1993.
2. U.S. EPA. 2008 EPA Budget in Brief, U.S. EPA, Office of the Chief Financial Officer, Washington D.C, www.epa.gov/ocfo
3. B. Yandle, *The Political Limitations of the Environmental Regulation*, Quorum Books, New York City, NY, 1989.
4. L. Theodore, Dissolve the USEPA...NOW, EM, Pittsburgh, 1995.
5. P. Samuel and P. Spencer, Facts Catch Up with 'Political Science.' *Consumers' Research*, 10 – 15, location unknown May 1993.
6. Adapted from: R. Dupont, L. Theodore, and K. Ganesan, *Pollution Prevention: The Waste Management Approach for the 21ˢᵗ Century*, CRC Press/Taylor & Francis Group, Boca Raton, FL, 2000.
7. L. Theodore, R. Dupont and J Reynolds, *Pollution Prevention*, Gordon and Breach Science Publishers, New York City, NY, originally published by the USEPA/APTI, RTP, NC, 1994.
8. R. Dupont, T. Baxter, and L. Theodore, *Environmental Management: Problems and Solutions*, CRC Press/Taylor & Francis Group, Boca Raton, FL, 1998.
9. M.K. Theodore and L. Theodore, *Introduction to Environmental Management*, CRC Press/ Taylor & Francis Group, Boca Raton, FL, 2010.
10. G. Burke, B. Singh, and L. Theodore, *Handbook of Environmental Management and Technology*, 2ⁿᵈ edition, John Wiley & Sons, Hoboken, NJ, 2005.
11. R. Dupont, T. Baxter, and L. Theodore, *Environmental Management: Problems and Solutions*, CRC Press/Taylor & Francis Group, Boca Raton, FL, 1998.
12. Adapted from: L. Stander and L. Theodore, *Environmental Regulatory Calculations Handbook*, CRC Press/Taylor & Francis Group, Boca Raton, FL, 2008.
13. L. Theodore and R. Dupont, *Environmental Health and Hazard Risk Assessment: Principles and Calculations*, CRC Press/Taylor & Francis Group, Boca Raton, FL, 2012.
14. Adapted from: L. Theodore, *Chemical Engineering: The Essential Reference*, McGraw-Hill, New York City, NY, 2014.

32

Term Projects (4): Health and Hazard Risk Assessment

32.1 Nuclear Waste Management
32.2 An Improved Risk Management Program
32.3 Bridge Rail Accident: Fault and Event Tree Analysis
32.4 HAZOP: Tank Car Loading Facility

Term Project 32.1

Nuclear Waste Management

As with many other types of waste disposal, radioactive waste disposal is no longer a function of technical feasibility but also a question of social or political acceptability. The present placement of facilities for the permanent disposal of municipal solid waste, hazardous chemical waste, and nuclear wastes alike has become an increasingly large part of waste management. Today, a large percentage of the money required to build a radioactive waste facility will be spent on the siting and licensing of the facility [1].

Nuclear or radioactive waste can be loosely defined as something that is no longer useful and that contains radioactive isotopes in varying concentration and forms. Radioactive waste is then further broken down into categories that describe waste by activity, by generation process, by molecular weight, and by volume [1].

Radioactive isotopes emit energy as they decay to more stable elements. The energy is emitted in various forms, including alpha particles, beta particles, neutrons, and gamma rays. The amount of energy that a particular radioactive isotope emits, the time frame over which it emits that energy, and the type of contact with humans, all help determine the hazard it poses to the environment. The major categories of radioactive waste that exist are high-level waste (HLW), low-level waste (LLW), transuranic waste (TRU), uranium mine and mill tailings, mixed wastes, and natural occurring radioactive materials (NORMs) [1,2].

The physical form of the waste is a critical factor in determining the probability that the waste will remain isolated from the biosphere. Many treatment processes can be employed to reduce the volume or increase the stability of waste that must ultimately be permanently disposed. Landfill fees for radioactive waste is assessed largely on the volume of the waste to be disposed. Current trends in the rising cost of waste disposal have led to the generators' implementing one or a number of waste minimization techniques.

You have been hired to develop an improved method of either treatment or disposal of nuclear waste, or both. Factors to be included in your analysis are:

1. Political concerns
2. Societal concerns
3. Economics
4. Environmental Concerns
5. Health and hazard risks [3]

Term Project 32.2

An Improved Risk Management Program

Developed under the *Clean Air Act's* (CAA's) Section 112(r), the *Risk Management Program* (RMP) rule (40 CFR Part 68) is designed to reduce the risk of accidental releases of acutely toxic, flammable, and explosive substances. A list of the regulated substances (138 chemicals) along with their threshold quantities is provided in the Code of Federal Regulations at 40 CFR 68.130.

In the RMP rule, EPA requires a *Risk Management Plan* that summarizes how a facility is to comply with EPA's RMP requirements. It details methods and results of hazard assessment, accident prevention, and emergency response programs instituted at the facility. The hazard assessment shows the area surrounding the facility and the population potentially affected by accidental releases.

EPA requirements include a three-tiered approach for affected facilities. A facility is affected if a process manufactures, processes, uses, stores, or otherwise handles any of the listed chemicals at or above the threshold quantities. The EPA approach is summarized in Table 32.1. For example, EPA defined Program 1 facilities as those processes that have not had an accidental release with offsite consequences in the five year prior to the submission date of the RMP and have no public receptors within the distance to a specified toxic or flammable endpoint associated with a worst-case release scenario. Program 1 facilities have to develop and submit a RMP and complete a registration that includes all processes that have a regulated substance present in more than a threshold quantity. They also have to: analyze the worst-case release scenario for the process or processes; document that the nearest public receptor is beyond the distance to a toxic or flammable endpoint; complete a five-year accident history for the process or processes; ensure that response actions are coordinated with local emergency planning and response agencies; and, certify that the source's worst-case release would not reach the nearest public receptors. Program 2 applies to facilities that are not Program 1 or Program 3 facilities. Program 2 facilities have to develop and submit the RMP as required for Program 1 facilities plus: develop and implement a management system; conduct a hazard assessment; implement certain prevention steps; develop and implement an emergency response program; and, submit data on prevention program elements for Program 2 processes. Program 3 applies to processes in Standard Industrial Classification (SIC) codes 2611 (pulp mills), 2812 (chloralkali), 2819 (industrial inorganics),

2821 (plastics and resins), 2865 (cyclic crudes), 2869 (industrial organics), 2873 (nitrogen fertilizers), 2879 (agricultural chemicals), and 2911 (petroleum refineries). These facilities belong to industrial categories identified by EPA as historically accounting for most industrial accidents resulting in off-site risk. Program 3 *also* (and this is important) applies to all processes subject to the OSHA Process Safety Management (PSM) standard (29 CFR 1910.119). Program 3 facilities have to develop and submit the RMP as required for Program 1 facilities plus: develop and implement a management system; conduct a hazard assessment; implement prevention requirements; develop and implement an emergency response program; and provide data on prevention program elements for the Program 3 processes [4].

As a recently assigned intern (to the health and hazard risk assessment group) at Region II EPA headquarters in New York City, NY, you have been requested by your immediate supervisor to develop an improved RMP for the Agency.

Comment: There are many State agencies that have developed RMPs (or the equivalent) that may be valuable in completing this project.

Table 32.1 RMP Approach [1]

Program	Description
1	Facilities submit RMP; complete registration of processes; analyze worst-case release scenarios; complete 5-year accident history; coordinate with local emergency planning and response agencies; and, certify that the source's worse-case release would not reach the nearest public receptors.
2	Facilities submit RMP; complete registration of processes; develop and implement a management system; conduct a hazard assessment; implement certain prevention steps; develop and implement an emergency response program; and, submit data on prevention program elements.
3	Facilities submit RMP; complete registration of processes; develop and implement a management system; conduct a hazard assessment; implement prevention requirements; develop and implement an emergency response program; and, provide data on prevention program elements.

Term Project 32.3

Bridge Rail Accident: Fault and Event Tree Analysis

As noted earlier in Part II-Chapter 19, a *fault tree* is a graphical technique used to analyze complex systems. The objective is to spotlight faulty conditions that can cause a system to fail. Fault tree analysis attempts to describe how and why an accident or other undesirable events have occurred. It may also be used to describe how and why an accident or other undesirable event could take place. A fault tree analysis also finds wide application in environmental management as it applies to hazard analysis and risk assessment of process and plant systems [5].

Fault tree analysis seeks to relate the occurrence of an undesired event, the so-called *top event*, to one or more antecedent events, called *basic events*. The top event may be, and usually is, related to the basic events via certain intermediate events. A fault tree diagram exhibits the casual chain linking the basic events to the intermediate events and the latter to the top event. In this chain, the logical connection between events is indicated by so-called *logic gates*. The principal logic gates are the "AND" gate, symbolized on the fault tree by a semi-circle shape, and the "OR" gate symbolized by a square bottom/triangle top shape [6].

An *event tree* provides a diagrammatic representation of event sequences that begin with a so-called initiating event and terminate in one or more undesirable consequences. In contrast to a fault tree, which works backward from an undesirable consequence to possible causes, an event tree works forward from the initiating event to possible undesirable consequences. The initiating event may be equipment failure, human error, power failure, or some other event that has the potential for adversely affecting an ongoing process or environment.

Summarizing, a fault tree works backward from an undesirable event or ultimate consequence to the possible causes and failures. Alternatively, an event tree works forward from an initial event, or an event that has the potential for adversely affecting an ongoing process, and ends at one or more undesirable consequences.

A natural gas pipeline is attached to a combined highway/railway bridge. You have been asked to investigate the risk of a fire or explosion at the bridge. Either a rail accident or an earthquake can rupture the natural gas line. Truck accidents on the bridge will not rupture the line. After a considerable amount of effort your staff has produced the following database [7].

1. The probability of rail accident at this bridge considering traffic, maintenance, and human error is 6 x 10⁻⁶/yr.
2. Pipeline rupture will occur in 60% of the rail accidents which are designated as "major".
3. The probability of an earthquake causing the pipeline to rupture is 1.5 x 10⁻⁸/yr.
4. The pipeline is inerted with nitrogen for maintenance work, one percent of the time.
5. Calm wind conditions and those with a westerly component will cause ignition from a rail accident 70% of the time. The wind will cause highway traffic to be the ignition source, thirty percent of the time.

For your calculations consider all the above events to be independent [7]. Your assignment is as follows:

1. Construct a fault tree
2. Find the cut sets
3. Identify the minimum cut sets
4. Calculate the probability of an earthquake causing a natural gas fire/explosion.
5. Calculate the probability of a rail accident causing a natural gas fire/explosion.
6. Based on your results how does the likelihood of the rail accident causing the fire/explosion compare to that of the earthquake?
7. Outline what steps can be taken to reduce the probability of an "accident".

Term Project 32.4

HAZOP: Tank Car Loading Facility

A. Hazard and Operability Study (HAZOP) is a structured, systematic "what if" analysis of a process, normally carried out by a team of five to ten people with expertise in various aspects of process design, operations, and safety. Combinations of *guide words* and process elements are examined. The guide words most commonly used are:

No or not
More

Less
As well as
Part of
Reverse
Other than

For each combination, the HAZOP team determines the possibility of deviations in flow, temperature, pressure, composition, etc. The team first determines whether the deviation is physically possible in the system. Then it determines what possible combination of circumstances would cause the deviation and what the consequences would be. Finally, the team recommends action to avoid any deviation which could lead to dangerous consequences. A recorder (secretary), takes careful notes of the discussion and recommendations on each combination of guide words and process step [8]

Carry out a HAZOP analysis of the tank car loading facility illustrated below in Figure 32.1. It is used for filling cars with sulfuric acid. Pertinent data follows [9].

When a tank car is lined up and connected, the feed pump is started by the operator. The pump shuts off automatically after a predetermined amount of sulfuric acid has been pumped. A low-level switch on the tank will shut off the pump if the tank inventory is too low.

An underline in an instrument bubble (e.g., on the LALL) in the figure indicates the instrument is in the control room. Otherwise instruments are located close to the equipment. Instrumentation symbols in the figure are:

1. FAL   Flow alarm-low
2. FIC   Flow indicator-controller
3. HS   Hand operated electrical switch
4. LAH   Level alarm-high
5. LAL   Level alarm-low
6. LALL   Level alarm-low,low
7. LI   Level indicator
8. LS   Electrical switch operated by tank-level signal
9. PAH   Pressure alarm, high
10. PAL   Pressure alarm, low
11. PI   Pressure indicator
12. TAH   Temperature alarm, high
13. TAL   Temperature alarm, low
14. S   Signal integrator

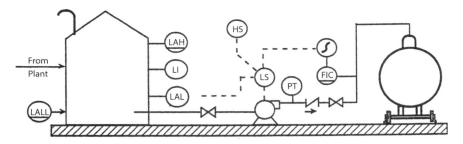

Figure 32.1 Tank car loading schematic

References

1. Adapted from: M.K. Theodore and L. Theodore, *Introduction to Environmental Management*, CRC Press/ Taylor & Francis Group, Boca Raton, FL, 2010.
2. G. Burke, B. Singh, and L. Theodore, *Handbook of Environmental Management and Technology*, 2nd edition, John Wiley & Sons, Hoboken, NJ, 2000.
3. L. Theodore and R. Dupont, *Environmental Health and Hazard Risk Assessment: Principles and Calculations*, CRC Press/Taylor & Francis Group, Boca Raton, FL, 2012.
4. Adapted from: L. Stander and L. Theodore, *Environmental Regulatory Calculations Handbook*, John Wiley & Sons, Hoboken, NJ, 2010.
5. Adapted from: L. Theodore, *Chemical Engineering: The Essential Reference*, McGraw-Hill, New York City, NY, 2014.
6. L. Theodore and R. Dupont, *Environmental Health and Hazard Risk Assessment: Principles and Calculations*, CRC Press/Taylor & Francis Group, Boca Raton, FL, 2012.
7. S. Shaefer and L. Theodore, *Probability and Statistics Applications in Environmental Science,* CRC Press/ Taylor & Francis Group, Boca Raton, FL, 2007.
8. L. Theodore and R. Dupont, *Environmental Health and Hazard Risk Assessment: Principles and Calculations*, CRC Press/Taylor & Francis Group, Boca Raton, FL, 2012.
9. D. Kauffman, *Process Synthesis and Design*, A Theodore Tutorial, Theodore Tutorials, East Williston, NY, originally published by the USEPA/APTI, RTP, NC, 1992.

33

Term Projects (3):
Unit Operations Laboratory
Design Projects

33.1 Hand Pump
33.2 Rooftop Garden Bed
33.3 Hydration Station Counter

Note: The three term projects presented were assigned in the undergraduate unit operations course at Manhattan college. A portion of their project grade is "Peformance Criteria", which evaluates how well their device meets specifications. The rubric for these criteria are included.

Term Project 33.1

Hand Pump

Design and build a hand pump that will transport 4 gallons of water from a bucket on the floor to an empty bucket at the top of a table (a height of approximately 29 inches). A schematic depicting the process is provided in Figure 33.1. Note: You will be connecting the pump to standard 5/8" garden hoses, so be sure to incorporate the appropriate fittings for the design.

The important performance variables are: (1) flow rate, (2) weight, and (3) cost. Your pump will be evaluated with these variables, and its performance in optimizing these variables will comprise 40% of the total grade as seen in Table 33.1.

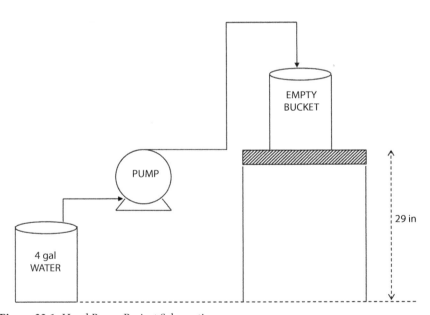

Figure 33.1 Hand Pump Project Schematic

Table 33.1 Hand Pump Project Performance Criteria

Flow Rate	
Flow Rate	**Grade**
> 1.2 gal/min	22/20
0.8 to 1.2 gal/min	20/20
0.6 to 0.8 gal/min	15/20
< 0.6 gal/min	10/20

Weight	
Weight	**Grade**
<2 lbs	12/10
2-4 lbs	10/10
2-6 lbs	8/10
>6 lbs	6/10

Cost	
Amount Spent	**Grade**
Less than 25 USD	10/10
Between 25 and 30 USD	8/10
Between 30 and 40 USD	6/10
Between 40 and 50 USD	4/10

Term Project 33.2

Rooftop Garden Bed

Design and build a miniature garden bed that is capable of supporting vegetable growth, and has a mechanism to collect, store, and deliver water. The bed must have an area of 1ft x 1ft, uses soil (or an alternative medium to support roots) and will ultimately be situated on a building rooftop. The minimum height must be 8 inches, and the maximum 12 inches. Accompanying the bed should be a mechanism to collect, store, and deliver water through active (i.e. requiring external energy) or passive (i.e. requiring no external

Table 33.2 Rooftop Garden Bed Project Performance Criteria

Weight of Bed, Support Material, and Water Collection System (not filled)	
Weight	**Grade**
<10 lbs	12/10
10-20 lbs	10/10
20-30 lbs	8/10
>30 lbs	6/10

Cost	
Amount Spent	**Grade**
Less than 25 USD	10/10
Between 25 and 30 USD	8/10
Between 30 and 40 USD	6/10
Between 40 and 50 USD	4/10

Water Distribution	
Flow Rate	**Grade**
Active System	Standard
Passive System	+5

energy) means. The bed must be EVENLY SATURATED (i.e. wet everywhere), and NOT FLOOD (i.e. water must not puddle at the top). Keep in mind the LOCATION of this bed when considering your design (e.g. wind, days of excessive rainfall, etc.) Performance of your design will comprise 20% of the total grade, and is detailed in Table 33.2.

Term Project 33.3

Hydration Station Counter

Bottled water has environmental consequences from both a material and energy perspective. Additionally, there are economic effects due to the production and waste management. One solution is to encourage the use of *hydration stations.* These devices: filter municipal water, and dispense it to reusable bottles that consumers possess. These stations reduce the garbage disposal of plastic, glass, and aluminum containers by encouraging the refilling of plastic containers at the station. A secondary purpose of the station will be to encourage individuals to replace their intake of soda with water.

In this project, you will be designing and building a device that translates the amount of water dispensed by a hydration station into the number of plastic bottles saved.

The overall project consists of three components:

1. An initial statistical analysis where the bottled water consumption of the institution is examined.
2. Development of a plan that would encourage members of the institution to use the Hydration Station.
3. Construction of the Hydration Station metering device, data collection and data dissemination hardware, to allow all members of the institution to have easy access to the hydration station's impact on reducing the container trash produced, as well as the impact of the Station on the reduction of non-water fluid intake. For the purposes of this project (and for evaluation), assume that we will be connecting your device to a standard 5/8" garden hose.

The final system should:

1. Have zero (or close to zero) impact on the environment
2. Collect all necessary data (possibilities include: size and type of container, amount of water received, may want to read bar code on container so that know who the user is)
3. Display data locally on a screen adjacent to the station, remotely & wirelessly to selected locations around the campus (e.g., the video screens in the cafeterias), and on a public website.
4. Develop a way to access and communicate the impact of the Hydration Station(s) on the trash output of institution and on the caloric intake of its members.

Performance of your design will comprise 20% of the total grade, and is detailed in Table 33.3.

Table 33.3 Hydration Station Project Performance Criteria

Device measures correct volume passing through	
Volume	**Grade**
< 1ml difference from actual	12/10
1 – 2 ml difference from actual	10/10
2 – 3 ml difference from actual	8/10
>3 difference from actual	6/10

Cost	
Amount Spent	Grade
Less than 200 USD	10/10
Between 200 and 250 USD	8/10
Between 250 and 300 USD	6/10
Greater than 300 USD	4/10

Reference

1. J.P. Abulencia – Unit Operations Laboratory Projects (Spring 2011 to Spring 2013)

34

Term Projects (4): Miscellaneous Topics

34.1 Standardizing Project Management
34.2 Monte Carlo Simulation: Bus Section Failures in Electrostatic Precipitators
34.3 Hurricane and Flooding Concerns
34.4 Meteorites

Term Project 34.1

Standardizing Project Management

Three problems in the project management chapter (Part II, Chapter 13) were concerned with the attempt to "standardize" project management. The two key players and corresponding publications are ISO 21500:2012 *Guidance on Project Management*, and ANSI's *A Guide to the Project Management Body of Knowledge* (PMBOK).

Elam [1] recently compared ISO 21500:2012 with the 5th edition of PMBOK. His comparison is presented in Table 34.1.

Table 34.1 Comparison of ISO 21500:2012 and PMBOK

	ISO 21500:2012	PMBOK 5th Edition
Process Groups	1. Initiating 2. Planning 3. Executing 4. Controlling 5. Closing	1. Initiating 2. Planning 3. Executing 4. Monitoring & Controlling 5. Closing
Knowledge Areas for PMBOK Subject Groups for ISO	1. Integration 2. Scope 3. Time 4. Cost 5. Quality 6. Human Resource 7. Communications 8. Risk 9. Procurement 10. Stakeholder	1. Integration Management 2. Scope Management 3. Time Management 4. Cost Management 5. Quality Management 6. Human Resource Management 7. Communications Management 8. Risk Management 9. Procurement Management 10. Stakeholder Management

After carefully reviewing both documents, provide the methodology employed by both groups, highlighting the differences. Also, speculate on what changes can be expected in the future due to recent developments in terrorism, health and hazard risk assessment, nanotechnology, energy policies (or lack thereof), water policies (or lack thereof), etc.

Term Project 34.2

Monte Carlo Simulation: Bus Section Failures in Electrostatic Precipitators

Theodore et al. [2] employed Monte Carlo methods in conjunction with the binomial and Weibull distributions discussed in the Probability and Statistics Chapter (Part II, Chapter 19) to estimate out-of-compliance probabilities for electrostatic precipitators (air pollution control particulate control devices) on the basis of observed bus section failures. The following definitions apply (see Figure 34.1):

> *Chamber.* One of many passages (M) for gas flow
> *Field.* One of several high voltage sections (N) for the removal of particulates; these fields are arranged in series (i.e., the gas passes from the first field into the second, etc.)
> *Bus section.* A region of the precipitator that is independently energized; a given bus section can be identified by a specific chamber and field.

Thus, an *M* x *N* electrostatic precipitator consists of *M* chambers and *N* fields. A precipitator is "out of compliance" when its overall collection efficiency falls below a designated minimum value because of bus section failures. When several bus sections fail, the effect of failures depends on where they are located. To determine directly whether a precipitator is out

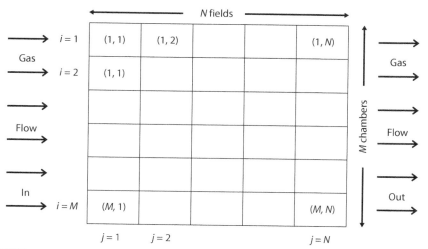

FIGURE 34.1 An *M* x *N* precipitator.

of compliance after a given number of bus sections have failed, it would be necessary to test all possible arrangement of the failure locations. The out-of-compliance probability is then given by the percent of arrangements that result in overall collection efficiencies less than the prescribed minimum standard. The number of arrangements to be tested often makes the direct calculation approach impractical. For example, Theodore et al. [2] were requested (as part of a consulting assignment) to investigate a precipitator unit consisting of 64 bus sections; if 4 of these were to fail, there would be 15,049,024, possible failure arrangements [3,4].

Instead of the direct calculation approach that would have required the evaluation of more than 15×10^6 potential failure arrangements, you have been requested to develop a similar procedure to perform this calculation, e.g., using a Monte Carlo technique [5] for an $M \times N$ bus section electrostatic precipitator.

Generate solutions for various combinations of $M \times N$ systems operating at a given (specified) overall collection efficiency [6]. Comment on the results.

Term Project 34.3

Hurricane and Flooding Concerns

Hurricane Sandy was the deadliest and most destructive tropical cyclone of the 2012 Atlantic hurricane season, as well as the second costliest hurricane in U.S. history. Sandy was a Category 3 storm at its peak intensity when it made landfall in Cuba and a Category 2 storm off the coast of the Northeastern United States. The storm became the largest Atlantic hurricane on record.

Hurricane Sandy affected 24 states, including the eastern seaboard from Florida to Maine and west across the Appalachian Mountains to Michigan and Wisconsin. There was particularly severe damage in New Jersey and New York. Its storm surge hit New York City on October 29, flooding streets, tunnels, and subway lines and cutting power in and around the city. Gasoline shortages, reminiscent of the Oil Embargo of 1973, lasted for two weeks. One of the authors (located on Long Island, New York) had no heat for 11 days and no TV power for 17 days.

Damage in the US was estimated at approximately $75 billion.

On October 28, New York City officials activated the coastal emergency plan, with subway closings and the evacuation of residents in areas hit earlier by Hurricane Irene in 2011. More than 76 evacuation shelters were open around the city. Mayor Michael Bloomberg ordered public schools closed the following day and called for a mandatory evacuation of areas near coastlines or waterways. Approximately 200 National Guard troops were deployed in the city. U.S. stock trading was suspended for two days. Amtrak canceled all Acela Express, Northeast Regional, Keystone, and Shuttle services for two days. More than 13,000 flights were cancelled across the U.S. on October 29, and more than 3,500 were called off October 30. Approximately 300 people were killed across the U.S., the Caribbean, and Canada, as a result of the storm.

Based on the above comments, obtain information on the following

1. What is the annual probability a Category 3, a Category 4, and a Category 5 hurricane will strike the New York City metropolitan area.
2. Estimate the damage (flooding, infrastructure, food and water shortages, and power failures) that can be expected with each Category hurricane.
3. Outline what steps can be taken to reduce and/or eliminate both the causes and consequences associated with each Category hurricane.

Note: The reader may choose to apply the above to any area of the U.S., including their own locale.

Comment: An urban planner [7] recently provided a real-world application that is of concern to administrators in the New York City metropolitan area. It was noted that Category 5 hurricanes strike the New York City metropolitan area with a 50 year recurrence interval. The analyses provided information on the risk that a Category 5 storm will strike in a 15 year period and investigated the recurrence interval that would be required to produce a 5% risk of a Category 5 hurricane striking in a 15 year period [8]. Interestingly, this "case study" has received little to no attention by city administrators. Discussion of this case study might explore what could be done to change this oversight.

Term Project 34.4

Meteorites

A *meteor* is defined as a small solid body entering the Earth's atmosphere from outer space. A meteor that reaches the Earth's surface before it is completely consumed is defined as a *meteorite*. A meteorite has also been defined as a natural object that has come from elsewhere in space. Some refer to large meteors/ meteorites as *asteroids*, while others refer to asteroids as small planets. Meteorites can be large or small. Most are produced by impacts of larger asteroids. When meteorites enter the atmosphere, frictional forces cause the body to heat up and emit light, thus forming a fireball, also known as a *shooting star* or *falling star*. Meteorites that are recovered after being observed as they enter the atmosphere or impact the Earth are called *falls*. All other meteorites are known as *finds*. Finally, meteorites are almost always named for the place where they land, for example, Tunguska (Siberia) [9].

Most meteoroids disintegrate when entering Earth's atmosphere. Only 5 or 6 a year are typically recovered and made known to scientists. Few meteorites are large enough to create large impact craters. Instead, they typically arrive at the surface at their terminal velocity [10] and, at most, create a small pit. Even so, falling meteorites have reportedly caused damage to property, and injuries to livestock and people.

Meteorites have traditionally been divided into three broad categories: stony meteorites are rocks, mainly composed of silicate minerals; iron meteorites are largely composed of metallic iron-nickel; and, stony meteorites contain large amounts of both metallic and rocky materials. Meteorites have also been classified according to their size; for example, those smaller than 2mm are classified as *micrometeorites*.

It's early morning on February 15, 2013 near Russia's Ural mountains. A large meteor has entered the Earth's atmosphere at a speed of approximately 30,000 mph and exploded into pieces about 25 miles above the ground, creating a sonic blast with the power of an atomic bomb. Over 1,000 people are injured.

The explosion occurred about 3,000 miles west of Tunguska, which in 1908 was the site of the largest recorded explosion of a meteor plunging to Earth. The present blast could have produced much more serious problems since the area contains nuclear and chemical weapons disposal facilities as well as some 6,000 tons of nerve agents.

Should there be concern? You be the judge. Scientists believe that a meteor explosion 6.5 million years ago in Mexico's Yucatan Peninsula may have been responsible for the extinction of the dinosaurs. The impact could have thrown up vast amounts of dust that blanketed the sky for decades and altered the climate of the Earth. Interestingly, at approximately the same time as the Tunguska explosion, an even larger meteor went sailing by Earth approximately 15,000 – 25,000 miles away. This meteor could have produced devastating effects had it approached Earth's atmosphere.

As a result of the recent 2013 meteor episode, NASA, in conjunction with PAT (Patrick Abulencia and Theodore) Consultants have initiated a joint project that will address three key issues.

1. Determine the annual probability with respect to size and frequency of meteorites striking/affecting the planet Earth.
2. Develop a plan to reduce or eliminate the likelihood of (1) occurring.
3. Develop an emergency plan of action in case (1) occurs.

References

1. Adapted from: D. Elam, *Standardized Project Management*, EM, Pittsburgh, PA, March 2013.
2. L. Theodore, J. Reynolds, F. Taylor, and S. Errico, Electrostatic Precipitator Bus Section Failure: Operation and Maintenance, Paper No. 84-96.10, presented at the 77th Annual Meeting of Air Pollution Control Association, June 24-29, San Francisco, CA, 1984.
3. A.M. Flynn and L. Theodore, *Health, Safety, and Accident Management in the Chemical Process Industries*, CRC Press/ Taylor & Francis Group, Boca Raton, FL, 2002.
4. L. Theodore and R. Dupont, *Environmental Health and Hazard Risk Assessment: Principles and Calculations*, CRC Press/Taylor & Francis Group, Boca Raton, FL, 2012.
5. S. Shaefer and L. Theodore, *Probability and Statistics Applications in Environmental Science*, CRC Press/ Taylor & Francis Group, Boca Raton, FL, 2007.
6. L. Theodore, *Air Pollution Control Equipment Calculations*, John Wiley & Sons, Hoboken, NJ, 2008.
7. G. Theodore, personal communication to L. Theodore, Interboro Partners, Brooklyn, NY, 2011.
8. L. Theodore and R. Dupont, "Environmental Health and Hazard Risk Assessment: Principles and Calculations," CRC Press/Taylor & Francis Group, Boca Raton, FL, 2012.
9. L. Theodore and R. Dupont, *Environmental Health and Hazard Risk Assessment: Principles and Calculations*, CRC Press/Taylor & Francis Group, Boca Raton, FL, 2012.
10. L. Theodore, *Air Pollution Control Equipment Calculations*, John Wiley & Sons, Hoboken, NJ, 2008.

Index

ABET, 454, 464, 468

Absorption, ix, 142–147, 164–165, 239

Acceptable risk, 327

Accident exposures, 318

Acid rain, xiii, 310, 352, 365, 370–371, 377

Actual number of theoretical stages, 521

Actual stage, mass transfer operation, 143

Adiabatic flame temperature, 68–69

Adsorbent, 76, 90–92, 149, 151, 519

Adsorber design procedure, xvi, 517, 519

Adsorption design, ix, 151

Adsorption equilibria, 51, 92, 149, 412

Adsorption factor, 144

Adsorption isotherm, 90–92, 149

Adsorption-desorption cycles, 152

Advanced wastewater treatment, 378

Air sterilizer, xiii, 391, 393

Alkaline chlorination, 379

Amendments to EPA legislation, 303–304, 338, 367–369, 374, 544–546

American national standards institute (ANSI), 271, 287, 299

American society for testing and materials (ASTM), 323, 336

Ammonium nitrate fertilizer, 374

Amortization, 226

Analysis of variance, xiv, 406, 412–413, 422

Annualized capital cost, 217–220

Anti-technology movement, ethics, 459, 472–475

Antoine equation coefficients, 491–492

Applied mathematics, vii, xv, 13, 39–41, 51–53, 57, 483, 485, 487

Applied thermodynamics, 80

Arrhenius equation, 174, 531

Art of questioning, 9

Artificial heart model, ethics, 473

Asteroids, term project, 570–573

Atmospheric lapse rate in heat transfer, 135

Atomic balance of equations, 65–66

Atomic particles, 16–17

Automatic control of reactor operation, 198–199, 211

Axial diffusion, 267

Baffles, heat exchanger design, 127–128

Balancing equations, stoichiometry, 64

Batch plants, 244

Batch reactors, x, 170, 174, 180, 184, 193

Batch sterilization, xiii, 391–392, 400

Benzene leaks, term project, 550

Bernoulli equation, 103

Binomial theorem, 420

Biochemical engineering,
    xiii, 14, 383–403
Biological oxygen demand (BOD), 400
Biot number, heat tranfer, 133
Black body, energy radiator, 126
Block diagrams, process
    control, 200, 210
Body centered cubic (BCC), 20, 26–27
Boltzmann constant, 35, 126
Bottoms, distillation column,
    153, 156, 493
Boundary conditions, transport
    phenomena, 257, 389
Bovine liver catalase, open ended, 397
Bragg's law of diffraction, 31–33, 35
Bravais lattice, 20–22
Breakpoint, adsorbers, 149–150
Breakthrough capacity, 150
Breakthrough curve, 149
Bridge rail accident, hazard risk,
    xvii, 553–554, 557
Building dams, water
    conservation, 372
Bulk motion, 254
Bus section failures in electrostatic
    precipitators, xvii,
    569–570, 573
Business related laws, etc., 220, 455
Calculating combustion
    temperatures, 496
Capital cost, plant design, 217–219,
    228, 487, 530, 538
Carbon nanotubes, 428, 430, 437
Cascade control, x, 199, 205–206
Cascade control, single input single
    output (SISO), 205
Cash pools, football pools, labor
    pools, etc., 423, 478
Catalysis, 181
Catalytic reactors, x, 109,
    170, 181, 183, 194
Category 5 hurricane, 571
Central atom in metallic
    crystal, 18–19, 31

Chemical engineering curriculum,
    1–2, 10, 160
Chemical kinetics, x,
    170–171, 173, 184
Chemical law of combining weights, 64
Chemical oxygen demand
    (COD), 372, 400
Chemical plant shipping
    facilities, xvi, 533–534
Chemical plant storage requirements,
    xvi, 533, 536
Chemical reaction equilibrium,
    viii, 81, 88–89, 97
Chemical reactors, x, xvi, 13, 169–195,
    211, 249, 483, 529, 531
Chen's correlation for tower
    daimeter, 161, 504
Clapeyron equation coefficients,
    491–492
Clean air act, 296, 437, 555
Clean water act, xiii, 338, 369, 377
Closed-loop system, process
    control, 199–200, 210
Coal-fired power plant, 70, 357
Colburn's equation for packing
    height, 144
Column quick-sizing, 521
Combustion systems, 239
Commensalism, 395
Commercial real estate, 348, 353
Composting, 312, 548
Comprehensive environmental
    response, 296, 302–303, 338
Concentration of a chemical in
    a mid-sized room, term
    project, 505, 523, 529–530
Condensation, ix, 120, 122,
    126–127, 136, 159, 308
Conduction, ix, 120–121, 123, 125, 128
Conduits, viii, 103, 106–107, 116,
    135, 265, 506–507
Conforming to moral standards, 464
Conservation law, viii, 60–62, 64, 80,
    101, 175, 184–186, 255–259

Conservation practices, xii, 345–347
Contaminant candidate list, 369
Continuous random variable,
    408–410, 417–419
Continuous stirred tank reactors
    (CSTR), x, 170, 176–178,
    183–185, 193–194, 199–201,
    205, 440, 530–531
Continuous-flow reactor, 70
Continuous-operation plants, 244
Contract law, xiv, 447–448, 459
Contributory infringement, 451
Control elements, process
    control, 198, 207–208
Control system, 199, 205,
    212, 240, 307, 320
Convection, ix, 120, 122–123, 125,
    127, 135, 254, 259, 266
Conversion, 88, 177, 188, 219, 227,
    246, 250, 344, 346, 358
Cooling towers, xvi, 129, 214, 243,
    249, 517, 525–526, 538
Coordination number, 18–19, 26
Copyright act of 1976, 452
Corrosion resistance, 27
Cost control, project management, 285
Cost of shipping materials, 534
Creativity and brainstorming, 3, 7
Criminal prosecution, open-ended
    problem, 456–457
Critical insulation thickness,
    131–133, 354
Critical thinking, 2–3, 6–7, 12
Criticism of current EPA regulatory
    approach, 452, 545
Cryogenics, ix, 120, 126–127, 139
Crystal lattice, 17, 23, 379
Crystal properties, 24
Crystallography of perfect
    crystals, vii, 16–17
Crystallography of real crystals, vii, 25
Cubic salts, 22
Cunningham correction factor,
    109, 117, 400, 441

Cyanide in water management,
    379, 467, 492
Decision regarding employment, 477
Defects in crystals, 30–31
Defense industry initiative in
    engineering ethics, 468
Delphi panel method, 281
Department of health and
    human services, 545
Dependent/independent variables,
    28, 45–49, 179, 238, 411, 486
Depriester chart, 87
Design and build, 296, 350,
    352, 562–565
Design approach, xi, 235, 240
Design of experiments, 414, 422
Design report in project management,
    xi, 235, 241–242
Desorption, 152
Deterministic systems, 46
Development of equations,
    xi, 254–255, 262
Diffusion effects, 254, 263
Dimensionless ratios, 261
Direct installation cost, 217
Directional crystal phenomena, 23–24
Discrete probability distributions, 410
Discrete random variable,
    408–409, 416–418
Dispersion model, 322
Dispose of waste out of state, 477–479
Dissociation constant, 385, 388
Distillate, 153–154, 156
Distillation, ix, xvi, 101, 109,
    142–143, 152–154, 156–158,
    164–165, 211, 229–230,
    245–246, 270, 308, 347,
    421, 517, 520, 522–523
Distribution function, 55,
    408–409, 416
Double-pipe heat exchanger,
    123, 127, 129
Drag coefficients, 314
Drinking water standards, 367–368

Driving force, 62–63, 121, 125, 127, 136, 144, 158, 353, 514

Drying, ix, 109, 142, 159, 164–166, 211

Due diligence process, 221

Effective argument, 7

Effective diffusivity, 389

Effective inquiry, 9

Effectiveness factors, 384, 488

Electrostatic precipitator, 214, 228, 407

Elliptical equation, 45

Emergency planning - meteorites, open-ended problems, xvii, 322

Emission source, 67

Emissivity, 126

Energy balances, 75, 90, 101, 248

Energy demand, 344

Energy independence, 351

Energy management, xii, 13, 343, 345, 347, 349–359, 361

Energy management program, matrix approach, 355

Energy quantity, xii, 80, 345–346

Energy resource options, 346

Energy resources, xii, 345–346, 352, 361, 487–488

Energy strategy, 351

Engineering ethics, xv, 464, 467, 470, 480

Enriching, 153–154

Enthalpy change, 68, 82, 497

Enthalpy effects, viii, 81, 83

Enthalpy of reaction, 75, 83–84, 89, 95–96, 185–187, 531

Entropy, xv, 75, 85–86, 94–96, 138, 356, 495–497, 513–514

Environmental effects of acid rain, 371

Environmental implication of nanotechnology, 433

Environmental justice, xv, 464, 468–469, 481

Environmental legislation, 295, 544–545

Environmental policy of the EPA, 468

Environmental risk assessment, xii, 319, 322, 337

Environmentalism, 296

Enzyme-substrate complex, 387–388

Enzymes, xiii, 384–389, 396–400

Equilibrium constant, K, 87, 96

Equilibrium curve, 144, 154, 156–157, 518

Equilibrium stage, 142–143, 154

Equipment description, 540

Equipment spacing, 539–540

Ergun equation for pressure drop, 184

Ethicists, 466

Eutrophication, 377

Evaporation, 90, 142, 159, 249, 298, 307, 507, 523

Event-tree diagram, 333

Executive summary, 242–243, 279

Expert witness, 465

Exposure assessment, 325–326

Extensive property, 23

Face-centered cubic (FCC), 20–22, 26–27

Factored method to determine capital cost, 216

Failure to report in employee ethics, 457

Falsifying, omitting info, lying, deceiving (FOLD), 464

Fanning friction factor, 105–106, 116, 180, 500–501

Fault tree analysis, 421

Feed-plate location, distillation column, 157

Feedforward and feedback control, x, 199–201, 204–205, 210–211

Fenske-underwood-giilliland method, 158, 520

Fermentation process, 394, 397

Fick's law, 254, 260

Film coefficient, 125, 137–138, 510

Filtration, ix, 109–110, 117, 379, 393, 547

Fins on a heat exchanger, 127–129, 136

First law of thermodynamics, 63, 80, 84
Fixed-bed adsorbers, 109, 149
Fixed-bed reactors, x, 109, 183–184, 191, 195
Flashing and boiling liquids, 507
Flocculation, 379
Flooding velocity in a packed column, 145, 503
Fluidization, ix, 109–110
Fluidized bed reactor, x, 110, 183, 401
Fossil fuels, 345, 370
Fourier's law, conduction, 121, 254, 260, 265
Free energy of reaction, 75, 89, 97
Free surface energy, 490
Freundlich model, vapor-solid equilibrium phenomena, 92
Fugitive emissions, 307–308
Fullerenes, 430
Fumes, solvents and exhaust, 307, 340
Future prospects of nanotechnology, xiv, 429, 434
Gantt chart, 275–276
Gauss methods, linear algebraic equations, 3, 42
Generalized correlation for flooding and pressure drop, 146
Generalized pressure-drop correlation, 161, 503–504, 508
Geometry of ionic unit cells, vii, 21
Geometry of metallic unit cells, vii, 20
Gilliland correlation, 521
Global warming, 135, 311, 339, 350, 352
Gold, nanomaterial/open-ended problem, 36, 117, 431
Gradients, 254, 259
Gray bodies, 126
Green chemistry, xii, 294, 300–301, 315
Green engineering, xii, 294, 300, 316
Hairpin design, heat exchanger, 128

Hazard risk assessment, xii, xvii, 13, 137, 139, 251, 291, 317–319, 321, 323, 325–329, 331–341, 382, 425, 444, 483
Hazardous and solid waste act, 299
Hazardous waste incineration, 77, 215, 230–231, 235, 239, 250–251, 361, 488
Health risk assessment, xii, 319, 323–324, 334–335
Heat exchanger design, approach temperature, 514
Heat flow, 120–122, 125
Heat transfer equation, 123, 127, 131, 135, 452
HEEL, residual adsorbate, 150
Height of transfer unit (HOG), 144–145, 518–519
Henry's law, 86–87, 144, 164, 338
Heterogeneous reactor, 182
Heterotrophic plate count, microbial growth, 373
Hexagonal close packed (HCC), 21
Hexavalent chromium, 378
Humidification, ix, 142, 159
Hurricane irene, term project, 570
Hydrologic cycle, xiii, 365–366, 377
Hyperbolic equation, 45
Identifying pollutants, 306
Implement a management system, environmental, 338, 355, 555–556
Implement an environmental regulation, 544
Implementation phase in project managmt, 277–278
Improved MSDS sheet, xvi, 543, 551
Incineration, 77, 215, 228–231, 235, 239, 250–251, 361, 471, 478, 488, 548
Incompressible fluid, 266
Indices, 24–25, 32, 216
Indoor air quality, 310–311
Inside battery limit plants, 538

Intensive property, 23
Interconnected heat exchangers, 512
Interference process, 451
Internal energy, 63, 75, 101
Internal reflux ratio, 155
International management
    standards, 271
International organization
    for standardization
    (ISO), 271, 298
Interphase diffusion, 146
Interstitial defect, 31
Interstitial impurities, vii, 25–26, 181
Interview for a new environmental
    management position, 477
Inventory tie-up costs, 536
Ionic crystal, cubic salt, cesium
    chloride, 22
Ionic unit cells, vii, 21
ISO 14000, xii, 294, 298–299,
    311–312, 315, 470–471, 476
Isothermal packed tower systems, 144
Isotope, 35
Kinetic energy, 31, 63, 490
Kirkbride equation, 522
L'Hopitals rule, critical insulation
    thickness, 132
Laminar flow, viii, 102,
    104–105, 263–264
Landfarming, 549
Landfills, 379, 478
Langmuir equation, 92
Laplace transform equations, 48–49,
    202, 207, 209, 212, 262
Latent heat, 122, 490–491, 525
Lattice strain, 31
Law of conservation of energy,
    62–63, 84, 513
Law of conservation of mass, 62
Leaching, ix, 142, 158–159, 165, 523
Least squares fit, regression
    analysis, 50, 411–412, 424
Legal considerations, xiv, 14, 445–449,
    451, 453, 455, 457, 459, 461

Lewis number, mass transfer
    equations, 262
Licensed professional engineer, 454
Limited monopolies, patents, 449
Linear algebraic equations,
    vii, 41–42, 52, 55
Linear system, 42
Liquid absorption process, 146
Liquid extraction, ix, xvi, 142, 158,
    166–167, 517, 523–525
Local emergency planning
    committee, 339, 375
Log-mean average heat transfer
    area, 122, 124
Log-mean temperature difference
    (LMTD), 136, 514
Managing project activities, xi, 271
Mass loading in wastewater
    treatment process, 371
Mass transfer operations, 152, 158, 165
Mass transfer zone, 149–150
Mass, energy and momentum,
    viii, 61–62, 238
Material balance calculation, 237
Material balance, macroscopic, 16, 28,
    61, 120, 173, 255, 262, 265, 430
Material science and
    engineering, 13, 34
Mathematical optimization,
    approximate measure, 48
Matrix formulation, energy
    managemnt, 351
Maximum contaminant levels, 368
Maximum thrust, fuselage, open-
    ended problem, 111
Maxwell-boltzmann distribution, 31
McCabe-Thiele diagram,
    distillation, 156
McCabe-Thiele method, distillation
    column design, 152
Mechanical energy equation, viii, 103
Membrane processes, ix, 142, 159–160
Metallic unit cells, vii, 20
Methyl-tert-butyl ether (MTBE), 359

Michaelis-Menten equtn, microbial kinetics, 385, 387–389, 397, 400

Microbial regrowth, 373

Micrometeorites, open-ended problem, 572

Microscopic approach, 61, 255–256, 262, 265–266, 430

Miller indices, 25, 32

Minimum number of theoretical stages, 520–521

Mixed cell cultures, 395

Mixed oxides, 431

Model describing the concentration of a nanochemical, 438

Modified Lang method, 217

Molecular balance, 65

Molecular diffusion, 254, 259–261, 393

Molecular level, 61, 254–256

Monod growth kinetics, 400–401

Monte Carlo method, xvii, 55, 424

Multicomponent distillation calculations, xvi, 520

Multicomponent separation, 158

Multiphase mixture, 25

Municipal wastewater treatment, 377, 380, 548

Mutualism, 395

Nanoimprinting, 435

Nanomachines, 428, 434

Nanomaterials, xiv, 428–430, 432, 437, 441, 444

Nanoparticles, 289, 428–429, 432–433, 442–443

Nanoregulations, 433–434

Nanotechnology environmental regulations, 436

Nanotechnology in actual commercial use, 432

Nanotechnology-related development, 434

National environmental policy act (NEPA), 295, 303

National institute of safety and health admin (NIOSH), 336

National pollutant discharge elimination system (NPDES), 369, 378, 388

Natural water bodies, 379

Natural water stabilizing organic matter, 379

Natural waters, 367, 377

Neutralism, open-ended problems, 395

Newton raphson method, 43

Newton's method of tangents, 44

Newton's second law, 254

Non-isothermal reactors, 185

Nonlinear algebraic equation, 43, 54

Nonpoint source water pollution, 378

Not-in-my-backyard (NIMBY), 472

NRTL method, 96

NRTL model, 88

Nuclear chemistry, 35

Nuclear power, 327, 356, 358

Nuclear waste, xvii, 310, 554

Number of transfer units, 144–145, 147, 518–519

Numerical method, 40, 45

Numerical methods, 40–41, 52–53, 56, 258, 262

Ohm's law, 28

Open-loop system, 199

Operating costs, x, 214–219, 227–231, 238, 272

Operational conditions, 242

Order of diffraction, 32

Order of magnitude estimate, 235

OSHA regulations, 321

Outside film coefficient, evaluating, 510

Overall heat transfer coefficient, 123–125, 127, 137, 185

Overall material balance, 154–155

Overall plate fractional efficiency, 157

Packed column, 143–145, 161, 503–504

Packed tower absorption, 147

Packing diameter versus hog, 518

Packing factors, vii, 23

Packing height, ix, 144–145
Parabolic equation, 45
Partial condenser, 154, 156
Partial derivative, 486
Partial differential equation,
	viii, 44, 263, 486
Partial reboiler, 154, 156, 520
Pasquel-Gifford atmospheric
	dispersion coefficients, 313
Patent rights, 450
Periodic table, 30, 35, 441
Permutations, 413–414
Personal liability suit, 455
Perturbation studies, x, 216, 219
Phase equilibrium, viii, 81, 86–87, 96
PID controller, 205
Pilot plants, 102, 242, 248
Piping and instrumentation, 237
Plagiarism, 474
Plant design, xi, xvi, 13, 214–215,
	233–251, 461, 483, 511,
	533–537, 539, 541
Plant tank farms, xvi, 533, 535
Plasmid instability, 397
Plate columns, ix, 146
Plate tower absorption, 147
Point sources, 369–370
Pollution prevention, xii, 294, 296,
	299–300, 303–304, 306, 311,
	315–316, 467, 487–488
Pollution prevention act, 299, 304
Polymorph, 26
Porous media, ix, 109, 117
Potential energy, 63, 101
Prandtl number, 261
Precipitation, 310, 366–367
Predicting, open-ended problems,
	xvi, 95, 138, 193, 195,
	266, 517, 525–526
Present-day atomic theory, 436
Present-value formula, 222
Present-worth analysis, 219
Pressure drop and flooding
	correlation, xv, 499, 503

Pressure drop variation with
	velocity, 500
Primary control loop, 206
Primary sludge solids, 547
Prime movers, viii, 101, 107
Probability and statistics, xiii, 14, 55,
	57, 74, 196, 291, 403–411,
	425, 494, 507–508
Probability distribution, 55,
	408–410, 416, 421
Problem definition, 7
Process control and instrumentation
	(PCI), x, 197, 199, 201,
	203, 205–207, 209, 211
Process environmental
	management, 293
Process feasibility, 241
Process flow diagram, 214, 236–237
Process schematic, 236
Process unit, 198–199, 238, 244
Producing nanoscaled
	particles, 431, 442
Progress reports, 279
Project control, 236, 241, 270, 275, 278
Project management, xi, xvii, 13, 215,
	247, 269–281, 283–289, 291
Project management institute, 287
Project management, matrix
	approach, 286
Project manager, 270, 272–274,
	277–278, 280, 285–286
Project planning, 273–275, 278
Project-related problems, 290
Properties of water, 377, 518
Proportional, 28, 63, 101, 126,
	138, 203–204, 211, 359
Proportional-integral
	derivative, 203–204
Public health service act, 367
Public water supply, 368
Purchasing equipment, 227, 288
Quality assurance, 276
Quality control, 276, 455
Radial diffusion, 264

Radiation, ix, 31, 33, 120, 123, 125–127, 137, 254, 322, 335–336, 364, 545
Radiation view factor, 137
Radioactive isotopes, 364
Radioactive waste disposal, 554
Radius ratio, 19–20
Raoult's law, 86–88, 96, 164
Rate of momentum, 100
Raw materials, 170, 235, 244, 248, 250, 360, 493, 534
Reaction velocity constant, 171, 173–174
Reactor size requirements, 530
Recombinant DNA, 396
Recuperator, 127
Recycling, 300, 312, 359–360, 477, 523
Reduce energy waste, 348
Reflux ratio, 153, 155, 158, 520–521
Refrigeration, ix, 120, 126–127
Regeneration, ix, 150–152, 181, 519
Regression analysis, xiv, 49, 398, 406, 411
Regulatory approach to solving environmtl problems, 545
Relative roughness, 106, 500–502
Relative volatilities, 152, 520–521
Renewable resources, 366
Resistance to heat flow, 121
Resistance to heat transfer, 124, 128
Resistivity, vii, 28–29
Resource conservation and recovery act (rcra), 296, 303, 338
Retardation factor, 550–551
Return bends, 128
Return on investment, 222–223
Reversible and irreversible reactions, 172, 180, 530
Reynolds number, viii, 102, 105–106, 116–117, 313–315, 393, 500–503
Risk evaluation, 325–326
Risk evaluation of accidents, 326
Risk management program, xvii

Risk-based corrective action, 322–323
Risk-based decision making, 322–323, 342
Rooftop garden bed project, xvii, 561, 563–564
Rotary dryer, 166
Rotary kiln, 122, 223–225, 239
Roughness of commercial pipes, 502
Roundoff error, 52
Runge Kutta, 44
Runoff, 366, 378
Russia's ural mountains, 572
Safe drinking water act (SDWA), xiii, 367, 377
Safety and accidents, xii, 319
Safety factors, 239, 326
Scale-up, xiii, 248, 391–394, 401, 434
Scatter diagram, 50–51, 411–412
Schmidt number, 261
SCORE, design process acronym, 234, 423
SDWA 1996 amendments, 368
Second law calculations, viii, 81, 84, 86
Second law of thermodynamics, viii, 80–81, 84–86, 94–95, 254, 512–514
Secondary control loop, 206
Selectivity, 173, 193
Semiconductors, matrix/solvent material, vii, 28–29
Sensible heat, 95, 122, 525
Sensible heating duties, 130
Service mark, 453
Shell-and-tube heat exchangers, 128, 136, 138, 511
Simple cubic, metallic unit cells, 20–23
Simpson's rule, 41
Simultaneous flow of two phases in pipes, 106, 506
Simultaneous linear algebraic equations, matrix, 42
Size distributions of nanoparticles, 442

Slip, two-phase flow, 106, 108, 506
Sludge, xvi, 298, 377–379,
    543, 547–549
Sludge disposal, 548–549
Sodium chloride, 20, 22, 497
Solution of equations, xi, 254, 258, 263
Solving differential equations, 485–486
Source reduction, 300, 304
Specific reaction rate, 171, 189
Specification, 129, 248, 273,
    288, 513–514
Spill prevention control and
    countermeasures, 338
Spray dryer, 159, 239
Stage efficiency, 143, 161
Staged column design, 154
Staged distillation, 156
Stagewise vs. continuous
    operation, 142
Standard deviation, 282–283,
    381, 423, 443
Standard enthalpy of formation, 83
Standardizing project
    management, xvii
Status reports, project
    management, 279
Steady-state, 65, 143, 202, 238,
    263–266, 306, 387–388,
    438–439
Stefan-boltzmann law, 126
Sterilization, xiii, 391–393, 400
Sterling silver, 26
Stochastic vs. deterministic
    system, 46
Stoichiometric ratios, 64
Stoke's law, 46, 108, 314, 393, 400
Storage tank problem, 54
Stripping gas, 147
Stripping in plate towers, 148
Stripping, mass transfer, ix,
    146–148, 152–155, 158, 523
Structural failure, 111
Study estimate, project
    management, 235

Sublimation, 90
Substitution impurity, 26
Superfund amendments and
    reauthorization act, 374
Surface water, 367, 373, 377, 382, 549
Surrogate regulators, open-
    ended problem, 227
Sustainability, xii, 294, 301–302,
    350, 467
Symbiosis, open-ended problem, 395
System boundary, 62
Temperature difference, 63,
    125, 127, 136, 514
Temperature profile, 266–267, 392
Terminal settling velocity, 108
Theoretical plate, 154
Thermal conductivity, 121–122,
    126, 128, 133, 136, 254,
    256, 266, 355, 430
Thermal death constant, 392
Thermal effects, x, 170, 184
Thermodynamics, viii, xv, xx, 3,
    13, 63, 70, 77–97, 138–139,
    173, 188, 196, 256, 483,
    494–498, 512–515
Thiele modulus, 390, 400
Threshold dose level, 331
Threshold planning quantity, 304, 374
Tort law, xiv, 448–449, 459
Total maximum daily load
    (TMDL), 370
Total organic carbon (TOC), 400
Total suspended solids (TSS),
    369, 380, 548
Tower diameter, ix, 145, 161,
    167, 504, 508, 519
Trade secrets, 447, 450
Trademark, 40, 453, 459–460
Transfer of momentum, 259
Transfer unit, 47, 144, 229, 518
Transport equations, xi, 254, 256, 258
Transport phenomena, xi, 13,
    61, 77, 118, 253–259,
    261–265, 267, 394, 488

Trapezoidal rule, 41
Treatment, storage and disposal
    facility, 478
Treybal, 158, 160, 163, 166–167,
    523, 526–527
Tube sheets, 128
Tubular flow reactor, x, 170,
    174, 178–180, 182–184,
    186–187, 193–194, 196
Turbulent flow, viii, 105–106,
    113, 266, 500, 502
Two-coefficient Clapeyron
    equation, 491
Two-phase flow, 116, 499
Two-phase pressure drop, 112, 507
Type of utility, 130
Type of waste and disposal
    method, 480
U.S. energy policy, xiii, 345, 350
U.S. forest service, 476
Underground storage tanks, 549–550
Understanding of patent law, 446
Unit cell, 21–23, 31, 36
Unit mole, 64
Unit process, 238
Urban planners, 350
Utility system, 198
Vacancies per unit cell, 36
Valves and fittings, viii, 107
Vapor pressure data, 491

Vapor pressure equation, 489–490, 492
Vapor recovery system, 250
Vaporization, 90, 122, 159,
    490–491, 497
Venturi scrubber, 228, 239
Volume requirements, reactors,
    xvi, 529–531
Waste load allocation, 378
Waste reduction technologies, 297
Wastewater standards, 369
Wastewater treatment plant/process,
    214, 298, 310, 369, 380
Water conservation, 372
Water management, xiii, 14, 361,
    363–367, 369, 371–373,
    375–377, 379, 381
Water quality standards, 370
Water usage, xiii, 365, 367
Weibull distribution, 410, 420, 424
Weighted-sum method, 56, 487–488
Wet-acid process, 493
Wilson's method, xvi, 88, 137, 510
Working charge, 150–151
X-ray diffraction, 20, 33–34
Yield, 24, 44, 47–48, 93, 173,
    182, 193, 198, 223, 229,
    399, 401–402, 424
Yucatan peninsula, open-
    ended problem, 572
Zeolites, 432